Genes, Behavior,

and the

DATE D'

Social Envi

D0642602

Moving Beyond the Nature/Nurture Debate

Committee on Assessing Interactions Among Social, Behavioral, and
Genetic Factors in Health

Board on Health Sciences Policy

Lyla M. Hernandez and Dan G. Blazer, Editors

INSTITUTE OF MEDICINE
OF THE NATIONAL ACADEMIES

THE NATIONAL ACADEMIES PRESS • 500 FIFTH STREET, N.W. • Washington, DC 20001

NOTICE: The project that is the subject of this report was approved by the Governing Board of the National Research Council, whose members are drawn from the councils of the National Academy of Sciences, the National Academy of Engineering, and the Institute of Medicine. The members of the committee responsible for the report were chosen for their special competences and with regard for appropriate balance.

This study was supported by Contract No. N01-OD-4-2139, TO 154 between the National Academy of Sciences and the National Institutes of Health's Office of Behavioral and Social Sciences Research, National Human Genome Research Institute, and the National Institute of General Medical Sciences. Any opinions, findings, conclusions, or recommendations expressed in this publication are those of the author(s) and do not necessarily reflect the view of the organizations or agencies that provided support for this project.

Library of Congress Cataloging-in-Publication Data

Genes, behavior, and the social environment : moving beyond the
 nature/nurture debate / Lyla M. Hernandez and Dan G. Blazer,
 editors ; Committee on Assessing Interactions, Among Social,
 Behavioral, and Genetic Factors in Health, Board on Health
 Sciences Policy.
   p. ; cm.
 Includes bibliographical references and index.
 ISBN 0-309-10196-4 (pbk.) — ISBN 0-309-66045-9 (PDFs)
 1. Behavior genetics. 2. Medical genetics. 3. Nature and nurture.
 4. Human genetics—Research.   I. Hernandez, Lyla M. II. Blazer,
 Dan G. (Dan German), 1944-   . III. Institute of Medicine (U.S.).
 Committee on Assessing Interactions, Among Social, Behavioral,
 and Genetic Factors in Health.
 [DNLM: 1. Genetics, Behavioral. 2. Sociobiology.   QU 450
 G3266 2006]
 QH457G458 2006
 616'.042—dc22
                                        2006023972

Additional copies of this report are available from the National Academies Press, 500 Fifth Street, N.W., Lockbox 285, Washington, DC 20055; (800) 624-6242 or (202) 334-3313 (in the Washington metropolitan area); Internet, http://www.nap.edu.

For more information about the Institute of Medicine, visit the IOM home page at: **www.iom.edu.**

"Knowing is not enough; we must apply.
Willing is not enough; we must do."
—Goethe

INSTITUTE OF MEDICINE
OF THE NATIONAL ACADEMIES

Advising the Nation. Improving Health.

THE NATIONAL ACADEMIES
Advisers to the Nation on Science, Engineering, and Medicine

The **National Academy of Sciences** is a private, nonprofit, self-perpetuating society of distinguished scholars engaged in scientific and engineering research, dedicated to the furtherance of science and technology and to their use for the general welfare. Upon the authority of the charter granted to it by the Congress in 1863, the Academy has a mandate that requires it to advise the federal government on scientific and technical matters. Dr. Ralph J. Cicerone is president of the National Academy of Sciences.

The **National Academy of Engineering** was established in 1964, under the charter of the National Academy of Sciences, as a parallel organization of outstanding engineers. It is autonomous in its administration and in the selection of its members, sharing with the National Academy of Sciences the responsibility for advising the federal government. The National Academy of Engineering also sponsors engineering programs aimed at meeting national needs, encourages education and research, and recognizes the superior achievements of engineers. Dr. Wm. A. Wulf is president of the National Academy of Engineering.

The **Institute of Medicine** was established in 1970 by the National Academy of Sciences to secure the services of eminent members of appropriate professions in the examination of policy matters pertaining to the health of the public. The Institute acts under the responsibility given to the National Academy of Sciences by its congressional charter to be an adviser to the federal government and, upon its own initiative, to identify issues of medical care, research, and education. Dr. Harvey V. Fineberg is president of the Institute of Medicine.

The **National Research Council** was organized by the National Academy of Sciences in 1916 to associate the broad community of science and technology with the Academy's purposes of furthering knowledge and advising the federal government. Functioning in accordance with general policies determined by the Academy, the Council has become the principal operating agency of both the National Academy of Sciences and the National Academy of Engineering in providing services to the government, the public, and the scientific and engineering communities. The Council is administered jointly by both Academies and the Institute of Medicine. Dr. Ralph J. Cicerone and Dr. Wm. A. Wulf are chair and vice chair, respectively, of the National Research Council.

www.national-academies.org

Independent Report Reviewers

Although the reviewers listed above have provided many constructive comments and suggestions, they were not asked to endorse the conclusions or recommendations nor did they see the final draft of the report before its release. The review of this report was overseen by **Jane E. Sisk**, Centers for Disease Control and Prevention and **Elena O. Nightingale**, Institute of Medicine. Appointed by the National Research Council and Institute of Medicine, they were responsible for making certain that an independent examination of this report was carried out in accordance with institutional procedures and that all review comments were carefully considered. Responsibility for the final content of this report rests entirely with the authoring committee and the institution.

Preface

Developing this report about facilitating integrated research on how the social environment and genetic function affect health outcomes has been tremendously rewarding, in large part because the effort was a collaboration among scientists from the social, behavioral, and biological sciences. Committee research and discussion illuminated associations among social factors and health, behaviors and health, and genetics and health. Committee collaboration resulted in a vision, described in this report, of how future research, transdisciplinary in nature, can contribute to the science of gene-social environment interactions and to explaining individual and population health and health disparities.

Yet, transdisciplinary research faces many challenges, not the least of which are those encountered when attempting to conduct collaborative research across disciplines. In a sense, the challenge of collaboration was illustrated in the work of this committee, whose scientists came from the fields of sociology, demography, psychology, psychiatry, research design, law, ethics, medicine, public health, epidemiology, biology, molecular virology, and genetics. Despite the fact that each committee member already had demonstrated a willingness to work with those from other disciplines on problems that crossed social, behavioral, and genetic lines, committee understanding and collaboration were not achieved effortlessly. Research conducted by different disciplines rests on different knowledge bases, often with different areas of focus—for example, the geneticist emphasizes individuals, while sociologists examine groups and societies. To form a group that could work collaboratively, it was necessary to devote meeting time to developing a common understanding of each others' definitions, terms, knowledge about what various disciplines have contributed to our understanding of disease risk, and an appreciation and value for the research designs and methods used by practitioners of the different disciplines. It was only after this had been accomplished that rapid progress could be made in developing an integrated approach to the task at hand—that of determining how researchers can begin to assess the impact on health of interactions among social, behavioral, and genetic factors.

In transdisciplinary research, investigators will be faced, on a broader scale, with the challenges that confronted this committee. Foremost among these challenges is the need to appreciate and value the contributions of other disciplines. Other challenges and approaches to addressing them are described in the body of the report, but the committee believes that the challenge of fostering true collaboration merited the emphasis that is provided in this preface. Successful transdisciplinary research that is conducted on gene-social environment interaction could provide a way for us to redefine how we think about health and disease. Such a redefinition, however, is not a short trip going forward with a specific goal in mind; rather, it is a journey that will require time and patience. This report and its recommendations are intended to launch us on that journey.

Dan G. Blazer, *Chair*
Committee on Assessing Interactions Among Social,
Behavioral, and Genetic Factors in Health

Acknowledgments

Over the course of this study, many individuals were willing to share their expertise, time, and thoughts with the committee. Their contributions invigorated committee deliberations and enhanced the quality of this report.

The study sponsors at the National Institutes of Health Office of Behavioral and Social Sciences Research, the National Human Genome Research Institute, and the National Institute of General Medical Sciences willingly provided information and responded to questions.

Invaluable information was provided by the authors of four commissioned papers: Steve W. Cole, Ph.D. (immunology), Myles S. Faith, Ph.D. and Tanja V.E. Kral, Ph.D. (obesity), Sharon Schwartz, Ph.D. (interactions), and Robert J. Thompson, Jr., Ph.D. (sickle cell disease).

The committee greatly appreciates the input of speakers whose presentations informed committee thinking, including Ronald Abeles, Arthur Beaudet, Sheldon Cohen, Eileen Crimmins, Anna Diez Roux, Ming D. Li, Colleen McBride, Margaret Locke, Brian Pike, and John Sheridan.

The committee was extremely fortunate in its staffing for this study. We wish to thank Lyla M. Hernandez, who did a remarkable job of directing the study. Thanks also go to Andrea Schultz who provided excellent research and administrative assistance to the committee. We are grateful to Christine Hartel for her insights and contributions in the writing of this report. We also wish to thank Andrew Pope for his guidance during the project, David Codrea for his handling of the financial accounting, Mark Chesnek for his cover design, and Sara Maddox for editing the draft document.

Contents

SUMMARY   1

1   INTRODUCTION   15
    Determinants of Health, 17
    Transdisciplinary Research, 19
    Commissioned Papers, 20
    Context, 21
    Goals of the Report, 23
    Conclusion, 23

2   THE IMPACT OF SOCIAL AND CULTRAL
    ENVIRONMENT ON HEALTH   25
    Defining the Social and Cultural Environment, 25
    The Influence of Social and Cultural Variables on Health:
        An Overview of Past Research, 26
    Aspects of Health Influenced by the Social Environment, 38
    Limitations of Current Research, 39
    Conclusion, 40

3   GENETICS AND HEALTH   44
    Genetic Susceptibility, 44
    Genetic Linkage Analysis and Genetic Association Studies, 47
    Gene-Environment and Gene-Gene Interactions, 50
    Mechanisms of Gene Expression, 52

Aspects of Health Influenced by Genetics, 56
Genetics of Populations as Related to Health and Disease, 60
Conclusion, 62

4   GENETIC, ENVIRONMENTAL, AND PERSONALITY
    DETERMINANTS OF HEALTH RISK BEHAVIORS   68
    Introduction and Overview, 68
    Definitions of Health Risk Behaviors, 69
    Tobacco Use, 70
    Unhealthy Eating Behaviors and Obesity, 72
    Physical Inactivity, 73
    Using Intermediate Phenotypes to Investigate the Effects of
        Gene-Environment Interactions, 74
    Conclusion, 82

5   SEX/GENDER, RACE/ETHNICITY, AND HEALTH   90
    Sex/Gender, 91
    Race/Ethnicity, 96
    Conclusion, 104

6   EMBEDDED RELATIONSHIPS AMONG SOCIAL,
    BEHAVIORAL, AND GENETIC FACTORS   109
    Thinking from the Bottom Up: Genomic Information
        Influencing Gene Expression, 110
    Thinking from the Bottom Up: Genomic Information
        Embedded in Biochemical Systems, 114
    Thinking from the Top Down: Social Factors Influencing
        Cells, Tissue, and Physiology, 116
    Molecular Mechanisms of Gene-Environment Interaction, 122
    The Need for Systems Approaches, 123

7   ANIMAL MODELS   132
    Role of Animal Models, 132
    Definitions from Animal Research, 136
    Identifying Gene-Social Environment Interactions Affecting Health
        and Disease, 139
    Future Issues, 149

8   STUDY DESIGN AND ANALYSIS FOR ASSESSMENT
    OF INTERACTIONS   161
    Definitions of Interactions, 161
    Research Designs for Evaluating Interactions, 168
    Statistical Issues Common to All Research Designs, 173
    Conclusion, 176

9 INFRASTRUCTURE   181
Education, 181
Mechanisms of Support, 186
Data, 187
Incentives and Rewards—NIH and Academe, 194
Peer Review, 198
Conclusion, 200

10 ETHICAL, LEGAL, AND SOCIAL IMPLICATIONS   202
Conveying Complex Scientific Findings Accurately, 203
Policy Does Not Inexorably Follow from Scientific
   Discoveries, 205
Ethical Implications for Research, 210

11 CONCLUSION   219

APPENDIXES
A   METHODOLOGY: DATA COLLECTION AND
     ANALYSIS   223

B   RECOMMENDATIONS FROM THE NATIONAL
     ACADEMY OF SCIENCES/NATIONAL ACADEMY OF
     ENGINEERING/INSTITUTE OF MEDICINE REPORT
     FACILITATING INTERDISCIPLINARY RESEARCH   232

C   SOCIAL, ENVIRONMENTAL, AND GENETIC
     INFLUENCES ON OBESITY AND OBESITY
     PROMOTING BEHAVIORS: FOSTERING
     RESEARCH INTEGRATION   236

D   THE INTERACTION OF SOCIAL, BEHAVIORAL, AND
     GENETICS FACTORS IN SICKLE CELL DISEASE   281

E   MODERN EPIDEMIOLOGIC APPROACHES TO
     INTERACTION: APPLICATIONS TO THE STUDY OF
     GENETIC INTERACTIONS   310

F   ACRONYMS   338

G   BIOGRAPHICAL SKETCHES   342

INDEX   351

Summary

During the twentieth century, great strides were made in reducing disease and improving the health of individuals and populations. Public health measures such as sanitation, improved hygiene, and vaccines led to major reductions in mortality and morbidity (Turnock, 2001). Increased attention to the hazards of the workplace resulted in reduced injuries and better health for workers (IOM, 2003a). Advances in biomedical research helped expand knowledge of disease and spurred the development of new clinical and pharmaceutical interventions. More recently, the sequencing of the human genome has provided information that holds the promise for further improving human health.

Over the years a large body of evidence has emerged indicating that social and behavioral factors such as socioeconomic status, smoking, diet, and alcohol use are important determinants of health (Berkman and Kawachi, 2000; IOM, 2000; Marmot and Wilkinson, 2006). Recent studies also suggest that examining interactions among genetic and social-environmental factors could greatly enhance understanding of health and illness. For example, Caspi and colleagues (2003) found "evidence of a gene-by-environment interaction, in which an individual's response to environmental insults is moderated by his or her genetic makeup." In a study showing how the social environment can influence biological response, Manuck et al. (2005) found that the socioeconomic status of communities is associated with variations in central nervous system serotonergic responsivity, which may have implications for the prevalence of psychological disorders and behaviors such as depression, impulsive aggression, and suicide.

1

As part of a strategy to determine how best to integrate research priorities to include an increased focus on the impact on health of interactions among social, behavioral, and genetic factors, the National Institutes of Health (NIH), Office of Behavioral and Social Sciences Research, in conjunction with the National Human Genome Research Institute and the National Institute of General Medical Sciences, requested that the Institute of Medicine undertake a study to examine the state of the science on gene-environment interactions that affect human health, with a focus on the social environment.[1] The goal of the study was to identify approaches and strategies to strengthen the integration of social, behavioral, and genetic research and to consider the relevant training and infrastructure needs. More specifically, NIH requested the following:

1. Review the state of the science on the interactions between the social environment and genetics that affect human health.
2. Develop case studies that will demonstrate how the interactions of the social environment and genetics affect health outcomes; illustrate the methodological issues involved in measuring the interactions; elucidate the research gaps; point to key areas necessary for integrating social, behavioral, and genetic research; and suggest mechanisms for overcoming barriers.
3. Identify gaps in the knowledge and barriers that exist to integrating social, behavioral, and genetic research in this area.
4. Recommend specific short- and long-term priorities for social and behavioral research on gene-social environment interactions; identify mechanisms that can be used to encourage interdisciplinary research in this area.
5. Assess workforce, resource, and infrastructure needs and make actionable recommendations on overcoming barriers and developing mechanisms to accelerate progress.

Chapter 2 of this report explores the impact of the social and cultural environment on health, examining what we know about the influences of these factors on health, and identifying the limitations of current research. Genetic factors and their impact on health are examined in Chapter 3, which focuses on what is known or theorized about the direct link between genes and health and what still must be explored to understand the environmental interactions and relative roles among genes that contribute to health and illness. The impact of behavioral factors on health is explored in Chap-

---

[1]For purposes of the study, sponsors clarified that the term *social environment* refers to the relations among people as individuals and in societies and not environmental conditions such as global warming and toxic waste, even if they result from human activities.

ter 4. While research on the impact of interactions has the potential to further the understanding of disease risk and aid in the development of effective interventions to improve the health of individuals and populations, there is a dearth of research that encompasses all three domains. Much remains to be learned about how these factors interact to impact health, including the most basic concept of defining interaction and how it can be characterized. Because greater etiological understanding is needed to identify future clinical research and develop effective interventions aimed at improving health outcomes, the committee focused its efforts on etiological research.

RECOMMENDATIONS

The recommendations discussed below are designed to explicate and facilitate research on the impact on health of interactions among social, behavioral, and genetic factors. Each recommendation is followed by a chapter number in which additional discussion related to the recommendation can be found.

Transdisciplinary Research

Contributions from research conducted over the past few decades, including the sequencing of the human genome, are pushing scientists to move beyond examining single agents of health and disease to a broader systems view, which is based on the understanding that health outcomes are the result of multiple determinants and their interactions (Lalonde, 1974; Evans and Stoddard, 1990; Kaplan et al., 2000; IOM, 2003a; IOM, 2003b). Understanding the associations between health and interactions among social, behavioral, and genetic factors requires research that embraces the systems view and includes an examination of the interactive pathways through which these factors operate to affect health.[2] Such research requires the participation of scientific investigators from a variety of different fields and a shift in focus from efforts that are dominated by single disciplines to research that involves collaborative participation of scientists with various expertise at all stages of the research process. While interdisciplinary research focuses on answering questions of mutual concern to those from various disciplines and multidisciplinary research involves research questions of both mutual and separate interest to participating investiga-

---

[2]Interactive physiological pathways pass information from the social world to genes and play a central role in understanding gene-environment interactions.

tors, transdisciplinary research "implies the conception of research questions that transcend the individual departments or specialized knowledge bases because they are intended to solve research questions that are, by definition, beyond the purview of the individual disciplines" (IOM, 2003b). Therefore, the committee makes the following recommendation:

> **Recommendation 1: Conduct Transdisciplinary, Collaborative Research.** *The NIH should develop Requests for Applications (RFAs) to study the impact on health of interactions among social, behavioral, and genetic factors and their interactive pathways (i.e., physiological). Such transdisciplinary research should involve the genuine collaboration of social, behavioral, and genetic scientists. Genuine collaboration is essential for the identification, incorporation, analysis, and interpretation of the multiple variables used.* (Chapter 6)

Key social variables which have been linked consistently and robustly to health outcomes include educational attainment, income and wealth, occupational status, social networks/social support, and work conditions. Well-established behavioral and psychological variables that affect health outcomes include tobacco/alcohol/drug use, eating behavior, physical activity, temperament, perceived stress and coping, perceived social support, emotional state, and motivation. Essential genetic factors affecting health include the DNA sequence variation, structural chromosomal changes, gene expression, epigenetic modifications, and downstream targets of gene expression.

In the search for a better understanding of genetic and environmental interactions as determinants of health, certain fundamental aspects of human identity (i.e., sex/gender and race/ethnicity) pose both a challenge and an opportunity for clarification. However because sex/gender and race/ethnicity are more complicated than they appear, they need to be considered and analyzed from a variety of perspectives, including social, cultural, psychological, historical, political, genetic, and geographic/ancestral.

Relevant physiological measurements and pathways should also be considered. Understanding the pathways through which interactions operate will aid in identifying links between major levels of organization of living systems: social groups of individuals, individuals composed of physiological systems, physiological systems composed of cells, and cells composed of molecules, especially DNA. Ultimately the results of such research may help to identify where to intervene along the causal chains and pathways between the social world and genes that cause disease in order to improve health outcomes.

Many determinants of health are not static, that is, they influence health in a variety of ways throughout the life course. For example, poverty may differentially and independently affect the health of an individual at different stages of life (e.g., in utero, during infancy and childhood, during pregnancy, or during old age). Personality traits and psychological status are also known to change over the lifespan and have potential to affect health. In addition to these well-established factors, a growing body of research has documented associations between cultural factors and health (Berkman and Kawachi, 2000). The influence of social, behavioral, and genetic factors on health involves dimensions of both time (critical stages in the life course and the effects of cumulative exposure) and the context or culture within which variables operate to influence health outcomes. Therefore, the committee makes the following recommendation:

> **Recommendation 2: Measure Key Variables Over the Life Course and Within the Context of Culture.** *The NIH should develop RFAs for studies of interactions that incorporate measurement, over the life course and within the context of culture, of key variables in the important domains of social, behavioral, and genetic factors.* (Chapter 6)

Modeling Strategies

For the most part research has taken a linear approach when examining the link between a particular set of variables (e.g., social-environmental or genetic variables) and health. Yet, there remains the need to connect and integrate knowledge across multiple determinants of health in order to understand the mechanisms of integration (for example, how social factors are translated into physiological effects on cellular responses, including changes in gene expression). Future studies should recognize that a linear approach does not reflect the integrated nature of how health outcomes are generated. Therefore, the committee recommends the following:

> **Recommendation 3: Develop and Implement New Modeling Strategies to Build More Comprehensive, Predictive Models of Etiologically Heterogeneous Disease.** *The NIH should emphasize research aimed at developing and implementing such models (e.g., pattern recognition, multivariate statistics, and systems-oriented approaches) for incorporating social, behavioral, and genetic factors and their interactive pathways (i.e., physiological) in testable models within populations, clinical settings, or animal studies.* (Chapter 6)

With approximately 30,000 genes in the human genome, most genes are likely to serve different functions at different times in different environments (McClintock et al., 2005). The ability to measure and evaluate differential gene expression has the potential to provide important insights into the study of health and disease. Alterations in DNA sequence and gene expression can be modified at different points throughout the life course, dictating variation in protein levels and functionality, as well as subsequent levels of metabolic products that are associated with those proteins. These factors can be measured through the use of genomic, transcriptomic, proteomic, and metabonomic technologies. However, further development of these technologies is needed to allow researchers to accurately study the molecular systems that interact with social and behavioral variables to influence health outcomes (see Chapter 6 for a more detailed discussion of these technologies). Thus, the committee recommends the following:

Recommendation 4: Investigate Biological Signatures. *Researchers should use genomic, transcriptomic, proteomic, metabonomic, and other high-dimensional molecular approaches to discover new constellations of genetic factors, biomarkers, and mediating systems through which interactions with social environment and behavior influence health.* (Chapter 6)

The context or culture within which individuals exist also is known to exert influence on health outcomes. Relevant social and cultural environments include not only an individual's immediate personal environment (e.g., his/her family), but also the broader social contexts such as the community in which a person resides. Health psychologists are increasingly calling attention to the critical role of sociocultural context, a necessary factor to consider if efforts to modify risk behaviors are to be effective. Different subgroups may have different genetic backgrounds, as well as varying cultural or socioeconomic characteristics that influence patterns of behavior, thereby creating a correlation between genotype and environmental exposure. Furthermore, it may be found that polymorphisms occurring in genotypes that act as destructive or protective factors for disease and health may be created, modified, or triggered by cultural and contextual factors. It is important to determine if research findings are applicable beyond a small population, and to capitalize on unique gene-environment interactions that could contribute to a broader understanding of factors, mechanisms, and processes. Therefore, the committee recommends the following:

Recommendation 5: Conduct Research in Diverse Groups and Settings. *The NIH should encourage research on the impact of interactions among social, behavioral, and genetic factors and their interactive pathways (i.e., physiological) on health that emphasizes diversity in groups and settings. Furthermore, NIH should support efforts to ensure that the findings of such research are validated by replication in independent studies, translated to patient-oriented research, conducted and applied in the context of* *public health, and used to design preventive and therapeutic approaches.* (Chapter 6)

Use of Animal Models

Animal research studies are an important complement to clinical and community-based research because they can serve as models for gene-environment interactions and pathways of human disease. Animal models can be used to conduct studies in which different aspects of social, behavioral, and genetic variables can be controlled, standardized, or manipulated to a significantly larger extent than can be accomplished in human studies. These models also allow for the invasive examination of organ-, tissue-, and region-specific mechanisms at the physiological, cellular, and molecular levels. Animals with short reproductive cycles and life spans provide an invaluable tool for conducting developmental and life-span studies, as well as breeding experiments and genetic manipulation that facilitate the elucidation of inherited traits and genetic effects. In some cases, animal models provide opportunities to establish causality through studies examining the temporal sequence of events, or studies involving removal followed by the add-back of hypothesized mediators at the genetic, protein, physiological, behavioral, or social environment level. Therefore, the committee makes the following recommendation:

Recommendation 6: Use Animal Models to Study Gene-Social Environment Interaction. *The NIH should develop RFAs that use carefully selected animal models for research on the impact on health of interactions among social, behavioral, and genetic factors and their interactive pathways (i.e., physiological).* (Chapter 7)

Research Design and Analysis

A clear formulation of the concept of "interaction" and an understanding of research designs that can be used to test for interactions are central to making progress in assessing the impact on health of interactions. Statistical tests for interaction are entirely dependent on the measurement scale (e.g., additive or multiplicative) used to evaluate the effects of different factors on health. Use of different measurement scales can lead to substantively different conclusions about whether or not interaction is present, and therefore, to different recommendations for intervention. Thus, determining the measurement scale is critical to the design of future studies and to the interpretation of their results. The choice of measurement scale should not be based on statistical convenience. Instead, it should be based on a theoretical model for disease causation that is more closely tied to biology.

Epidemiologists have built a conceptual framework for interaction based on the counterfactual model and the sufficient-component cause model (see Chapter 8 for discussion of these models). Beginning with this conceptual framework, defined at the level of an individual, it is possible to predict patterns of risk in the population when interaction is or is not present. Such an analysis leads to the conclusion that an additive scale for testing interaction more closely reflects the underlying biology than a multiplicative scale. That is, when two factors participate in the same sufficient cause (interaction defined conceptually), disease risks in individuals with both risk factors will be greater than expected from the additive effects of each risk factor alone.

Currently, most of the statistical software commonly used for epidemiologic analysis includes tests for interaction on a multiplicative scale but not on an additive scale. Thus testing for interactions on an additive scale requires the development of new, accessible statistical software. Also, tests for interaction require extremely large sample sizes; hence multisite collaborations may be required in order to assemble databases of sufficient size that are needed to assure adequate statistical power. Additionally, given the complexity in defining interaction and testing for it, new, efficient study designs should be developed for testing interaction. Therefore, the committee recommends the following:

Recommendation 7: Advance the Science of the Study of Interactions. *Researchers should base testing for interaction on a conceptual framework rather than simply the testing of a statistical model, and they must specify the scale (e.g., additive or multiplicative) used to evaluate whether or not interactions are present. If a multiplicative scale is used, consistency with an additive relation between the effects of different factors also should be evaluated. The NIH should develop RFAs for research on developing study designs that are efficient at testing interactions, including variations in interactions over time and development.* (Chapter 8)

Infrastructure

Research conducted to elaborate the impact of interactions among social, behavioral, and genetic factors on human health places several demands on the research infrastructure. This infrastructure includes the human infrastructure (e.g., education and training), data, and incentives and rewards.

The foundation of the research enterprise is the education of its researchers. Given that advances in genomics have been recent—and the challenge of incorporating genetic research with behavior and social factors is even more recent—it is likely that there are many current researchers who have gaps in their scientific training. Furthermore, training is needed for pre- and postdoctoral students. While universities (and high schools), NIH, and other funders of research training share responsibility for educating researchers, NIH, as the major funder of biomedical and behavioral research, is poised to make major contributions to training a cadre of researchers to conduct transdisciplinary research. Several existing mechanisms could be used as is or modified to facilitate the education of investigators in transdisciplinary research. Therefore, the committee recommends the following:

Recommendation 8: Expand and Enhance Training for Transdisciplinary Researchers. *The NIH should use existing and modified training tools both to reach the next generation of researchers and to enhance the training of current researchers. Approaches include individual fellowships (F31, F32) and senior fellowships (F33), transdisciplinary institutional grants (T32, T90), and short courses.* (Chapter 9)

The study of interactions presents a significant need for datasets that provide information across multiple disciplines, thus allowing the evaluation of gene-environment interactions. Datasets to study such interactions are typically large, difficult to collect, and costly. Therefore, it is important to support the development and use of datasets that can be shared among a wide audience of researchers.

Datasets that already include biological and genetic measures could be augmented to include social and behavioral variables. However, these additions must not only be feasible, but more importantly, they must be scientifically compelling. Alternatively, new datasets with the necessary variables could be developed. For example, health conditions or diseases could be identified for which there is a suspected or known genetic contribution, behavioral factors are likely to be involved, and hypotheses have been formed regarding the role of social factors.

Because there is a significant need for datasets that provide information for the three domains discussed (social, behavioral, and genetic factors), the committee recommends the following:

Recommendation 9: Enhance Existing and Develop New Datasets. *The NIH should support datasets that can be used by investigators to address complex levels of social, behavioral, and genetic variables and their interactive pathways (i.e., physiological). This should include the enhancement of existing datasets that already provide many, but not all, of the needed measures (e.g., the National Longitudinal Survey of Youth, ADDHealth) and the encouragement of their use. Furthermore, NIH should develop new* *datasets that address specific topics that have high potential for showing genetic contribution, social variability, and behavioral contributions—topics such as obesity, diabetes, and smoking.* (Chapter 9)

The report *Facilitating Interdisciplinary Research* (NAS/NAE/IOM, 2004) outlined several key conditions for effective interdisciplinary research, including "sustained and intense communication, talented leadership, appropriate reward and incentive mechanisms (including career and financial rewards), adequate time, seed funding for initial exploration, and willingness to support risky research." The committee believes that these same conditions apply to transdisciplinary research. Although aspects of university functioning, such as rewards and incentives, are not within the purview of NIH, they may ultimately affect the ability of NIH to find researchers who can conduct the kind of transdisciplinary research that is envisioned here.

One major challenge is acknowledging multiple investigators on team projects. The recent NIH announcement of plans to recognize multiple Principal Investigators represents a significant advancement in providing external recognition for members of research teams. As NIH explores such new approaches, the next step would be for universities to use that information in ways that would ensure that the impact of the incentives and rewards are felt at the campus level, such as the credit toward promotion and tenure that accrues to those who participate in such projects.

Scientific peer review of research applications also is a key step involved in the support of any area of research. It is not uncommon to hear investigators lament that transdisciplinary projects have difficulty in undergoing the peer review process. An important goal, therefore, is to ensure that transdisciplinary work is fairly reviewed and truly valued throughout the review process. It is not enough to simply place people from different disciplines on a review group. Specific steps need to be taken to ensure that reviewers will be able to appreciate the transdisciplinary nature or goals of a proposal. These steps include selecting reviewers who have engaged in transdisciplinary work and training reviewers about its importance and the differences between transdisciplinary research and other types of research.

Approaches to advance the field of transdisciplinary research need to be systematically applied toward the goal of fostering a type of research that has inherent scientific challenges and that faces specific institutional hurdles. Therefore, the committe makes the following recommendation:

Recommendation 10: Create Incentives to Foster Transdisciplinary Research. *The NIH and universities should explore ways to create incentives for the kinds of team science needed to support transdisciplinary research. Areas to address include (1) hiring, promotion, and tenure policies that acknowledge the contributions of* *collaborators on transdisciplinary teams; (2) peer review that includes reviewers who have experience with inter- or transdisciplinary research and are educated about the complexity and challenges involved in such research; (3) mechanisms for peer review of research grants that ensure the appropriate evaluation of transdisciplinary research projects; and (4) credit for collaborators in teams, such as NIH acknowledgement of co-investigators and university sharing of incentive funds.* (Chapter 9)

Ethical, Legal, and Social Implications

Several important ethical and legal issues need to be addressed when considering information produced by research assessing the impact on health of interactions among social, behavioral, and genetic factors. Although these issues apply to all types of research, they are especially sensitive when considering the transdisciplinary research discussed in this report. First is the issue of conveying complex scientific findings accurately to the public, policymakers, and other researchers. Claims about scientific findings are at times simplified and even exaggerated, sometimes because of the complexity of the concepts or because of economic and social pressures to emphasize the significance of findings in easily understandable terms. These difficulties are compounded by the fact that the media, understandably, prefers straightforward, easy-to-deliver messages. However, failure to convey the limitations and complexity of scientific findings has a significant impact, because beliefs about causation of health and disease affect the allocation of responsibility and resources, which has ethical and social implications.

Another issue of concern is the development of policy based on scientific findings. The array of factors that must be considered in deciding how to use the knowledge gained from research on gene-environment interactions in developing social policy is very broad and extends far beyond the science itself and into a variety of social and ethical considerations. To address difficulties in how individuals and groups understand complex scientific findings, as well as the potential impact such findings could have on policy development, the committee makes the following recommendations:

Recommendation 11: Communicate with Policymakers and the Public. *Researchers should (1) be mindful of public and policymakers' concerns, (2) develop mechanisms to involve and inform these constituencies, (3) avoid overstating their scientific findings, and (4) give careful consideration to the appropriate level of community involvement and the level of community oversight needed for such studies.* (Chapter 10)

In addition to research assessing the impact on health of interactions among social, behavioral, and genetic factors and their interactive pathways (i.e., physiological), improving health also requires individuals to act upon research findings. Therefore, the committee recommends the following:

Recommendation 12: Expand the Research Focus. *The NIH should develop RFAs for research that elucidate how best to encourage people to engage in health-promoting behaviors that are informed by a greater understanding of these interactions, how best to effectively communicate research results to the public and other stakeholders, and how best to inform research participants about the nature of the investigation (gene-environment interactions) and the uses of data following the study.* (Chapter 10)

According to the Criteria for IRB Approval of Research (45 CFR § 46.111(a)(7) (2006)), Institutional Review Boards (IRBs) are responsible for ensuring, where appropriate, the protection of research participants' privacy and the data regarding them. Studying the impact of interactions among social, behavioral, and genetic factors on health requires the collection of information about relevant DNA variants as well as clinical or other phenotypic information. This often includes sensitive personal behavior information and social factors. The risk to research participants could be substantial if such information is accessed by people and institutions outside the study. Given the sensitivity of such research and its implications, it is of primary importance to address the issues of data sharing and informed consent. Therefore, the committee makes the following recommendations:

Recommendation 13: Establish Data-Sharing Policies That Ensure Privacy. *IRBs and investigators should establish policies regarding the collection, sharing, and use of data that include information about (1) whether and to what extent data will be shared; (2) the level of security to be provided by all members of the research team as well as the research and administrative process; (3) the use of state-of-the-art security for collected data, including, but not limited to, NIH's Certificates of Confidentiality; (4) the use of formal criteria for identifying the circumstances under which individual research results will be revealed; and (5) how, before sharing data with others, recipients must agree to use data only in ways that are consistent with those agreed to by the research participants. Furthermore, if a mechanism to identify individual research participants is retained in the database, IRBs and investigators should consider whether to contact participants prior to initiating research on new hypotheses or other new research. (Chapter 10)*

Recommendation 14: Improve the Informed Consent Process. *Researchers should ensure that informed consent includes the following: (1) descriptions of the individual and social risks and benefits of the research; (2) the identification of which individual results participants will and will not receive; (3) the definition of the procedural protections that will be provided, including access policies and scientific and lay oversight; and (4) specific security, privacy, and confidentiality protections for protect the data and samples of research participants.* (Chapter 10)

CONCLUDING REMARKS

This report is intended to encourage and facilitate the growth of transdisciplinary research on the impact on health of interactions among social, behavioral, and genetic factors. Such research could further understanding of disease risk and aid in the development of effective interventions to improve the health of individuals and populations. Yet, achieving such understanding is not a short-term effort. Immediate priorities for action include training investigators in transdisciplinary research, expanding and developing datasets that include social, behavioral, and genetic variables (measured over the life course), developing new research strategies, and attending to the important ethical, legal, and social implications of such research. Such steps will facilitate the conduct of hypothesis-generating research to identify high-priority areas for study, which will then lead to targeted studies of interactions focused on specific health outcomes.

Health outcomes are multidetermined and result from complex interactions of many factors over time. Yet, the study of health outcomes has been driven primarily by disciplines that focus upon their own unique areas of expertise. If the study of health outcomes is to advance, investigators must break out of these disciplinary "silos" and attack the determinants of health in concert.

REFERENCES

Berkman L, Kawachi I, editors. 2000. *Social Epidemiology.* New York: Oxford University Press.

Caspi A, Sugden K, Moffitt TE, Taylor A, Craig IW, Harrington H, McClay J, Mill J, Martin J, Braithwaite A, Poulton R. 2003. Influence of life stress on depression: Moderation by a polymorphism in the 5-HTT gene. *Science* 301(5631):386-389.

Evans R, Stoddard G. 1990. Producing health, consuming health care. *Social Science Medicine* 31(12):1347-1363.

IOM (Institute of Medicine). 2000. *Promoting Health: Intervention Strategies from Social and Behavioral Research.* Washington, DC: National Academy Press.

IOM. 2003a. *The Future of the Public's Health in the 21st Century.* Washington, DC: The National Academies Press.

IOM. 2003b. *Who Will Keep the Public Healthy? Educating Health Professionals for the 21st Century.* Washington, DC: The National Academies Press.

Kaplan G, Everson S, Lynch J. 2000. The contribution of social and behavioral research to an understanding of the distribution of disease: A multilevel approach. In: Smedley B, Syme S, editors. *Promoting Health: Intervention Strategies from Social and Behavioral Research.* Washington, DC: National Academy Press. Pp. 37-80.

Lalonde M. 1974. *Perspective on the Health of Canadians.* Ottawa, ON: Ministry of Supply and Services.

Manuck SB, Bleil ME, Petersen KL, Flory JD, Mann JJ, Ferrell RE, Muldoon MF. 2005. The socio-economic status of communities predicts variation in brain serotonergic responsivity. *Psychological Medicine* 35(4):519-528.

Marmot MG, Wilkinson RD, editors. 2006. *Social Determinants of Health.* 2nd edition. Oxford, England: Oxford University Press.

McClintock MK, Conzen SD, Gehlert S, Masi C, Olopade F. 2005. Mammary cancer and social interactions: Identifying multiple environments that regulate gene expression throughout the life span. *Journals of Gerontology: Series B 60B*(Special Issue 1):32-41.

NAS/NAE/IOM (National Academy of Sciences/National Academy of Engineering/Institute of Medicine). 2004. *Facilitating Interdisciplinary Research.* Washington, DC: The National Academies Press.

Turnock BJ. 2001. *Public Health: What It Is and How It Works.* Gaithersburg, MD: Aspen Publishers, Inc.

1

Introduction

Why are some people healthy and others not? It seems a simple question. The answers, however, are complex and have to do not only with disease and illness, but also with who we are, where we live and work, and the social and economic policies of our government, all of which play a role in determining our health.

Institute of Medicine. *The Future of the Public's Health in the 21st Century, 2003.*

In recent years, attempts to determine why some people are healthy while others experience pain and illness most often have focused on the biological aspects of health. Biomedical research has contributed enormously to our knowledge of disease and to the development of new medical technologies and clinical and pharmaceutical interventions that improve the lives of so many. Most recently, the mapping of the human genome has uncovered new information about the association of genomics with disease. As Guttmacher and Collins have noted, "Genomics, which has quickly emerged as the central basic science of biomedical research, is poised to take center stage in clinical medicine as well" (Guttmacher and Collins, 2004).

Yet, over the years a large body of evidence also emerged that indicates that "almost half of all causes of mortality in the United States are linked to social and behavioral factors such as smoking, diet, alcohol use, sedentary life-style, and accidents" (IOM, 2000). Few diseases or conditions are caused purely by genetic factors; most are the result of interactions between genetic and environmental factors. Therefore, in order to continue to expand our knowledge of how to improve the health of individuals and populations, it becomes imperative to conduct research that explores how the interactions among social, behavioral, and genetic factors affect health. As a result, many are now engaged in attempts to determine how best to integrate research priorities to include a greater focus on these factors.

The National Institutes of Health (NIH), Office of Behavioral and Social Sciences Research, in conjunction with the National Human Genome

Research Institute and the National Institute of General Medical Sciences requested that the Institute of Medicine (IOM) undertake a study to examine the state of the science on gene-environment interactions that affect human health, with a focus on the social environment.[1] The study was to identify approaches and strategies to strengthen the integration of social, behavioral, and genetic research and to consider the relevant training and infrastructure needs. More specifically, the study was to:

1. Review the state of the science on the interactions between the social environment and genetics that affect human health.
2. Develop case studies that will demonstrate how the interactions of the social environment and genetics affect health outcomes; illustrate the methodological issues involved in measuring the interactions; elucidate the research gaps; point to key areas necessary for integrating social, behavioral, and genetic research; and suggest mechanisms for overcoming barriers.
3. Identify gaps in the knowledge and barriers that exist to integrating social, behavioral, and genetic research in this area.
4. Recommend specific short- and long-term priorities for social and behavioral research on gene-social environment interactions; identify mechanisms that can be used to encourage interdisciplinary research in this area.
5. Assess workforce, resource, and infrastructure needs and make actionable recommendations on overcoming barriers and developing mechanisms to accelerate progress.

In response to the NIH request, IOM established the Committee on Assessing Interactions Among Social, Behavioral, and Genetic Factors in Health. (See Appendix A for a discussion of committee methodology and Appendix G for biographical sketches of committee members.)

Assessing the impact on health of interactions among social, behavioral, and genetic factors is an emerging and complex field. Much remains to be learned about how these factors interact to impact health, including the most basic concept of defining interaction and how it can be characterized. Because there is a need for greater etiological understanding in order to identify future clinical research or develop effective interventions aimed at improving health outcomes, the committee has focused its efforts and this report on etiological research.

Of primary importance to the work of the committee was the recogni-

---

[1]For purposes of the study, sponsors clarified that the term *social environment* refers to the relations among people as individuals and in societies and not to environmental conditions such as global warming and toxic waste even if they result from human activities.

tion that multiple determinants contribute to the health of individuals and populations. Furthermore, the committee emphasized the development of a common understanding of concepts and terms crucial to advancing our understanding of the interaction of multiple determinants of health.

DETERMINANTS OF HEALTH

The definition of health has evolved over time, as has its measurement. Because health cannot be measured directly, a number of variables have been used as indicators of the concept of health. Prior to the mid-1900s, negative indicators such as mortality and disease rates were used—with the idea that the lower the rate, the healthier the population. Mortality or disease rates continue to be used as broad indicators when comparing populations—such as infant mortality rates or rates of specific diseases. However, a view of health as something much broader than the mere absence of disease has led to an evolution in thinking about the framework for health determinants.

One such framework was developed by Lalonde (1974) and includes environment, lifestyle, human biology, medical care, and health care organization as major determinants of health. The Lalonde framework recognized the importance of individual risk factors to health and led to further analysis and exploration of these factors' impact on health. A more complex model developed by Evans and Stoddart (1990) suggested a new framework for health determinants:

> It should accommodate distinctions among disease, as defined and treated by the health care system, health and functioning, as perceived and experienced by individuals, and well-being, a still broader concept to which health is an important but not the only, contributor. It should . . . permit and encourage a more subtle and more complex consideration of both behavioural and biological responses to social and physical environments.

Since the work of Evans and Stoddart, a number of models of health determinants have been developed. A 1999 IOM report explored core concepts of health, proposing a model of determinants that illustrated how *individual characteristics* (biology and life course, lifestyle and health behavior, illness behavior, personality and motivation, and values and preferences) and *environmental characteristics* (social and cultural, economic and political, physical and geographic, and health and social care) influence *health-related quality of life* (symptoms, functional status, health perceptions, and opportunity) (IOM, 1999).

Kaplan and colleagues (2000) proposed a framework that "builds bridges between levels rather than attributing primary importance to one level or another." Their multilevel approach to health determinants includes pathophysiological pathways, genetic/constitutional factors, individual risk factors, social relationships, living conditions, neighborhoods and communities, institutions, and social and economic policies as the major forces that affect health. *The Future of the Public's Health* (IOM, 2003a) and *Who Will Keep the Public Healthy?* (IOM, 2003b) emphasized that improving the health of populations requires understanding the ecology of health and the interconnectedness of the biological, behavioral, physical, and socioenvironmental spheres.

For the purposes of developing this report, the committee has focused its examination and analysis of factors on three major domains: social factors, behavioral factors, and genetic factors. Furthermore, the committee found it most useful to embrace a model that includes multiple determinants of health that are related and linked in many ways. Such a model is frequently referred to as an *ecological model*, because it emphasizes the linkages and relationships among multiple factors (or determinants) affecting health. As noted by IOM (2003b):

> An ecological model assumes that health and well-being are affected by interaction among multiple determinants including biology, behavior, and the environment. Interaction unfolds over the life course of individuals, families, and communities, and evidence is emerging that societal-level factors are critical to understanding and improving the health of the public.

An ecological model, therefore, provides the appropriate framework for assessing the impact on health of interactions among social, behavioral, and genetic factors.

Despite the fact that a complex interplay of factors influences vulnerability and resistance to disease, "the vast majority of the nation's health research resources have been directed toward biomedical research endeavors" (IOM, 2000). However, recent studies suggest that research on interactions of genetics with social-environmental factors is essential to understanding health and illness. For example, Caspi and colleagues (2003) found "evidence of a gene-by-environment interaction, in which an individual's response to environmental insults is moderated by his or her genetic makeup." In another study, Manuck et al. (2005) found that the socioeconomic status of communities is associated with variations in central nervous system serotonergic responsivity, which may have implications on the prevalence of psychological disorders and behaviors such as depression, impulsive aggression, and suicide.

Various pathway diagrams have been developed to represent the many ways in which social, behavioral, and genetic factors influence health. However, it is the committee's hope that, no matter which model one chooses, the discussion and recommendations set forth in this report will facilitate efforts to examine the interaction of multiple determinants on health, with a specific emphasis upon the interaction of the social environment and behavior with the genome.

TRANSDISCIPLINARY RESEARCH

To conduct research on interactions requires a shift in focus from research that is dominated by single disciplines—even when their work is complementary—to transdisciplinary research. *"Transdisciplinary research involves broadly constituted teams of researchers that work across disciplines in the development of the research questions to be addressed"* (IOM, 2003b). While interdisciplinary research focuses on answering a question of mutual concern to those of various disciplines, and multidisciplinary research involves research on questions of both mutual and separate interest to participating investigators, transdisciplinary research "implies the conception of research questions that transcend the individual departments or specialized knowledge bases because they are intended to solve . . . research questions that are, by definition, beyond the purview of the individual disciplines" (IOM, 2003b). Transdisciplinary research calls for the various disciplines involved to work together as a team to define the nature of the problem to be resolved. As stated in the above mentioned IOM report (2003b):

> *The practical ramifications of such an approach are that the disciplines will no longer function like "silos" that exist side-by-side, deeply rooted in their respective traditions. Rather, these disciplines will involve more broadly constituted and integrated "teams."*

In other words, in transdisciplinary research all of the disciplines involved are forced to change the ways in which they think about the problem, which in turn requires a transformation in the training of a cadre of new investigators, as well as new training for experienced investigators. Making this shift may be the most difficult challenge that is involved in studying the interaction of the genome and the social environment.

Developing teams of scientists who can engage in and conduct the necessary transdisciplinary research presents several practical difficulties. For example, researchers from different disciplines must be able to understand and value one another's language, concepts, and methods. Additionally, sources of data that support such transdisciplinary efforts must be developed or enhanced. Nowhere are the difficulties of transdisciplinary research better illustrated than when attempting to develop research efforts that addresses the impact of interactions among social, behavioral, and genetic factors. Such research could include research scientists from the fields of anthropology, sociology, psychology, genetics, molecular biology, biostatistics, and epidemiology.

Unfortunately, barriers to conducting effective transdisciplinary research exist within the institutions that prepare researchers, which are, for the most part, organized along single discipline departmental lines. In this system, promotion and rewards within the institution flow from the departments, each of which tends to value most highly the research and teaching that is conducted within its particular sphere. In addition, faculty members within these institutions do not have the knowledge and skills that are needed to engage in transdisciplinary research or teaching. Even when the results from such research emerge, there are not enough "peers" available to evaluate those results because most scientists and reviewers, while firmly grounded in their respective disciplines, are not sufficiently grounded in the other disciplines that may be involved in the research.

The report by the National Academy of Sciences/National Academy of Engineering/Institute of Medicine *Facilitating Interdisciplinary Research* (NAS/NAE/IOM, 2004) outlines several changes that are needed to foster interdisciplinary research, many of which also could be applied to attempts to facilitate transdisciplinary research, such as overcoming institutional barriers related to policies that govern hiring, promotion, tenure, and resource allocation. Furthermore, that report suggests that much can be learned from industry and national laboratories that organize research efforts around the problems they wish to address rather than by discipline. (See Appendix B for a complete list of recommendations from the report on interdisciplinary research.)

COMMISSIONED PAPERS

The committee commissioned papers to examine areas that might prove fruitful for investigation of the impact on health of the interaction among social, behavioral, and genetic factors. In the paper on obesity (see Appendix C), Myles S. Faith and Tanya V.E. Kral present evidence that genetic and social-environmental factors promote obesity through their independent influences on intermediary behavioral variables. Robert J. Thompson, Jr., in his study of sickle cell anemia (a Mendelian single-gene disorder) found that the severity of the symptoms of this disease is influenced by social and behavioral factors. He found that stress (primary related to dealing with daily hassles) and stress processing (primarily in relation to cognitive appraisals and attributions, coping methods, and family support) is associated with variability in the manifestation of sickle cell disease (see Appendix D). The paper on interactions prepared by Sharon Schwartz

(Appendix E) explores new ways of thinking about biologic interaction. A paper on immunology prepared by Steve W. Cole discusses what is known about the interaction between genes and the social environment in the context of immune system function.

Each of these papers, as well as additional analysis and synthesis of information conducted by committee members, points to the need for research on the interactions among social, behavioral, and genetic factors and their impact on health.

CONTEXT

In its charge to the committee, NIH sponsors defined *social environment* as the relations among people as both individuals and in societies. The term was not defined in a way that included environmental conditions such as global warming and toxic waste, even if they result from human activities. The committee has chosen to emphasize certain variables of the social environment as having high potential for research about interactions, both because there is a large body of evidence that examines the impact of these variables on health and because there exist well-established and well-accepted measures for the investigation of these variables. These variables are *socioeconomic status, race/ethnicity, social networks/social support,* and the *psychosocial work environment.* Furthermore, the committee determined that the life course perspective is crucial when studying interactions because, as stated in Chapter 2, "the influence of social and cultural variables on health involves dimensions of both *time* (critical stages in the life course and the effects of cumulative exposure) as well as *place* (multiple levels of exposure)." Chapter 2 explores the impact of the social and cultural environment on health, providing definitions, examining what we know about the influences of these factors on health, and identifying the limitations of current research.

Genetic factors and their impact on health are examined in Chapter 3, which focuses on what is known or theorized about the direct link between genes and health and what still must be explored to understand the environmental interactions and relative roles among genes that contribute to health and illness. This chapter describes simple Mendelian patterns of disease inheritance and also explores genetic susceptibility to disease as the consequence of the joint effects of many genes, each with small to moderate effects and often with interaction among themselves and the environment that give rise to the distribution of disease risk that is seen in a population. It also includes a discussion of epigenetic phenomenon, mechanisms of gene expression and regulation, aspects of health influenced by genetics, and the limitations of current research on the interactions of genetic factors with social and behavioral factors.

The impact of *behavioral factors* on health is explored in Chapter 4. The term *behavior* includes two components. First are observable behaviors that influence health, including smoking, drinking, drug use, diet, and exercise. Such factors are frequently referred to as risk factors. The second component includes certain psychological characteristics, including cognitive and emotional function and resilience. The discussion of behavioral and psychological states includes an examination of stress and coping, and it identifies the limits of current research on the interactions of behavioral factors with genetics and social factors.

The search for a better understanding of genetic and environmental interactions as determinants of health has revealed some fundamental yet complex aspects, or traits, of human identity that pose a challenge to researchers, but also provide an opportunity for clarification. Chapter 5 discusses two such complex traits: sex/gender and race/ethnicity. These traits are particularly useful and important because they have clear social dimensions that need to be taken into account in order to understand their impact on health, and each has genetic underpinnings to varying degrees. As discussed earlier, the committee believes that research on the impact of interactions among social, behavioral, and genetic factors on health must be conducted from a *life course* perspective:

> As a concept, a life course is defined as "a sequence of socially defined events and roles that the individual enacts over time" (Giele and Elder, 1998).

> These events and roles do not necessarily proceed in a given sequence, but rather constitute the sum total of the person's actual experience. Thus the concept of life course implies age-differentiated social phenomena distinct from uniform life-cycle states and the life span (Families.com, 2003).

Chapter 6 discusses how future research needs to reflect the integrated nature of the social and physical environment and gene function that is the salient feature of biological systems, describing the variety of models needed in light of the fact that rarely is there a one-to-one relationship between genes and a trait.

The use of animal models for understanding interactions is explored in Chapter 7, which describes what can be learned from animal models about how social systems regulate physiological systems and gene functions, presents criteria for the conduct of animal models, and describes the limitations of and power for generalizations from animal studies.

Chapter 8 explores research design and analysis approaches for the study of interactions. This chapter defines types of interactions, provides an example of the systems or pathways through which the social environment affects health—including behavioral, physiological, cellular, and genetic paths—and discusses models through which genes and the social environment could affect health. It also describes a progression of studies that could be used to study each of these models and discusses statistical issues related to testing gene-environment interactions.

Chapter 9 addresses infrastructure needs. It examines three aspects of infrastructure: education, data, and incentives and rewards. The discussion explores ways in which existing mechanisms can be focused to strengthen the infrastructure and also examines potential new mechanisms that could be developed.

Chapter 10 addresses ethical and social implications, focusing on such factors as the need for transparency in research and exploring the level of general public understanding of research on interactions, the disclosure of research results to participants, the social meaning of the research that is conducted, and the challenges of data privacy and availability. Chapter 11, the final chapter, briefly summarizes the main points of the study and presents the conclusions of the report.

GOALS OF THE REPORT

The primary goals of this report are to provide a research framework for assessing the impact of interactions among social, behavioral, and genetic factors on health, to identify and make recommendations about infrastructure needs and options, and to emphasize the importance of integrating the ethical and social implications into all research involving these interactions.

CONCLUSION

Research assessing the impact of interactions among social, behavioral, and genetic factors on health holds great promise for helping explicate some of the complex relationships between health outcomes and the myriad multiple determinants of health. The findings of such research may well assist us in devising interventions that will benefit both individuals and the larger populations and groups within society.

REFERENCES

Caspi A, Sugden K, Moffitt TE, Taylor A, Craig IW, Harrington H, McClay J, Mill J, Martin J, Braithwaite A, Poulton R. 2003. Influence of life stress on depression: Moderation by a polymorphism in the 5-HTT gene. *Science* 301(5631):386-389.

Evans R, Stoddart G. 1990. Producing health, consuming health care. *Social Science Medicine* 31(12):1347-1363.

Families.com. 2003. *Family Issues Encyclopedia (L): Life Course Theory.* [Online]. Available: issues.families.com/life-course-theory-1051-1055-iemf [accessed October 17, 2005].

Giele JZ, Elder GH Jr. 1998. *Methods of Life Course Research: Qualitative and Quantitative Approaches.* Thousand Oaks, CA: Sage.

Guttmacher AE, Collins FS. 2004. Genomic medicine: A primer. In: Guttmacher A, Collins FS, Drazen JM, editors. *Genomic Medicine: Articles from the New England Journal of Medicine.* Baltimore, MD: The Johns Hopkins University Press. Pp. 3-13.

IOM (Institute of Medicine). 1999. *Gulf War Veterans: Measuring Health.* Washington, DC: National Academy Press.

IOM. 2000. *Promoting Health: Intervention Strategies from Social and Behavioral Research.* Washington, DC: National Academy Press.

IOM. 2003a. *The Future of the Public's Health.* Washington, DC: The National Academies Press.

IOM. 2003b. *Who Will Keep the Public Healthy: Educating Health Professionals for the 21st Century.* Washington, DC: The National Academies Press.

Kaplan G, Everson S, Lynch J. 2000. The contribution of social and behavioral research to an understanding of the distribution of disease: A multilevel approach. In: Smedley B, Syme S, editors. *Promoting Health: Intervention Strategies from Social and Behavioral Research.* Washington, DC: National Academy Press. Pp. 37-80.

Lalonde M. 1974. *A New Perspective on the Health of Canadians.* Ottawa, ON: Ministry of Supply and Services.

Manuck SB, Bleil ME, Petersen KL, Flory JD, Mann JJ, Ferrell RE, Muldoon MF. 2005. The socio-economic status of communities predicts variation in brain serotonergic responsivity. *Psychological Medicine* 35(4):519-528.

NAS/NAE/IOM (National Academy of Sciences/National Academy of Engineering/Institute of Medicine). 2004. *Facilitating Interdisciplinary Research.* Washington, DC: The National Academies Press.

2

The Impact of Social and Cultural Environment on Health

DEFINING THE SOCIAL AND CULTURAL ENVIRONMENT

Health is determined by several factors including genetic inheritance, personal behaviors, access to quality health care, and the general external environment (such as the quality of air, water, and housing conditions). In addition, a growing body of research has documented associations between social and cultural factors and health (Berkman and Kawachi, 2000; Marmot and Wilkinson, 2006). For some types of social variables, such as socioeconomic status (SES) or poverty, robust evidence of their links to health has existed since the beginning of official record keeping. For other kinds of variables—such as social networks and social support or job stress—evidence of their links to health has accumulated over the past 30 years. The purpose of this chapter is to provide an overview of the social variables that have been researched as inputs to health (the so-called social determinants of health), as well as to describe approaches to their measurement and the empirical evidence linking each variable to health outcomes.

It should be emphasized at the outset that the social determinants of health can be conceptualized as influencing health at *multiple levels* throughout the *life course*. Thus, for example, poverty can be conceptualized as an exposure influencing the health of individuals at different levels of organization—within families or within the neighborhoods in which individuals reside. Moreover, these different levels of influence may co-occur and interact with one another to produce health. For example, the detrimental health impact of growing up in a poor family may be potentiated if that family also happens to reside in a disadvantaged community (where other families are poor) rather than in a middle-class community. Furthermore, poverty may differentially and independently affect the health of an individual at different stages of the life course (e.g., in utero, during infancy and childhood, during pregnancy, or during old age).

In short, the influence of social and cultural variables on health involves dimensions of both *time* (critical stages in the life course and the effects of cumulative exposure) as well as *place* (multiple levels of exposure). The contexts in which social and cultural variables operate to influence health outcomes are called, generically, the *social and cultural environment*.

THE INFLUENCE OF SOCIAL AND CULTURAL VARIABLES ON HEALTH: AN OVERVIEW OF PAST RESEARCH

In recent years, social scientists and social epidemiologists have turned their attention to a growing range of social and cultural variables as antecedents of health. These variables include SES, race/ethnicity, gender and sex roles, immigration status and acculturation, poverty and deprivation, social networks and social support, and the psychosocial work environment, in addition to aggregate characteristics of the social environments such as the distribution of income, social cohesion, social capital, and collective efficacy. Comprehensive surveys of current areas of research in the social determinants of health can be found in existing textbooks (Marmot and Wilkinson, 2006; Berkman and Kawachi, 2000). This chapter focuses on presenting the key research findings for a few selected social variables—SES, the psychosocial work environment, and social networks/ social support. These variables are highlighted because of their robust associations with health status and their well-documented and reliable methods of measuring these variables, and because there are good reasons to believe that these variables interact with both behavioral as well as inherited characteristics to influence health. Race/ethnicity, another set of important variables with robust associations to health, is addressed in Chapter 5.

SES and Health

An association between SES and health has been recognized for centuries (Antonovsky, 1967). Socioeconomic differences in health are large, persistent, and widespread across different societies and for a diverse range of health outcomes. In the social sciences, SES has been measured by three different indicators, taken either separately or in combination: educational attainment, income, and occupational status. Although these measures are moderately correlated, each captures distinctive aspects of social position, and each potentially is related to health and health behaviors through distinct mechanisms.

Educational Attainment

Education is usually assessed by the use of two standard questions that ask about the number of years of schooling completed and the educational credentials gained. The quality of education also may be relevant to health, but it is more difficult to assess accurately. An extensive literature has linked education to health outcomes, including mortality, morbidity, health behaviors, and functional limitations. The relationship between lower educational attainment and worse health outcomes occurs throughout the life course. For example, infants born to Caucasian mothers with fewer than 12 years of schooling are 2.4 times more likely to die before their first birthday than infants born to mothers with 16 or more years of education (NCHS, 1998). The pattern of association between maternal education and infant mortality has been described as a "gradient," with higher mortality risk occurring with successively lower levels of educational attainment (NCHS, 1998). A similar pattern of educational disparities is apparent for all racial/ethnic groups, including African American, Hispanic, American Indian, and Asian/Pacific Islander infants (NCHS, 1998). Steep educational gradients also are observed for children's health (e.g., cigarette smoking, sedentarism and obesity, elevated blood lead levels), health in midlife (e.g., mortality rates between the ages of 25 and 64), and at older ages (the prevalence of activity limitations resulting from chronic conditions such as diabetes and hypertension) (NCHS, 1998).

An association between education and health in observational data does not necessarily imply *causation*. For example, an association between lower educational attainment and an increased risk of premature mortality during midlife (even in longitudinal study designs) may partly reflect the influence of *reverse causation*—that is, lower educational attainment in adulthood may have been the consequence of serious childhood illness that truncated the ability of a given individual to complete his/her desired years of schooling (and which independently placed that person at higher risk of premature mortality). Alternatively, the association between education and health may partly reflect *confounding* by a third variable, such as ability, which is a prior common cause of both educational attainment and health status. Although highly unlikely, in the extreme case, if the association between education and health is entirely accounted for by confounding bias, then improving the individual's level of schooling would do nothing to improve his/her health chances.

The totality of the evidence suggests, nonetheless, that education is a causal variable in improving health. Natural policy experiments—such as the passage of compulsory schooling legislation at different times in different localities within the United States—suggest that higher levels of education are associated with better health (lower mortality) (Lleras-Muney, 2002). In addition, randomized trials of preschool education, such as the High/Scope Perry Preschool Project, indicate beneficial outcomes even in adolescence and adulthood, such as fewer teenage pregnancies, lower rates of high-school drop-out, and better earnings and employments prospects (which may independently improve health chances) (Parks, 2000; Reynolds et al., 2001). It is therefore likely that the association between schooling and health reflects both a causal effect of education on health, as well as an interaction between the level of schooling and inherited characteristics.

Several causal pathways have been hypothesized through which higher levels of schooling can improve health outcomes. They include the acquisition of knowledge and skills that promote health (e.g., the adoption of healthier behaviors); improved "health literacy" and the ability to navigate the health care system; higher status and prestige, as well as a greater sense of mastery and control, associated with a higher level of schooling (a psychosocial mechanism); as well as the indirect effects of education on earnings and employment prospects (Cutler and Lleras-Muney, 2006). Although it is not established which of these pathways matter more for health, they each are likely to contribute to the overall pattern of higher years of schooling being associated with better health status. Moreover, the evidence points to the importance of improving access to preschool education as a means of enhancing the health prospects of disadvantaged children (Acheson, 1998).

Income

The measurement of income is more complex than assessing educational attainment. Survey-based questions inquiring about income must minimally specify the following components: (a) time frame—for example monthly, annually, or over a lifetime (in general, the shorter the time frame for the assessment of income, the greater the measurement error); (b) sources, such as wages and salary, self-employment income, rent, interest and dividends, pensions and social security, unemployment benefits, alimony and near-cash sources such as food stamps; (c) unit of measurement, that is, whether income is assessed for the individual or the household (with appropriate adjustments for household size in the latter case); and (d) whether it is gross or disposable income (i.e., taking account of taxes and transfer payments). In addition to the higher rate of measurement error for income (as compared to educational attainment), this variable also is associated with higher refusal rates in surveys that are administered to the general population.

As with education, an extensive literature has documented the association between income and health. For example, even after controlling for educational attainment and occupational status, post-tax family income was associated with a 3.6-fold mortality risk among working-age adults in the Panel Study of Income Dynamics, comparing the top (>$70,000 in 1984 dollars) to the bottom (<$15,000) categories of income (Duncan et al., 2002). The association between income and mortality also has been described as a "gradient" (Adler et al., 1994). That is, the excess risks of poor health are not confined simply to individuals below the official poverty threshold of income. Rather, an individual's chances of having good health (e.g., avoiding premature mortality) improve with each incremental rise in income (although the relationship is also steepest at lower levels of income and tends to flatten out beyond incomes that are about twice the median level).

Also, as with education, the causal direction of an association between income and health does not entirely run from income → health. That is, the relationship between the two variables is acknowledged to be dynamic and reciprocal. Ill health is a potent cause of job loss and reduction in income. Indeed, income as an indicator of SES is more susceptible to reverse causation than education, which tends to be completed in early adult life prior to the onset of major causes of morbidity and functional limitations.

Nonetheless, tests of the income/health relationship in different datasets suggest that lower income is likely to be a cause of worse health status. For example, children do not normally contribute to household incomes, yet their health is strongly associated with levels of household income in both the Panel Study of Income Dynamics and the National Health Interview Surveys (Case et al., 2002). Furthermore, the adverse health effects of lower income accumulate over children's lives, so that the relationship between income and children's health becomes more pronounced as children grow older (Case et al., 2002).

An alternative possibility is that the relationship between income and health is explained by a third variable—such as inherited ability—that is associated with both socioeconomic mobility and the adoption of health maintenance behaviors. However, even inherited ability is unlikely to entirely account for the income/health association. If inherited ability is the sole explanation for the income/health relationship, we would not expect to find any association between family income and health among children who are adopted soon after birth by nonbiological parents (assuming that adoptive parents do not get to choose the children they will adopt based on their background, including their socioeconomic circumstances). Yet, in the National Health Interview Survey, the impact of family income on child health has been found to be similar among children who were adopted by nonbiological parents compared to children who were reared by their biological parents (Case et al., 2002). Other types of tests of the income/health association—such as the use of instrumental variable estimation (Ettner,

1996) and the observation of natural experiments that resulted in exogenous increases in income (Costello et al., 2003)—similarly have led to the conclusion that the effect of higher incomes on improved health status is likely to be causal.

The causal pathways linking income to health are likely to be different from those linking education to health. Most obviously, income enables individuals to purchase various goods and services (e.g., nutrition, heating, health insurance) that are necessary for maintaining health. Additionally, secure incomes may provide individuals with a psychological sense of control and mastery over their environment. (See Chapter 4 for a detailed discussion of psychological factors and health.) That said, it has also been observed that higher incomes are associated with healthier behaviors (such as wearing seatbelts and refraining from smoking in homes) that do not, in themselves, cost money (Case and Paxson, 2002). Although the causal mechanisms underlying these relationships are not clear, it has been speculated that "the lack of adequate resources strips parents of the energy necessary to wrestle children into seat belts. Poorer parents may also smoke to buffer themselves from poverty-related stress and depression" (Case and Paxson, 2002).

Debate also exists in the literature concerning whether it is *absolute* income or *relative* income that matters for health (Kawachi and Kennedy, 2002). The absolute income theory posits that an individual's level of wellbeing is determined by his/her own (absolute) level of income, and only his/her own income. Many definitions of poverty, for example, are based upon the concept of the failure to meet a minimal standard of living defined in absolute terms (e.g., the inability to afford food). By contrast, the relative income theory posits that individual health is determined by the relative distance (or gap) between a given individual's income and that of others around him/her (Kawachi and Kennedy, 2002).

The concept of relative income has been operationalized in empirical research by measures of *relative deprivation* (at the individual level) as well as by aggregate measures of income inequality (at the community level). Measures of relative deprivation involve assessments of the income distance between individuals and their comparison (or reference) group—that is defined by others who are alike with respect to age group, occupational class, or community of residence. The causal mechanisms underlying the relationship between absolute income and health are linked to the ability to access material goods and services necessary for the maintenance of health. Relative income is hypothesized to be linked to health through psychosocial stresses generated by invidious social comparisons as well as by the inability to participate fully in society because of the failure to attain normative standards of consumption. Growing evidence has suggested an association between relative deprivation (measured among individuals) and poor health

outcomes (Aberg Yngwe et al., 2003; Eibner et al., 2004). A related literature has attempted to link the societal distribution of income (as an aggregate index of relative deprivation) to individual health outcomes, although
the findings in this area remain contested (Subramanian and Kawachi,
2004; Lynch et al., 2004).

Variables other than household income also may be useful for health
research—such as assets including inherited wealth, savings, or ownership
of homes or motor vehicles (Berkman and Macintyre, 1997). While *income*
represents the *flow of resources* over a defined period, *wealth* captures the
stock of assets (minus liabilities) at a given point in time, and thus indicates
economic reserves. Measuring wealth is particularly salient for studies that
involve subjects towards the end of the life course, a time when many
individuals have retired and depend on their savings. In the Panel Study of
Income Dynamics, for example, only a weak association was seen between
post-tax family income and mortality among post-retirement-age subjects,
while measures of wealth continued to indicate a strong association with
mortality risk (Duncan et al., 2002).

Finally, measures of income, poverty, and deprivation have been extended to incorporate the dimension of *place*. Growing research, utilizing
multilevel study designs, has conceptualized economic status as an attribute
of neighborhoods (Kawachi and Berkman, 2003). These studies have revealed that residing in a disadvantaged (or high-poverty) neighborhood
imposes an additional risk to health beyond the effects of individual SES. A
recent Department of Housing and Urban Development randomized experiment in neighborhood mobility, the so-called Moving To Opportunity
study, found results consistent with observational data: Moving from a
poor to a wealthier neighborhood was associated with significant improvements in adult mental health and rates of obesity (Kling et al., 2004).
Disadvantaged neighborhoods are often characterized by adverse physical,
social, and service environments, including exposure to more air pollution
via proximity to heavy traffic, a lack of local amenities such as grocery
stores, health clinics, and safe venues for physical activity, and exposure to
signs of social disorder (Kawachi and Berkman, 2003). In other words, the
relevant social and cultural "environments" for the production of health
include not only an individual's immediate personal environment (e.g., his/
her family), but also the broader social contexts such as the community in
which a person resides.

Occupational Status

The third standard component of SES that typically is measured by
social scientists is occupational status, which summarizes the levels of prestige, authority, power, and other resources that are associated with differ-

ent positions in the labor market. Occupational status has the advantage over income of being a more permanent marker of access to economic resources.

Three main traditions can be discerned in the way in which different disciplines have approached the measurement of aspects of occupations relevant to health. In the traditional occupational health field, researchers have focused on the physical aspects of the job, such as exposure to chemical toxins or physical hazards of injury (Slote, 1987). In the fields of occupational health psychology and social epidemiology, researchers have focused on characterizing the psychosocial work environment, including measures of job security, psychological job demands and stress, and decision latitude (control over the work process) (Karasek and Theorell, 1990). Finally, the sociological tradition has tended to focus on occupational status, which includes both objective indicators (e.g., educational requirements associated with different jobs) as well as subjective indicators (e.g., the level of prestige associated with different jobs in the occupational hierarchy) (Berkman and Macintyre, 1997).

Several alternative approaches currently exist for the measurement of occupational status. For a detailed description, see Berkman and Macintyre (1997) as well as Lynch and Kaplan (2000). For example, the Edwards classification (U.S. Census Bureau, 1963) is a scheme based upon the conceptual distinction between manual and nonmanual occupations. The Edwards classification was used to demonstrate that individuals who grew up in manual (as compared to nonmanual) households during childhood and adolescence were at increased risk of developing heart disease in later adult life, independently of the individual's own attained SES (Gliksman et al., 1995). An alternative and commonly used measure of occupational status is the Duncan Socioeconomic Index (SEI), which combines subjective ratings of occupational prestige with objective measures of education and incomes associated with each occupation. SEI scores, which range from 0 to 100, were originally constructed by Duncan (1961) using data from the 1947 National Opinion Research Center study, which provided public opinions about the relative prestige rankings of representative occupations. These prestige rankings were then combined with U.S. Census information on the levels of education and incomes associated with each Census-defined occupation. The resulting SEI scores have been updated several times (Burgard et al., 2003). In the Wisconsin Longitudinal Survey of men and women who graduated from Wisconsin high schools in 1957 (53 or 54 years old in 1992-1993), Duncan SEI scores were inversely associated with self-reported health, depression, psychological well-being, and smoking status (Marmot et al., 1997).

As is the case with both education and income, an association between occupational status and health may partly reflect reverse causation. That is, ill health (e.g., depression or alcoholism) is a major cause of downward occupational mobility, as well as a constraint on upward social mobility. An individual's choice of occupation also may reflect unmeasured variables (such as ability) that simultaneously influence health status. Although the adverse health impact of job loss (e.g., through factory closure studies) is widely accepted (Kasl and Jones, 2000), fewer studies have convincingly demonstrated a causal effect of variables such as occupational prestige on health outcomes. As noted above, existing measures of occupational status such as the Duncan SEI combine measures of prestige with indicators of education and income that are thought to affect health independently. In addition, there are uncertainties regarding the optimal time point for measuring occupational status, especially since individuals change occupations over their life course. Job changes that occur earlier in people's careers are often associated with upward social mobility, while late-career changes may be related to a diminished capacity to function within demanding occupations (Burgard et al., 2003). For this reason, the frequently used "final occupation"—that is the occupation of an individual at the time of death or at the onset of disease—may not be an optimal indicator of the occupational conditions experienced over the individual's life course. Few studies have examined the health effects of occupational status over an individual's entire life course (Burgard et al., 2003), although some evidence suggests that persistently low occupational status measured at multiple time points or downward status mobility over time may be associated with worse health outcomes (Williams, 1990).

The potential pathways linking occupational status to health outcomes are again distinct from those linking either education or income to health. First, higher status (and nonmanual) occupations are less likely to be associated with hazardous exposures to chemicals, toxins, and risks of physical injury. Higher status jobs also are more likely to be associated with a healthier psychosocial work environment (Karasek and Theorell, 1990), including higher levels of control (decision latitude) as well as a greater range of skill utilization (lack of monotony). A greater sense of control in turn implies improved ability to cope with daily stress, including a reduced likelihood of deleterious coping behaviors such as smoking or alcohol abuse. Undoubtedly, a major intervening pathway between occupational status and health is through the indirect effects of higher incomes and access to a wider range of resources such as powerful social connections.

In summary, there is good evidence linking each of the major indicators of SES to health outcomes. Together, education, income, and occupation mutually influence and interact with one another over the life course to shape the health outcomes of individuals at multiple levels of social organization (the family, neighborhoods, and beyond).

Social Networks, Social Support, and Health

An independent social determinant of health is the extent, strength, and quality of our social connections with others. Recognition of the importance of social connections for health dates back as far as the work of Emile Durkheim. More recently John Bowlby (1969) maintained that secure attachments are not only necessary for food, warmth, and other material resources, but also because they provide love, security, and other *nonmaterial* resources that are necessary for normal human development (Berkman and Glass, 2000). Certain periods during the life course may be critical for the development of bonds and attachment (Fonagy, 1996). According to attachment theory, secure attachments during infancy satisfy a universal human need to form close affective bonds (Bowlby, 1969).

Two social variables are of particular interest in characterizing social relationships: social networks and social support. *Social networks* are defined as the web of person-centered social ties (Berkman and Glass, 2000). Its assessment includes the structural aspects of social relationships, such as size (the number of network members), density (the extent to which members are connected to one another), boundedness (the degree to which ties are based on group structures such as work and neighborhood), and homogeneity (the extent to which individuals are similar to one another). Its assessment also may extend to aspects including frequency of contact, extent of reciprocity, and duration. *Social support* refers to the various types of assistance that people receive from their social networks and can be further differentiated into three types: instrumental, emotional, and informational support. *Instrumental support* refers to the tangible resources (such as cash loans, labor in kind) that people receive from their social networks, while *emotional support* includes less tangible (but equally important) forms of assistance that make people feel cared for and loved (such as sharing confidences, talking over problems). *Informational support* refers to the social support that people receive in the form of valuable information, such as advice about healthy diets or tips about a new cancer screening test.

A variety of pencil-and-paper instruments exist to measure both social networks and social support; for a detailed guide, see Cohen et al. (2000). Several of these instruments have been psychometrically validated and indicate good internal consistency and test-retest reliability. However, one criticism of measurement in this area has been the lack of an established "gold standard." The variety of different measures currently in use makes it difficult to compare results across studies (Seeman, 1998).

A substantial body of epidemiological evidence has linked social networks and social support to positive physical and mental health outcomes throughout the life course (Stansfeld, 1999). Social connectedness is believed to confer generalized host resistance to a broad range of health outcomes, ranging from morbidity and mortality to functional outcomes (Cassel, 1976). Prospective epidemiological studies in adult populations have found consistently that social networks predict the risk of all-cause and cause-specific mortality (including cardiovascular disease, cancer, and traumatic causes of death) (Berkman and Glass, 2000). For mental health outcomes, a wealth of evidence indicates that social support buffers the effects of stressful life events and helps to prevent the onset of psychiatric disorders, particularly depression (Kawachi and Berkman, 2001). Both social networks and social support have been linked to better prognoses and survival following major illnesses, such as myocardial infarction, stroke, and certain types of cancer, including melanoma (Berkman and Glass, 2000). Some experimental evidence in the field of psychoneuroimmunology has suggested that social connectedness may confer host resistance against the development of infections (Cohen et al., 2000). In addition, a growing body of research has linked social support to neuroendocrine regulation. For example, the presence of a supportive caregiver among children has been shown to lower hypothalamic-pituitary-adrenal (HPA) reactivity (as measured by salivary cortisol levels) to maternal separation (Gunnar et al., 1992). Among adults, social support predicts lower levels of HPA axis and sympathetic nervous system reactivity in laboratory-based challenge paradigms (Seeman and McEwen, 1996).

The relationship between social networks/social support and health is bidirectional in two ways. First, major illnesses (such as a diagnosis of depression or HIV) can be a potent trigger of changes in social networks and social support. Depression typically results in social withdrawal, while newly diagnosed patients with HIV may find that members of their social network either avoid them (because of the associated stigma) or rally to their support. Second, social networks/social support can be both a positive and negative influence on health outcomes simultaneously. For example, it may not be health promoting to belong to one's intimate network if that network happens to be one of injection drug users. Similarly, abusive partners or abusive parents are sources of negative social support. The association between social networks/social support and health also may reflect confounding by a third variable, such as temperament or personality. (See Chapter 4 for a detailed discussion of personality and temperament.)

The most rigorous approach to overcoming the threats to causal inference (caused by endogeneity or omitted variable bias) is to conduct a randomized controlled trial. To date, however, the results of randomized trials of social support provision have been mixed. For example, recent large-scale randomized trials following major illnesses, such as myocardial infarction (Writing Committee for the ENRICHD Investigators, 2003), stroke (Glass et al., 2004), and metastatic breast cancer (Goodwin et al., 2001), have not found beneficial effects on clinical outcomes (improved survival or functional recovery). However, it is premature to conclude on the basis of these intervention trials that social support has no causal effect on health. For example, it has been pointed out that most of the observational evidence on social support has focused on support received from naturally occurring networks, while most interventions have attempted to bolster social support through strangers (e.g., patient support groups) (Cohen et al., 2000). The typical "treatment" in intervention studies also may have been of insufficient "dose" or duration to affect clinical outcomes. The bottom line seems to be that effective interventions to strengthen social support (to affect clinical outcomes) have yet to be devised (Cohen et al., 2000).

From the standpoint of mechanisms, recent research suggests that affiliative behavior has a basis in biology. Animal models point to the role of the neuropeptide oxytocin in facilitating various social behaviors such as maternal attachment and pair bonding (Zak et al., 2004). Social support and the administration of oxytocin have been shown to reduce stress responses during a public speaking task (Heinrichs et al., 2003). In the emerging field of neuroeconomics, it was recently demonstrated that the intranasal administration of oxytocin causes a substantial increase in trust among humans, thereby greatly increasing the benefits from social interactions (Kosfeld et al., 2005). If oxytocin is indeed the biological substrate for prosocial behavior, these preliminary findings suggest promising experimental and laboratory-based approaches for investigating gene-environment interactions in the association of social support and health.

The investigation of the health effects of social networks/social support can be further extended to the community level. The concept of *social capital* has been defined as the resources that are available to members of communities and other social contexts (e.g., workplaces) by virtue of the existence of a rich network of social interactions (Kawachi et al., 2004). Measures of social capital typically emphasize two components, both measured (or aggregated) to the community level. The *structural* component of social capital includes the extent and intensity of associational links and activity in society (e.g., density of civic associations; measures of informal sociability; indicators of civic engagement). The *cognitive* component assesses people's perceptions of trust, sharing, and reciprocity (Harpham et al., 2002). A growing number of multilevel studies have found an association between community stocks of social capital and individual health outcomes (e.g., mortality, self-rated health, some health behaviors) *net* of the influence of individual socioeconomic characteristics (Kawachi et al., 2004). Although causality in this area is still contested (Pearce and Smith, 2003), there are plausible grounds for supposing that a more socially cohesive community (evidenced by higher stocks of social capital) would be better able to protect the health of its members. For example, higher stocks of social capital are associated with the improved ability of communities to exercise informal social control over deviant behaviors (such as smoking and drinking by minors), as well as to undertake collective action for mutual benefit (e.g., passage of local ordinances to restrict smoking in public places). Social capital and social cohesion are therefore potentially important characteristics of the "social and cultural environment" that ultimately influence patterns of health achievement.

The Psychosocial Work Environment and Health

The psychosocial work environment—particularly exposure to job stress—has been linked to the onset of several conditions, including cardiovascular disease, musculoskeletal disorders, and mental illness (Marmot and Wilkinson, 2006). Two models of job stress have received particular attention in the literature: the job demand-control model (Karasek and Theorell, 1990) and the effort-reward imbalance model (Siegrist et al., 1986). The demand-control model posits that it is the combination of high psychological demands and low level of control (low decision authority and skill utilization) that leads to high physiological strain among workers and hence to the onset of disease (such as hypertension and cardiovascular disease) (Marmot and Wilkinson, 2006). A pencil-and-paper questionnaire to measure job demands and job control has been developed and validated for use in population-based studies (and can be accessed at www.uml.edu/Dept/WE/research/jcq).

In contrast to the demand-control model of job stress, the effort-reward imbalance model developed by Siegrist maintains that working conditions produce adverse health outcomes when the costs associated with the job (e.g., high level of effort) exceed its rewards (money, esteem, and career opportunities) (Siegrist et al, 1986). As with the demand-control model, a self-administered questionnaire has been developed and validated. Both the demand-control model and the effort-reward imbalance model have been shown to predict the incidence of cardiovascular disease and other health outcomes in longitudinal observational studies (Marmot and Wilkinson, 2006).

The relationship between job stress and health is likely to be reciprocal, however. For example, the onset of subtle illness symptoms may result in the worker switching to a less demanding job. In theory, this issue could be addressed in longitudinal studies through careful and repeated assessments of workers' health symptoms over time. On the other hand, other problems, such as omitted variable bias, can present formidable challenges to causal inference in this field. For example, some individuals may "select into" certain occupations based on temperament, personality, and innate "hardiness;" while others may "select out" of stressful jobs for the same reasons. If these third variables (temperament, hardiness) remain unmeasured, their omission may result in biased estimates of the effect of psychosocial working conditions on health outcomes. Future research in psychosocial work environment should therefore attempt to control for these variables and investigate the potential interactions between inherited individual characteristics and the psychosocial work environment in producing differential patterns of health and disease.

ASPECTS OF HEALTH INFLUENCED BY
THE SOCIAL ENVIRONMENT

Social variables potentially affect health outcomes throughout the entire spectrum of etiology: from disease onset (beginning prenatally and accumulating in their effects throughout the life course) to disease progression and survival. During each stage of the disease continuum, social-environmental variables can influence outcomes in a variety of different ways. Prior to the onset of disease, social variables might influence the risk of prenatal infections, the adoption of risky or health-promoting behaviors, or the ability to cope with adverse circumstances. Subsequent to the development of illness, social variables may determine the rate of progression of disease (or recovery) through differential rates of access to treatment, treatment adherence, coping behaviors, or "direct" effects on immune surveillance and tissue repair.

It is important to note, however, that the relevance and magnitude of the associations between social-environmental variables and health outcomes can *vary* at different points of the disease process. For example, the incidence of some cancers, notably breast cancer and melanoma, is *higher* among more advantaged SES groups, reflecting in part the underlying socioeconomic distribution of their risk factors. For breast cancer, the increased incidence among higher SES women is in part explained by reproductive factors, including earlier age at menarche, later age at first birth, and lower fertility.[1] On the other hand, *survival* following the diagnosis of breast cancer consistently favors higher SES women, due, among other things, to earlier detection and better access to effective treatment (Lochner and Kawachi, 2000). Likewise, observational evidence suggests the strong

---

[1]It should be noted that genetic factors also may apparently vary by socioeconomic group. For example, the prevalence of the BRCA1 gene mutations is higher among women of Ashkenazi Jewish descent than among other women. In turn, Americans of Ashkenazi Jewish origin tend to have a higher than average socioeconomic position than the average. Disentangling the various contributions of genes and social factors is therefore challenging (McClain et al., 2005).

role of social support in improving survival and functional recovery following major diseases (such as stroke or heart attack), but the evidence is less consistent for preventing the *incidence* of disease (where social *networks* appear to have a stronger role) (Seeman, 1998).

There also may be critical stages in the life course during which the social environment has a stronger impact on later life health outcomes. For example, the Barker hypothesis implicates the prenatal period as being particularly relevant for the later development of coronary heart disease and some cancers (Barker and Bagby, 2005). In addition, social-environmental conditions often *cumulate* over the life course, so that for example, persistent poverty may be more detrimental to health than transient poverty, and studying the *dynamic trajectories* of social variables is likely to be of additional interest in explaining patterns of health. Finally, social-environmental conditions may be *reproduced* across generations, because parents "pass on" their disadvantage to their children. For example, poor households are more likely to have sick children (Cutler and Lleras-Muney, 2006). Childhood illness can in turn truncate the educational and occupational mobility of the affected individuals. This constitutes a social mechanism—separate from a genetic mechanism—for the inheritance or transmission of disease risk. There may, of course, be gene-environment interactions involved in the ways in which these two separate influences shape the patterns of health across the life course.

LIMITATIONS OF CURRENT RESEARCH

The current state of research on social variables demonstrates incredible potential for improving our understanding of health. It also provides an excellent backdrop for contributing to the development research and the research agenda on gene-environment interactions. Specifically, benefits may result from the increased interest in understanding gene-environment interactions that may include insights into the social variables that represent important sources of variance and increased understanding about how physiological pathways for some disease processes might be modified, constrained, or moderated by environmental influences. For example, if one were interested in how stress is related to drug abuse, given the higher levels of chronic social stress, an ethnically diverse sample would be of great benefit to drawing conclusions about extremes of the stress continuum by studying African Americans who have experienced psychosocial sources such as racism and discrimination (e.g., Clark et al., 1999). Additionally, how the accumulation of stressful experiences over a lifetime impacts the relationship between stress, SES, and drug abuse would provide important additional information about how genetic mechanisms work.

CONCLUSION

There remain important unanswered questions in understanding the contribution of the social and cultural environment to health. Given the burgeoning interest in examining gene-environment interactions in health, there exists an opportunity to make a major investment in new research initiatives—parallel to current investments in genetics and molecular science—to expand our understanding of social and cultural influences on health. A research agenda for expanding the scope of such research has already been outlined by previous National Research Council reports.[2] This chapter has presented an overview of the state of the field in the measurement of social-environmental variables and our empirical understanding of the mechanisms by which these variables influence disease onset and progression. Significant opportunities are at hand to bridge the gaps in our understanding of how social and genetic factors interact and mutually influence health outcomes. The next chapter discusses the relationship of genetics and health.

REFERENCES

Aberg Yngwe M, Fritzell J, Lundberg O, Diderichsen F, Burstrom B. 2003. Exploring relative deprivation: Is social comparison a mechanism in the relation between income and health? *Social Science & Medicine* 57(8):1463-1473.

Acheson D (Chair). 1998. *Independent Inquiry into Inequalities in Health Report*. London: The Stationery Office.

Adler N, Boyce T, Chesney M, Cohen S, Folkman S, Kahn R, Syme S. 1994. Socioeconomic status and health: The challenge of the gradient. *American Psychologist* 49(1):15-24.

Antonovsky A. 1967. Social class, life expectancy and overall mortality. *Milbank Memorial Fund Quarterly* 45(2):31-73.

Barker DJ, Bagby SP. 2005. Developmental antecedents of cardiovascular disease: A historical perspective. *Journal of the American Society of Nephrology* 16(9):2537-2544.

Berkman L, Glass T. 2000. Social integration, social networks, social support, and health. In: Berkman L, Kawachi I, editors. *Social Epidemiology*. New York: Oxford University Press. Pp. 137-173.

Berkman L, Kawachi I, editors. 2000. *Social Epidemiology*. New York: Oxford University Press.

Berkman L, Macintyre S. 1997. The measurement of social class in health studies: Old measures and new formulations. In: Kogevinas M, Pearce N, Susser M, Boffetta P, editors. *Social Inequalities and Cancer*. Lyon, France: IARC Scientific Publication Number 138. Pp. 31-64.

Bowlby, J. 1969. *Attachment and Loss. Vol. 1. Attachment*. London: Hogarth Press.

---

[2]*Promoting Health: Intervention Strategies from Social and Behavioral Research*, 2000; *New Horizons in Health: An Integrative Approach*, 2001; and *Understanding Racial and Ethnic Differences in Health and Late Life*, 2004.

Burgard S, Stewart J, Schwartz J. 2003. *Occupational Status.* San Francisco, CA: MacArthur Network on SES and Health.

Case A, Paxson C. 2002. Parental behavior and child health. *Health Affairs* 21(2):164-178.

Case A, Lubotsky D, Paxson C. 2002. Economic status and health in childhood: The origins of the gradient. *American Economic Review* 92(5):1308-1334.

Cassel J. 1976. The contribution of the social environment to host resistance: The fourth Wade Hampton Frost lecture. *American Journal of Epidemiology* 104(2):107-123.

Clark R, Anderson NB, Clark VR, Williams DR. 1999. Racism as a stressor for African Americans: A biopsychosocial model. *American Psychologist* 54(10):805-816.

Cohen S, Underwood LG, Gottlieb BH. 2000. *Social Support Measurement and Intervention.* New York: Oxford University Press.

Costello EJ, Compton SN, Keeler G, Angold A. 2003. Relationships between poverty and psychopathology: A natural experiment. *Journal of the American Medical Association* 290(15):2023-2029.

Cutler D, Lleras-Muney A. 2006. *Education and Health: Evaluating Theories and Evidence.* Ann Arbor, MI: National Poverty Center.

Duncan GJ, Daly MC, McDonough P, Williams DR. 2002. Optimal indicators of socioeconomic status for health research. *American Journal of Public Health* 92(7):1151-1157.

Duncan OD. 1961. A socioeconomic index for all occupation. In: Reiss A Jr., editor. *Occupations and Social Status.* New York: Free Press. Pp. 109-138.

Eibner C, Sturn R, Gresenz CR. 2004. Does relative deprivation predict the need for mental health services? *Journal of Mental Health Policy and Economics* 7(4):167-175.

Ettner SL. 1996. New evidence on the relationship between income and health. *Journal of Health Economics* 15(1):67-85.

Fonagy P. 1996. Patterns of attachment, interpersonal relationships and health. In: Blane D, Brunner E, Wilkinson R, editors. *Health and Social Organization: Towards Health Policy for the Twenty-First Century.* London: Routledge Press. Pp. 125-151.

Glass TA, Berkman LF, Hiltunen EF, Furie K, Glymour MM, Fay ME, Ware J. 2004. The Families in Recovery from Stroke Trial (FIRST): Primary study results. *Psychosomatic Medicine* 66(6):889-897.

Gliksman MD, Kawachi I, Hunter D, Colditz GA, Manson JE, Stampfer MJ, Speizer FE, Willett WC, Hennekens CH. 1995. Childhood socioeconomic status and risk of cardiovascular disease in middle aged U.S. women: A prospective study. *Journal of Epidemiology and Community Health* 49(1):10-15.

Goodwin PJ, Leszcz M, Ennis M, Koopmans J, Vincent L, Guther H, Drysdale E, Hundleby M, Chochinov HM, Navarro M, Speca M, Hunter J. 2001. The effect of group psychosocial support on survival in metastatic breast cancer. *New England Journal of Medicine* 345(24):1719-1726.

Gunnar MR, Larson MC, Hertsgaard L, Harris ML, Brodersen L. 1992. The stressfulness of separation among nine-month-old infants: Effects of social context variables and infant temperament. *Child Development* 63(2):290-303.

Harpham T, Grant E, Thomas E. 2002. Measuring social capital within health surveys: Key issues. *Health Policy and Planning* 17(1):106-111.

Heinrichs M, Baumgartner T, Kirschbaum C, Ehlert U. 2003. Social support and oxytocin interact to suppress cortisol and subjective responses to psychosocial stress. *Biological Psychiatry* 54(12):1389-1398.

Karasek RA, Theorell T. 1990. *Healthy Work: Stress, Productivity, and the Reconstruction of Working Life.* New York: Basic Books.

Kasl S, Jones B. 2000. The impact of job loss and retirement on health. In: Berkman L, Kawachi I, editors. *Social Epidemiology.* New York: Oxford University Press. Pp. 118-136.

Kawachi I, Berkman L. 2001. Social ties and mental health. *Journal of Urban Health* 78(3):458-467.

Kawachi I, Berkman LF. 2003. *Neighborhoods and Health*. New York: Oxford University Press.

Kawachi I, Kennedy BP. 2002. *The Health of Nations*. New York: The New Press.

Kawachi I, Kim D, Coutts A, Subramanian S. 2004. Commentary: Reconciling the three accounts of social capital. *International Journal of Epidemiology* 33(4):682-690.

Kling J, Liebman J, Katz L, Sanbonmatsu L. 2004. *Moving to Opportunity and Tranquility: Neighborhood Effects on Adult Economic Self-Sufficiency and Health from a Randomized Housing Voucher Experiment*. Princeton IRS Working Paper 481. Princeton, NJ: Princeton University.

Kosfeld M, Heinrichs M, Zak PJ, Fischbacher U, Fehr E. 2005. Oxytocin increases trust in humans. *Nature* 435(7042):673-676.

Lleras-Muney A. 2002. *The Relationship Between Education and Adult Mortality in the United States*. Working Paper 8986. Cambridge, MA: NBER (National Bureau of Economic Research).

Lochner K, Kawachi I. 2000. Socioeconomic status. In: Hunter D, Colditz G, editors. *Cancer Prevention: The Causes and Prevention of Cancer*. Vol. 1. Dordrecht, Netherlands: Kluwer Academic Publishers.

Lynch J, Kaplan G. 2000. Socioeconomic position. In: Berkman L, Kawachi I, editors. *Social Epidemiology*. New York: Oxford University Press. Pp. 13-35.

Lynch J, Smith G, Harper S, Hillemeier M, Ross N, Kaplan GA, Wolfson M. 2004. Is income inequality a determinant of population health? Part 1. A systematic review. *Milbank Quarterly* 82(1):5-99.

Marmot M, Ryff CD, Bumpass LL, Shipley M, Marks NF. 1997. Social inequalities in health: Next questions and converging evidence. *Social Science & Medicine* 44(6):901-910.

Marmot MG, Wilkinson RD, editors. 2006. *Social Determinants of Health*. 2nd edition. Oxford, England: Oxford University Press.

McClain MR, Nathanson KL, Palomaki GE, Haddow JE. 2005. An evaluation of BRCA1 and BRCA2 founder mutations penetrance estimates for breast cancer among Ashkenazi Jewish women. *Genetics in Medicine: Official Journal of the American College of Medical Genetics* 7(1):34-39.

NCHS (National Center for Health Statistics). 1998. *Health, United States, 1998 with Socioeconomic Status and Health Chartbook*. Hyattsville, MD: NCHS.

Parks G. 2000. The High/Scope Perry Preschool Project. *Juvenile Justice Bulletin* 1-8.

Pearce N, Smith GD. 2003. Is social capital the key to inequalities in health? *American Journal of Public Health* 93(1):122-129.

Reynolds AJ, Temple JA, Robertson DL, Mann EA. 2001. Long-term effects of an early childhood intervention on educational achievement and juvenile arrest: A 15-year follow-up of low-income children in public schools. *Journal of the American Medical Association* 285(18):2339-2346.

Seeman T. 1998. *Social Support*. San Francisco: MacArthur Network on SES and Health.

Seeman TE, McEwen BS. 1996. Impact of social environment characteristics on neuroendocrine regulation. *Psychosomatic Medicine* 58(5):459-471.

Siegrist J, Siegrist K, Weber I. 1986. Sociological concepts in the etiology of chronic disease: The case of ischemic heart disease. *Social Science and Medicine* 22(2):247-253.

Slote, L. 1987. *Handbook of Occupational Safety and Health*. New York: Wiley.

Stansfeld S. 1999. Social support and social cohesion. In: Marmot M, Wilkinson R, editors. *Social Determinants of Health*. Oxford, England: Oxford University Press. Pp. 155-178.

Subramanian SV, Kawachi I. 2004. Income inequality and health: What have we learned so far? *Epidemiologic Reviews* 26:78-91.

U.S. Census Bureau. 1963. *Methodology and Scores of Socioeconomic Status.* Working Paper No. 15. Washington, DC: U.S. Government Printing Office.

Williams D. 1990. Socioeconomic differentials in health: A review and redirection. *Social Psychology Quarterly* 53(2):81-99.

Writing Committee for the ENRICHD Investigators. 2003. Effects of treating depression and low perceived social support on clinical events after myocardial infarction: The Enhancing Recovery in Coronary Heart Disease Patients (ENRICHD) randomized trial. *Journal of the American Medical Association* 289(23):3106-3116.

Zak PJ, Kurzban R, Matzner WT. 2004. The neurobiology of trust. *Annals of the New York Academy of Sciences* 1032:224-227.

3

Genetics and Health

Although there are many possible causes of human disease, family history is often one of the strongest risk factors for common disease complexes such as cancer, cardiovascular disease (CVD), diabetes, autoimmune disorders, and psychiatric illnesses. A person inherits a complete set of genes from each parent, as well as a vast array of cultural and socioeconomic experiences from his/her family. Family history is thought to be a good predictor of an individual's disease risk because family members most closely represent the unique genomic and environmental interactions that an individual experiences (Kardia et al., 2003). Inherited genetic variation within families clearly contributes both directly and indirectly to the pathogenesis of disease. This chapter focuses on what is known or theorized about the direct link between genes and health and what still must be explored in order to understand the environmental interactions and relative roles among genes that contribute to health and illness.

GENETIC SUSCEPTIBILITY

For more than 100 years, human geneticists have been studying how variations in genes contribute to variations in disease risk. These studies have taken two approaches. The first approach focuses on identifying the individual genes with variations that give rise to simple Mendelian patterns of disease inheritance (e.g., autosomal dominant, autosomal recessive, and X-linked) (see Table 3-1; Mendelian Inheritance in Man). The second approach seeks to understand the genetic susceptibility to disease as the con-

TABLE 3-1 Online Mendelian Inheritance in Man (OMIM) Statistics (as of May 15, 2006), Number of Entries

	Autosomal	X-Linked	Y-Linked	Mitochondrial	Total
Gene with known sequence	10,215	472	48	37	10,772
Gene with known sequence and phenotype	349	31	0	0	380
Phenotype description molecular basis known	1,710	153	2	26	1,891
Mendelian phenotype or locus, molecular basis unknown	1,384	134	4	0	1,522
Other, mainly phenotypes with suspected Mendelian basis	2,065	145	2	0	2,212
Total	15,723	9,353	56	63	16,777

SOURCE: OMIM, www.ncbi.nlm.nih.gov/Omim/mimstats.html, accessed May 15, 2006.

sequence of the joint effects of many genes. Each of these approaches will be discussed below.

In general, diseases with simple Mendelian patterns of inheritance tend to be relatively uncommon or frequently rare, with early ages of onset, such as phenylketonuria, sickle cell anemia, Tay-Sachs disease, and cystic fibrosis. In addition, some of these genes have been associated with extreme forms of common diseases, such as familial hypercholesterolemia, which is caused by mutations in the low-density lipoprotein (LDL) receptor that predispose individuals to early onset of heart disease (Brown and Goldstein, 1981).

Another example of Mendelian inheritance is familial forms of breast cancer associated with mutations in the BRCA1 and BRCA2 genes that predispose women to early onset breast cancer and often ovarian cancer. The genes identified have mutations that often are highly penetrant—that is, the probability of developing the disease in someone carrying the disease susceptibility genotype is relatively high (greater than 50 percent). These genetic diseases often exhibit a genetic phenomenon known as *allelic heterogeneity,* in which multiple mutations within the same gene (i.e., alleles) are found to be associated with the same disease. This allelic heterogeneity often is population specific and can represent the unique demographic and mutational history of the population.

In some cases, genetic diseases also are associated with *locus heterogeneity*, meaning that a deleterious mutation in any one of several genes can give rise to an increased risk of the disease. This is a finding common to many human diseases including Alzheimer's disease and polycystic kidney disease. Both allelic heterogeneity and locus heterogeneity are sources of variation in these disease phenotypes since they can have varying effects on the disease initiation, progression, and clinical severity.

Environmental factors also vary across individuals and the combined effect of environmental and genetic heterogeneity is *etiologic heterogeneity*. Etiologic heterogeneity refers to a phenomenon that occurs in the general population when multiple groups of disease cases, such as breast cancer clusters, exhibit similar clinical features, but are in fact the result of differing events or exposures. Insight into the etiology of specific diseases as well as identification of possible causative agents is facilitated by discovery and examination of disease cases demonstrating etiologic heterogeneity. The results of these studies may also highlight possible gene-gene interactions and gene-environment interactions important in the disease process. Identifying etiologic heterogeneity can be an important step toward analysis of diseases using molecular epidemiology techniques and may eventually lead to improved disease prevention strategies (Rebbeck et al., 1997).

As opposed to the Mendelian approach, the second approach to studying how variations in genes contribute to variations in disease risk focuses on understanding the genetic susceptibility to diseases as the consequence of the joint effects of many genes, each with small to moderate effects (i.e., polygenic models of disease) and often interacting among themselves and with the environment to give rise to the distribution of disease risk seen in a population (i.e., multifactorial models of disease). This approach has been used primarily for understanding the genetics of birth defects and common diseases and their risk factors. As described below, several steps are involved in developing such an understanding.

As a first step, study participants are asked to provide a detailed family history to assess the presence of familial aggregation. If individuals with the disease in question have more relatives affected by the disease than individuals without the disease, familial aggregation is identified. While familial aggregation may be accounted for through genetic etiology, it may also represent an exposure (e.g., pesticides, contaminated drinking water, or diet) common to all family members due to the likelihood of shared environment.

When there is evidence of familial aggregation, the second step is to focus research studies on estimating the *heritability* of the disease and/or its risk factors. Heritability is defined as the proportion of variation in disease risk in a population that is attributable to unmeasured genetic variations inferred through familial patterns of disease. It is a broad population-based measure of genetic influence that is used to determine whether further genetic studies are warranted, since it allows investigators to test the overarching null hypothesis that no genes are involved in determining disease risk. Twin studies and family studies are frequently used in the study of heritability.

Twin studies comparing the disease and risk factor variability of monozygotic and dizygotic twins have been a common study design used to easily estimate both genetic and cultural inheritance. Studies of monozygotic twins reared together versus those reared apart also have been important in estimating both genetic and environmental contributions to patterns of inheritance. The modeling of the sources of phenotypic variation using family studies has become quite sophisticated, allowing the inclusion of model parameters to represent the additive genetic component (i.e., polygenes), the nonadditive genetic component (i.e., genetic dominance, as well as gene-environment and gene-gene interactions), shared family environment, and individual environments. The contributions of these factors have been shown to vary by age and population.

When significant evidence of genetic involvement is established, the next step is to identify the responsible genes and the mutations that are associated with increased or decreased risk, using either genetic linkage analysis or genetic association studies. For example, in the study of birth defects, this often involves the search for chromosomal deletions, insertions, duplications, or translocations.

GENETIC LINKAGE ANALYSIS AND GENETIC ASSOCIATION STUDIES

The human genome is made up of tens of thousands of genes. With approximately 30,000 genes to choose from, assigning a specific gene or group of genes to a corresponding human disease demands a methodical approach consisting of many steps. Traditionally, the process of gene discovery begins with a linkage analysis that assesses disease within families. Linkage analyses are typically followed by genetic association studies that assess disease across families or across unrelated individuals.

Genetic Linkage Analysis

The term *linkage* refers to the tendency of genes proximally located on the same chromosome to be inherited together. Linkage analysis is one step in the search for a disease susceptibility gene. The goal of this analysis is to approximate the location of the disease gene in relation to a known genetic marker, applying an understanding of the patterns of linkage. Traditional linkage analysis that traces patterns of heredity of both the disease phenotype and genetic markers in large, high-risk families have been used to locate disease-causing gene mutations such as the breast cancer gene (BRCA1) on chromosome 17 (Hall et al., 1990).

Because the mode of inheritance is often not clear for common diseases, an alternative approach to classic linkage analysis was developed to capitalize on the basic genetic principle that siblings share half of their alleles on average. By investigating the degree of allelic sharing across their genomes, pairs of affected siblings (i.e., two or more siblings with the same disease) can be used to identify chromosomal regions that may contain genes whose variations are related to the disease being studied. If numerous sibling pairs affected by the disease of interest exhibit a greater than expected sharing of the known alleles of the polymorphic genetic marker being used, then the genetic marker is likely to be linked (that is, within close proximity along the chromosome) to the susceptibility gene responsible for the disease being studied. To find chromosomal regions that show evidence for linkage using this affected sibling pair method typically requires typing numerous affected sibships with hundreds of highly polymorphic markers uniformly positioned along the human genome (Mathew, 2001).

This approach has been widely used to identify regions of the genome thought to contribute to common chronic diseases. However, results of linkage analyses have not been consistently replicated. The inability to successfully replicate linkage findings may be a result of insufficient statistical power (that is, including an inadequate number of sibling pairs with the disease of interest) or results that included false positives in the original study. An alternate explanation could be that different populations are affected by different susceptibility genes than those that were studied originally (Mathew, 2001). Without consistent replication of results it is premature to draw conclusions about the contribution of a gene locus to a specific disease.

Upon the confirmation of a linkage, researchers can begin to search the region for the candidate susceptibility gene. The search for a single susceptibility gene for common diseases often involves examination of very large linkage regions, containing 20 to 30 million base pairs and potentially hundreds of genes (Mathew, 2001). It is also important to note, however, that while linkage mapping is a powerful tool for finding Mendelian disease genes, it often produces weak and sometimes inconsistent signals in studies of complex diseases that may be multifactorial. Linkage studies perform best when there is a single susceptibility allele at any given disease locus and generally performs poorly when there is substantial genetic heterogeneity.

Genetic Association Studies

Technological advances in high-throughput genotyping have allowed the direct examination of specific genetic differences among sizable numbers of people. Genetic association techniques are often the most efficient approach for assessing how specific genetic variation can affect disease risk. Genetic association studies, which have been used for decades, have perpetually progressed in terms of the development of new study designs (such as case-only and family-based association designs), new genotyping systems (such as array-based genotyping and multiplexing assays), and new methods used for addressing biases such as population (Haines and Pericak-Vance, 1998).

Analysis of the effects of genetic variation typically involves first the discovery of single nucleotide polymorphisms (SNPs)[1] and then the analysis of these variations in samples from populations. SNPs occur on average approximately every 500 to 2,000 bases in the human genome. The most common approach to SNP discovery is to sequence the gene of interest in a representative sample of individuals. Currently, sequencing of entire genes on small numbers of individuals (~25 to 50) can detect polymorphisms occurring in 1 to 3 percent of the population with approximately 95 percent confidence. The Human DNA Polymorphism Discovery Program of the National Institute of Environmental Health Sciences' Environmental Genome Project is one example of the application of automated DNA sequencing technologies to identify SNPs in human genes that may be associated with disease susceptibility and response to environment (Livingston et al., 2004). The National Heart, Lung, and Blood Institute's Programs in Genomic Applications also has led to important increases in our knowledge about the distribution of SNPs in key genes thought to be already biologically implicated in disease risk (i.e., biological candidate genes[2]).

Impressive and rapid advances in SNP analysis technology are rapidly redefining the scope of SNP discovery, mapping, and genotyping. New array-based genotyping technology enables "whole genome association" analyses of SNPs between individuals or between strains of laboratory animal species (Syvanen, 2005). Arrays used for these analyses can represent hundreds of thousands of SNPs mapped across a genome (Klein et al.,

---

[1]An SNP is the DNA sequence variation that occurs when a single nucleotide (A, T, C, or G) in the genome sequence is altered (Smith, 2005).

[2]A candidate gene is a gene whose protein product is involved in the metabolic or physiological pathways associated with a particular disease (IOM, 2005).

2005; Hinds et al., 2005; Gunderson et al., 2005). This approach allows rapid identification of SNPs associated with disease and susceptibility to environmental factors. The strength of this technology is the massive amount of easily measurable genetic variation it puts in the hands of researchers in a cost-effective manner ($500 to $1,000 per chip). The criteria for the selection of SNPs to be included on these arrays are a critical consideration, since they affect the inferences that can be drawn from using these platforms. Of course, the ultimate tool for SNP discovery and genotyping is individual whole genome sequencing. Although not currently feasible, the rapid advancement of technology now being stimulated by the National Human Genome Research Institute's "$1,000 genome" project likely will make this approach the optimal one for SNP discovery and genotyping in the future.

With the ability to examine large quantities of genetic variations, researchers are moving from investigations of single genes, one at a time, to consideration of entire pathways or physiological systems that include information from genomic, transcriptomic, proteomic, and metabonomic levels that are all subject to different environmental factors (Haines and Pericak-Vance, 1998). However, these genome- and pathway-driven study designs and analytic techniques are still in the early stages of development and will require the joint efforts of multiple disciplines, ranging from molecular biologists to clinicians to social scientists to bioinformaticians, in order to make the most effective use of these vast amounts of data.

GENE-ENVIRONMENT AND GENE-GENE INTERACTIONS

The study of gene-environment and gene-gene interactions represents a broad class of genetic association studies focused on understanding how human genetic variability is associated with differential responses to environmental exposures and with differential effects depending on variations in other genes. To illustrate the concept of gene-environment interactions, recent studies that identify genetic mutations that appear to be associated with differential response to cigarette smoke and its association with lung cancer are reviewed below. Tobacco smoke contains a broad array of chemical carcinogens that may cause DNA damage. There are several DNA repair pathways that operate to repair this damage, and the genes within this pathway are prime biological candidates for understanding why some smokers develop lung cancers but others do not. In a study by Zhou et al. (2003), variations in two genes responsible for DNA repair were examined for their potential interaction with the level of cigarette smoking and concomitant association with lung cancer. Briefly, one putatively functional mutation in the XRCC1 (X-ray cross-complementing group 1) gene and two putatively functional mutations in the ERCC2 (excision repair cross-complementing group 2) gene were genotyped in 1,091 lung cancer cases and 1,240 controls. When the cases and controls were stratified into heavy smokers versus nonsmokers, Zhou et al. (2003) found that nonsmokers with the mutant XRCCI genotype had a 2.4 times greater risk of lung cancer than nonsmokers with the normal genotype. In contrast, heavy smokers with the mutant XRCCI genotype had a 50 percent reduction in lung cancer risk compared to their counterparts with the more frequent normal genotype. When the three mutations from these two genes were examined together in the extreme genotype combination (individual with five or six mutations present in his/her genotype) there was a 5.2 time greater risk of lung cancer in nonsmokers and a 70 percent reduction of risk in the heavy smokers compared to individuals with no mutations. The protective effect of these genetic variations in heavy smokers may be caused by the differential increase in the activity of these protective genes stimulated by heavy smoking. Similar types of gene-smoking interactions also have been found for other genes in this pathway, such as ERCC1. These studies illustrate the importance of identifying the genetic variations that are associated with the differential risk of disease related to human behaviors. Note that this type of research also raises many different kinds of ethical and social issues, since it identifies susceptible subgroups and protected subgroups of subjects by both genetic and human behavior strata (see Chapter 10).

The study by Zhou et al. (2003) also demonstrates the increased information provided by jointly examining the effects of multiple mutations on toxicity-related disease. Other studies of mutations in genes involved in the Phase II metabolism (GSTM1, GSTT1, GSTP1) also have demonstrated the importance of investigating the joint effects of mutations (Miller et al., 2002) on cancer risk. Although these two studies focused on the additive effects of multiple genes, gene-gene interactions are another important component to develop a better understanding of human susceptibility to disease and to interactions with the environment.

To adequately understand the continuum of genomic susceptibility to environmental agents that influences the public's health, more studies of the joint effects of multiple mutations need to be conducted. Advances in bioinformatics can play a key role in this endeavor. For example, methods to screen SNP databases for mutations in transcriptional regulatory regions can be used for both discovery and functional validation of polymorphic regulatory elements, such as the antioxidant regulatory element found in the promoter regions of many genes encoding antioxidative and Phase II detoxification enzymes (Wang et al., 2005). Comparative sequence analysis methods also are becoming increasingly valuable to human genetic studies, because they provide a means to rank order SNPs in terms of their potential deleterious effects on protein function or gene regulation (Wang et al., 2004). Methods of performing large-scale analysis of nonsynonymous SNPs to predict whether a particular mutation impairs protein function (Clifford et al., 2004) can help in SNP selection for genetic epidemiological studies and can be used to streamline functional analysis of mutations that are found to be statistically associated with differential response to environmental factors such as diet, stress, and socioeconomic factors.

MECHANISMS OF GENE EXPRESSION

Identifying genes whose variations are associated with disease is just the first step in linking genetics and health. Understanding the mechanisms by which the gene is expressed and how it is influenced by other genes, proteins, and the environment is becoming increasingly important to the development of preventive, diagnostic, and therapeutic strategies.

When genes are expressed, the chromosomal DNA must be transcribed into RNA and the RNA is then processed and transported to be translated into protein. Regulating the expression of genes is a vital process in the cell and involves the organization of the chromosomal DNA into an appropriate higher-order chromatin structure. It also involves the action of a host of specific protein factors (to either encourage or suppress gene expression), which can act at different steps in the gene expression pathway.

In all organisms, networks of biochemical reactions and feedback signals organize developmental pathways, cellular metabolism, and progression through the cell cycle. Overall coordination of the cell cycle and cellular metabolism results from feed-forward and feedback controls arising from sets of dependent pathways in which the initiation of events is dependent on earlier events. Within these networks, gene expression is controlled by molecular signals that regulate when, where, and how often a given gene is transcribed. These signals often are stimulated by environmental influences or by signals from other cells that affect the gene expression of many genes through a single regulatory pathway. Since a regulatory gene can act in combination with other signals to control many other genes, complex branching networks of interactions are possible (McAdams and Arkin, 1997).

Gene regulation is critical because by switching genes on or off when needed, cells can be responsive to changes in environment (e.g., changes in diet or activity) and can prevent resources from being wasted. Variation in the DNA sequences associated with the regulation of a gene's expression are therefore likely candidates for understanding gene-environment interactions at the molecular level, since these variations will affect whether an environmental signal transduced to the nucleus will successfully bind to the promoter sequence in the gene and stimulate or repress gene expression. Combining genomic technologies for SNP genotyping with high-density gene expression arrays in human studies has only recently elucidated the extent to which this type of molecular gene-environment interaction may be occurring. Cells also regulate gene expression by post-transcriptional modification; by allowing only a subset of the mRNAs to go on to translation; or by restricting translation of specific mRNAs to only when and where the product is needed. The genetic factors that influence post-transcriptional control are much more difficult to study because they often involve multiprotein complexes not easily retrieved or assayed from cells. At other levels, cells regulate gene expression through *epigenetic* mechanisms, including DNA folding, histone acetylation, and methylation (i.e., chemical modification) of the nucleotide bases. These mechanisms are likely to be influenced by genetic variations in the target genes as well as variations manifested in translated cellular regulatory proteins. Gene regulation occurs throughout life at all levels of organismal development and aging.

A classic example of developmental control of gene expression is the differential expression of embryonic, fetal, and adult hemoglobin genes (see Box 3-1). The regulation of the epsilon, delta, gamma, alpha, and beta genes occurs through DNA methylation that is tightly controlled through developmental signals. During development a large number of genes are turned on and off through epigenetic regulation. One of the fastest growing fields in genetics is the study of the developmental consequences of environmental exposures on gene expression patterns and the impact of genetic variations on these developmental trajectories.

An Example of a Single-Gene Disorder with Significant Clinical Variability: Sickle Cell Disease[3]

Sickle cell disease refers to an autosomal recessive blood disorder caused by a variant of the β-globin gene called sickle hemoglobin (Hb S). A single nucleotide substitution (T→A) in the sixth codon of the β-globin gene results in the substitution of valine for glutamic acid (GTG→GAG), which can cause Hb S to polymerize (form long chains) when deoxygenated (Stuart and Nagel, 2004). An individual inheriting two copies of Hb S (Hb SS) is considered to have sickle cell *anemia*, while an individual inheriting one copy of Hb S plus another deleterious β-globin variant (e.g., Hb C or Hb β-thalassemia) is considered to have sickle cell *disease*. An individual is considered to be a carrier of the sickle cell trait if he/she has one copy of the

---

[3]The sickle cell example is abstracted from a commissioned paper prepared by Robert J. Thompson, Jr., Ph.D. (Appendix D).

BOX 3-1
Gene Expression and Globin

The production of hemoglobin is regulated by a number of transcriptional controls, such as switching, that dictate the expression of a different set of globin genes in different parts of the body throughout the various stages of the development process. This transcriptional regulation of globin genes is a result of many different DNA sequences and methylation of those sequences. The process begins shortly after conception when the yolk of the egg sac expresses genes that are responsible for the embryonic hemoglobin are deactivated, while the genes responsible for producing fetal hemoglobin in the liver are activated. Upon birth, the adult globin genes are activated and the bone marrow stem cells begin to produce adult hemoglobin and red blood cells (Rimoin et al., 2002).

A group of diseases that are the result of defective switching among the globin genes during the development process are called thalassemias. This class of diseases results in the decreased capacity to carry oxygen due to the complete absence of hemoglobin or the production of abnormal hemoglobin. Two types of thalassemias, alpha and beta, are the product of ineffective gene regulation. The

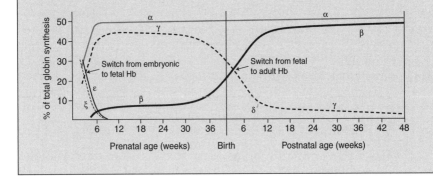

normal β-globin gene and one copy of the sickle variant (Hb AS) (Ashley-Koch et al., 2000).

Four major β-globin gene haplotypes have been identified. Three are named for the regions in Africa where the mutations first appeared: BEN (Benin), SEN (Senegal), and CAR (Central African Republic). The fourth haplotype, Arabic-India, occurs in India and the Arabic peninsula (Quinn and Miller, 2004).

Disease severity is associated with several genetic factors (Ashley-Koch et al., 2000). The highest degree of severity is associated with Hb SS, followed by Hb s/β0-thalassemia, and Hb SC. Hb S/β⁺-thalassemia is associated with a more benign course of the disease (Ashley-Koch et al., 2000). Disease severity also is related to β-globin haplotypes, probably due to globin genes activated during fetal stages of development are often not completely deactivated following the birth of individuals affected by thalassemia. Although it is not nearly as effective as the hemoglobin produced by the bone marrow, the remaining globin activation in the liver cells offers an additional source of necessary oxygen suppliers to the cells (Rimoin et al., 2002).

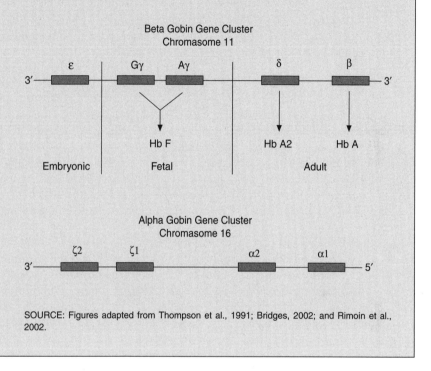

SOURCE: Figures adapted from Thompson et al., 1991; Bridges, 2002; and Rimoin et al., 2002.

variations in hemoglobin level and fetal hemoglobin concentrations. The Senegal haplotype is the most benign form, followed by the Benin, and the Central African Republic haplotype is the most severe form (Ashley-Koch et al., 2000).

Thus, although sickle cell disease is a monogenetic disorder, its phenotypic expression is multigenic (see Appendix D). There are two cardinal pathophysiologic features of sickle cell disease—chronic hemolytic anemia and vasoocclusion. Two primary consequences of hypoxia secondary to vasoocclusive crisis are pain and damage to organ systems. The organs at greatest risk are those in which blood flow is slow, such as the spleen and bone marrow, or those that have a limited terminal arterial blood supply, including the eye, the head of the femur and the humerus, and the lung as the recipient of deoxygenated sickle cells that escape the spleen or bone marrow. Major clinical manifestations of sickle cell disease include painful events, acute chest syndrome, splenic dysfunction, and cerebrovascular accidents.

Efforts to enhance clinical care are focusing on increasing our understanding of the pathophysiology of sickle cell disease in order to facilitate a precise prognosis and individualized treatment. Required is knowledge about which genes are associated with the hemolytic and vascular complications of sickle cell disease and how variants of these genes interact among themselves and with their environment (Steinberg, 2005).

ASPECTS OF HEALTH INFLUENCED BY GENETICS

Because every cell in the body, with rare exception, carries an entire genome full of variation as the template for the development of its protein machinery, it can be argued that genetic variation impacts all cellular, biochemical, physiological, and morphological aspects of a human being. How that genetic variation is associated with particular disease risk is the focus of much current research. For common diseases such as CVD, hypertension, cancer, diabetes, and many mental illnesses, there is a growing appreciation that different genes and different genetic variations can be involved in different aspects of their natural history. For example, there are likely to be genes whose variations are associated with a predisposition toward the initiation of disease and other genes or gene variations that are involved in the progression of a disease to a clinically defined endpoint. Furthermore, an entirely different set of genes may be involved in how an individual responds to pharmaceutical treatments for that disease. There also are likely to be genes whose variability controls how much or how little a person is likely to be responsive to the environmental risk factors that are associated with disease risk. Finally, there are thought to be genes that affect a person's overall longevity that may counteract or interact with genes that may otherwise predispose that person to a particular disease outcome and thus may have an additional impact on survivorship.

In many ways, we are only at the beginning the process of developing a true understanding of how genomic variations give rise to disease susceptibility. Indeed many would argue that, without incorporating the equally important role of the environment, we will never fully understand the role of genetics in health. As progress is made through utilizing the new technologies for measuring biological variation in the genome, transcriptome, proteome, and metabonome, we are likely to have to make large shifts in our conceptual frameworks about the roles of genes in disease. Global patterns of genomic susceptibility are likely to emerge only when we consider the influence of the many interacting components working simultaneously that are dependent on contexts such as age, sex, diet, and physical activity that modify the relationship with risk. For the most part, we are still at the stage of documenting the complexity, finding examples and types of genetic susceptibility genes, understanding disease heterogeneity, and postulating ways to develop models of risk that use the totality of what we know about human biology, from our genomes to our ecologies to model risk.

Cardiovascular Disease (CVD)

The study of CVD can be used to illustrate the issues that are encountered in using genetic information in order to understand the etiology of the most common chronic diseases as well as in identifying those at highest risk of developing these diseases. The majority of CVD cases have a complex multifactorial etiology, and even full knowledge of an individual's genetic makeup cannot predict with certainty the onset, progression, or severity of disease (Sing et al., 2003). Disease develops as a consequence of interactions between a person's genotype and exposures to environmental agents, which influence cardiovascular phenotypes beginning at conception and continuing throughout adulthood. CVD research has found many high-risk environmental agents and hundreds of genes, each with many variations that are thought to influence disease risk. As the number of interacting agents involved increases, a smaller number of cases of disease will be found to have the same etiology and be associated with a particular genotype (Sing et al., 2003). The many feedback mechanisms and interactions of agents from the genome through intermediate biochemical and physiological subsystems with exposure to environmental agents contribute to the emergence of a given individual's clinical phenotype. In attempting to sort out the relative contributions of genes and environment to CVD, a large array of factors must be considered, from the influence of genes on cholesterol (e.g., LDL levels) to psychosocial factors such as stress and anger. Although hundreds of genes have been implicated in the initiation, progression, and clinical manifestation of CVD, relatively little is known about how a person's environment interacts with these genes to tip the balance between the atherogenic and anti-atherogenic processes that result in clinically manifested CVD. Please see Chapters 4 and 6 for further discussion of effects of social environment on CVD.

It is well known that many social and behavioral factors ranging from socioeconomic status, job stress, and depression, to smoking, exercise, and diet affect cardiovascular disease risk (see Chapters 2, 3, and 6 for more detailed discussion of these factors). As more studies of gene-environment interaction consider these factors as part of the "environment," which are examined in conjunction with genetic variations, multiple intellectual and methodological challenges arise. First, how are the social factors embodied such that an interaction with a particular genotype can be associated with differential risk? Second, how can we handle complex interactions to address questions, such as how does an individual's genotype influence his/her behavior? For example, one's genetic susceptibility to nicotine addiction is actually a risk factor for CVD and its effect on CVD risk may be contingent on interactions with other genetic factors.

Pharmacogenetics

It has been well established that individuals often respond differently to the same drug therapy. The drug disposition process is a complex set of physiological reactions that begin immediately upon administration. The drug is absorbed and distributed to the targeted areas of the body where it interacts with cellular components, such as receptors and enzymes, that further metabolize the drug, and ultimately the drug is excreted from the body (Weinshilboum, 2003). At any point during this process, genetic variation may alter the therapeutic response of an individual and cause an adverse drug reaction (ADR) (Evans and McLeod, 2003). It has been estimated that 20 to 95 percent of variations in drug disposition, such as ADRs, can be attributed to genetic variation (Kalow et al., 1998; Evans and McLeod, 2003).

Sensitivity to both dose-dependent and dose-independent ADRs can have roots in genetic variation. Polymorphisms in kinetic and dynamic factors, such as cytochrome P450 and specific drug targets can cause these individuals susceptibilities to ADRs. While the characteristics of the ADR dictate the true significance of these factors, in most cases, multiple genes are involved (Pirmohamed and Park, 2001). Future analyses using genomewide SNP profiling could provide a technique for assessing several genetic susceptibility factors for ADRs and ascertaining their joint effects. One of the challenges to the study of the relationship between genetic variation and ADRs is an inadequate number of patient samples. To remedy this problem, Pirmohamed and Park (2001) have proposed that prospective randomized controlled clinical trials become a part of standardized practice to ultimately prove the clinical utility of genotyping all patients as a measure to prevent ADRs.

Here we review some of the current work in pharmacogenetics as an example of what might be expected to arise from rigorous study of the interaction between social, behavioral, and genetic factors. Researchers have provided a few well-established examples of differences in individual drug response that have been ascribed to genetic variations in a variety of cellular drug disposition machinery, such as drug transporters or enzymes responsible for drug metabolism (Evans and McLeod, 2003). For example:

• With the knowledge that the HER2 gene is overexpressed in approximately one fourth of breast cancer cases, researchers developed a
humanized monoclonal antibody against the HER2 receptor in hopes of
inhibiting the tumor growth associated with the receptor. Genotyping advanced breast cancer patients to identify those with tumors that overexpress
the HER2 receptor has produced promising results in improving the clinical
outcomes for these breast cancer patients (Cobleigh et al., 1999).

• A therapeutic class of drugs called thiopurines is used as part of the
treatment regimen for childhood acute lymphoblastic leukemia. One in 300
Caucasians has a genetic variation that results in low or nonexistent levels
of thiopurine methyltransferase (TPMT), an enzyme that is responsible for
the metabolism of the thiopurine drugs. If patients with this genetic variation are given thiopurines, the drug accumulates to toxic levels in their body
causing life-threatening myelosuppression. Assessing the TPMT phenotype
and genotype of the patient can be used to determine the individualized
dosage of the drug (Armstrong et al., 2004).

• The family of liver enzymes called cytochrome P450s plays a major
role in the metabolism of as many as 40 different types of drugs. Genetic
variants in these enzymes may diminish their ability to effectively break
down certain drugs, thus creating the potential for overdose in patients
with less active or inactive forms of the cytochrome P450 enzyme. Varying
levels of reduced cytochrome P450 activity is also a concern for patients
taking multiple drugs that may interact if they are not properly metabolized
by well-functioning enzymes. Strategies to evaluate the activity level of
cytochrome P450 enzymes have been devised and are valuable in planning
and monitoring successful drug therapy. Some pharmaceutical drug trials
are now incorporating early tests that evaluate the ability of differing forms
of cytochrome P450 to metabolize the new drug compound (Obach et al.,
2006).

Some pharmacogenetics research has focused on the treatment of psychiatric disorders. With the introduction of a class of drugs known as
selective serotonin re-uptake inhibitors (SSRIs), pharmacological treatment
of many psychiatric disorders changed drastically. SSRIs offer significant
improvements over the previous generation of treatments, including improved efficacy and tolerance for many patients. However, not all patients
respond positively to SSRI treatment and many experience ADRs. New
pharmacogenetic studies have indicated that these ADRs may be the result
of genetic variations in serotonin transporter genes and cytochrome P450
genes. Further study and replication of these findings are necessary. If the
characterization of the genetic variations is completed and is fully understood it would be possible to screen and monitor patients using genotyping

techniques to create individualized drug therapies similar to those discussed above (Mancama and Kerwin, 2003).

A significant challenge to the development of individualized drug therapies is the often polygenic or multifactorial inherited component of drug responses. Isolating the polygenic determinants of the drug responses is a sizable task. A good understanding of the drug's mechanism of action and metabolic and disposition pathways should be the basis of all investigations. This knowledge can aid in directing genome-wide searches for gene variations associated with drug effects and subsequent candidate-gene approaches of investigation. Additionally, proteomic and gene-expression profiling studies are also important ways to substantiate and understand the pathways by which the gene of interest operates to affect the individual's response to the drug (Evans and McLeod, 2003). It is not enough to show an association; characterization of the underlying biological mechanisms is an essential component of moving genetic findings into the area of risk reduction. Another key component of utilizing genetics to improve prevention and reduce disease is an understanding of the distribution of the genetic variations in the populations being served.

GENETICS OF POPULATONS AS RELATED TO
HEALTH AND DISEASE

Human populations differ in their distribution of genetic variations. This is a consequence of their historical patterns of mutation, migration, reproduction, mating, selection, and genetic drift. Inherited mutations typically occur during gametogenesis within a single individual and then can be passed on to offspring for many generations. Whether that mutation goes on to become a prevalent polymorphism (i.e., a mutation with a population frequency of greater than 1 percent) is determined by both evolutionary forces and chance events. For example, it depends on whether the original child who inherited the mutation survives to adulthood and reproduces and whether that child's children survive to reproduce, and so on. The number of children in a family also influences the prevalence of the mutation, and this is often tied to environmental factors that impact fertility and mating patterns that influence the speed with which a private mutation becomes a public polymorphism. There are well-known examples of what are called founder mutations in which this trajectory can be documented. For example, one particular district in what is Quebec (Canada) today was originally founded by only a few families from a particular French province. One of the founding fathers carried a 10kb deletion in his LDL receptor (LDL-R) gene that was passed down through the generations quickly and today is carried by 1 in 154 French Canadians in northeastern Quebec. This mutation is associated with familial hypercholesterolemia, and French Canadians have one of the highest prevalences of this disease in the world because of the small founding populations followed by population expansion (Moorjani et al., 1989).

There are also a number of examples where mutations that arise in an individual become more prevalent because of the selective advantage they impart on their carriers. The best known example is the mutation associated with sickle cell anemia. The geographical pattern of this mutation strongly mirrors the geographical pattern of malarial infection. It has been molecularly demonstrated that individuals carrying the sickle cell mutation have a resistance to malarial infection. Because many of the selection pressures that may have given rise to the current distribution of mutations in particular populations are in our evolutionary past, it is difficult to assess how much variation within or among populations is due to these types of selection forces.

Another major force in determining the distribution of genetic variations within and among human populations is their migration and reproductive isolation. According to our best knowledge, one of the most important periods in human evolution occurred approximately 100,000 years ago, when some humans migrated to other continents from the African basin and established new communities with relative reproductive isolation. Genetic differences among people in different geographical areas have been associated with the concept of race for hundreds of years. Although race is still used as a label, the original concept of race as genetically distinct subspecies of humans has been rejected through modern genetic information. For numerous reasons, discussed in the section below, it is more appropriate to reconceptualize the old genetics of race into a more accurate genetics of ancestry.

In addition to distant evolutionary patterns of migration, more modern migration patterns also have had a profound effect on the genetics of populations. For example, the current population of the United States and much of North America is very diverse genetically as a consequence of the mixing of many people from many different countries and continents.

A central reason for studying the origins and nature of human genetic variation is that the similarities and differences in the type and frequencies of genetic variations within and among populations can have a profound impact on studies that attempt to understand the influence of genes on disease risk. For example, some genetic variations, such as the apolipoprotein E protein polymorphisms, are found in every population and have very similar genotype frequencies around the world (Wu et al., 2002; Deniz Naranjo et al., 2004). The variation's association with increased heart disease and Alzheimer's disease could be and has been tested in many of the world's populations. Other mutations such as the 10kb deletion in the LDL-R gene described above are more population-specific variations.

Furthermore, from a statistical point of view, the effect of a genetic variation on the continuum of risk found in any population is correlated with its frequency. For example, common genetic polymorphisms with frequencies near 50 percent cannot be associated with large phenotypic effects within a population because the genotype classes each represent a large fraction of the population and, since most risk is normally distributed, the average risk for a highly prevalent genotype class cannot deviate from the overall risk of the population to any large degree. This correlation between genotype frequency and effect does not mean that common variations cannot be significant in their effects. The statistical significance of an association between a genetic variant and a disease is a joint function of sample size and the size of the effect. In addition, genetic research among populations that differ in their genotype frequencies can differ in their inferences about which polymorphisms have significant effects even if the absolute phenotypic effect is the same. See Cheverud and Routman (1995) for a more formal statistical explanation of this phenomenon and its impact on assessing gene-gene interactions.

Another key consideration in understanding the relationship between genetic variations and measures of disease risk is the population differences in the correlations between genotype frequencies at different SNP locations. There are two common reasons why the frequency of an allele or genotype at a particular SNP could be correlated with the frequency of an allele or genotype for a different SNP. First, a phenomenon known as *linkage disequilibrium* creates correlations among SNPs as a consequence of the mutation's history. When mutations arise, they occur on a particular genetic background, which creates a correlation with the other SNPs on the chromosome. Second, the mixing of populations known as admixture that occurs typically through migration means that SNPs with population-specific frequencies will be correlated in a larger mixed sample. In this case, *population stratification* is the cause of the correlation, and there has been much genetic epidemiological research on this phenomenon and how to control for it. Population stratification is thought to be a possible source of spurious genetic associations with disease (see Box 3-2).

CONCLUSION

In large part, the twentieth century was dominated by studies of human health and disease that focused on identifying single genetic and environmental agents that could explain variation in disease susceptibility. This new century has been characterized by huge advances in our understanding of Mendelian disorders with severe clinical outcomes. However, the Men-

BOX 3-2
Population Stratification (Confounding)

When the risk of disease varies between two ethnic groups, any genetic or environmental factor that also varies between the groups will appear to be related to disease. This phenomenon is called "population stratification" in epidemiologic studies investigating the effect of a genetic factor on disease, and it is a form of confounding. Population stratification refers to the presence of subgroups—for example ethnic groups—in the sample, which could potentially cause a spurious association between genetic variations and trait. Concerns about population stratification have raised doubts about the credibility of some reported findings in candidate gene studies and have led to calls for the routine use of related controls in case-control studies of genetic factors to eliminate the possibility of population stratification (Lander and Schork, 1994; Altshuler et al., 1998). In fact, although population stratification is frequently used as an explanation for nonreplicable associations in the literature, there are few actual examples to support this assumption (Risch, 2000) and many agree that the problem probably has been overstated (Cardon and Bell, 2001). For example, Wacholder et al. (2000) argued that population stratification to an extent large enough to distort results is unlikely to occur in many realistic situations. Population stratification is a manifestation of confounding—that is, the distortion of the relationship between the exposure of interest and disease due to the effect of a true risk factor that is related to the exposure (Wacholder et al., 2000). Thus, in population stratification ethnicity acts as a surrogate for the true risk factor, which may be environmental or genetic. This means that controlling for ethnicity can reduce the confounding bias.

Ardlie et al. (2002) evaluated four moderately sized case-control studies for the presence of population structure and concluded that carefully matched case-control samples in U.S. and European populations are unlikely to contain levels of population stratification that would result in significantly inflated numbers of false positive associations. However, methods have been developed by which unlinked genetic markers can be used to detect stratification and even correct for it when it is present (Pritchard and Rosenberg, 1999; Satten et al., 2001).

delian paradigm has failed to elucidate the genetic contribution to susceptibility to most common chronic diseases, which researchers know have a substantial genetic component because of their familial aggregation and studies that demonstrate significant heritabilities for these diseases. Likewise, environmental and social epidemiological studies have been wildly successful in illuminating the role of many environmental factors such as diet, exercise, and stress on disease risk. However, these environmental factors still do not, by themselves, fully explain the variance in the prevalence of several diseases in different populations. Researchers are only now beginning to study in earnest the potential interactions between the genetic and environmental factors that are likely to be contributing to a large fraction of disease in most populations. There is much that can be done to incorporate measures of social environment into genetic studies and to also incorporate genetic measures into social epidemiological studies.

Over the last two decades, progress in identifying specific genes and mutations that explain genetic susceptibility to common conditions has been relatively slow, for a variety of reasons. First, the diseases being studied tend to be complex in their etiology, meaning that different people in a population will develop disease for different genetic and/or environmental reasons. Any single genetic or environmental factor is expected to explain only a very small fraction of disease risk in a population. Moreover, these factors are expected to interact, and other biological processes (e.g., epigenetic modifications) are likely to be contributors to the complex puzzle of susceptibility. An accurate phenotypic definition of disease and its subtypes is crucial to identifying and understanding the complexities of disease-specific genetic and environmental causes.

Second, geneticists only recently have developed the knowledge base or methods needed to measure genetic variations and their metabolic consequences with sufficient ease and cost-effectiveness so that the large number of genes thought to be involved can be studied. With the completion of the Human Genome Project in 2003, many different scientific entities (e.g., the Environmental Genome Project and the International HapMap Consortium) have been working to identify the mutational spectra in human populations, and genetic epidemiologists are just now beginning to understand the extensive nature of common variations (>1 percent population frequency) within the human genome that could be affecting people's risk of disease. The SNP data generated by these initiatives are now centrally located in a number of public databases, including the National Center for Biotechnology Information's dbSNPs database, the National Cancer Institute's CGAP Genetic Annotation Initiative SNP Database, and the Karolinska Institute Human Genic Bi-Allelic Sequences Database. At present, the largest dataset on human variation is being generated by the International HapMap Project,[4] which is genotyping millions of SNPs on 270 individuals from 4 geographically separated sites from around the world. The International HapMap Project has greatly increased the number of validated SNPs available to the research community to be used to study human variation and is producing a map of genomic haplotypes in four populations with ancestry from parts of Africa, Asia, and Europe. In addition, high-throughput methods of genotyping large numbers of SNPs (thousands) in large epidemiological cohorts are only now becoming available

---

[4]See www.hapmap.org.

(see above). Unfortunately, high-throughput methods of measuring the environment have not kept a similar pace. For many studies of common disease, a rate-limiting step to increasing our understanding will continue to be the difficult and costly measurement of environmental factors.

Finally, progress also has been hampered because of a lack of adequate investment in developing new methods of analysis that can incorporate the high-dimensional biological reality that we can now measure. The complex genetic and environmental architecture of multifactorial diseases is not easily detected or deciphered using the traditional statistical modeling methods that are focused on the estimation of a single overall model of disease for a population. For example, using traditional logistic regression methods it would be simply impossible to enter all the hundreds of genetic variations that are thought to be involved in CVD risk or in any of the other common disease complexes currently being studied. Beyond the obvious issues of power and overdetermination in such a large-scale model, we also do not know how to model or interpret interactions among many factors simultaneously or how to incorporate the rare, large effects of some genes relative to the common, small effects of others. New modeling strategies that take advantage of advances in pattern recognition, machine learning, and systems analysis (e.g., scale-free networks, Bayesian belief networks, random forest methods) are going to be needed in order to build more comprehensive, predictive models of these etiologically heterogeneous diseases.

The field of human genetics, like many other disciplines, is in transition, and there is much to be gained by joining forces with a wide range of other disciplines that are focused on improving prevention and reducing the disease burden in our populations.

REFERENCES

Altshuler D, Kruglyak L, Lander E. 1998. Genetic polymorphisms and disease. *New England Journal of Medicine* 338(22):1626.

Ardlie KG, Lunetta KL, Seielstad M. 2002. Testing for population subdivision and association in four case-control studies. *American Journal of Human Genetics* 71(2):304-311.

Armstrong VW, Shipkova M, von Ahsen N, Oellerich M. 2004. Analytic aspects of monitoring therapy with thiopurine medications. *Therapeutic Drug Monitoring* 26(2):220-226.

Ashley-Koch A, Yang Q, Olney R. 2000. Sickle hemoglobin (Hb S) allele and sickle cell disease: A HuGE review. *American Journal of Epidemiology* 151(9):839-845.

Bridges, K. 2002. *Hemoglobinopathies (Hemoglobin Disorders)*. [Online]. Available: sickle. bwh.harvard.edu/hemoglobinopathy.html [accessed May 15, 2006].

Brown MS, Goldstein JL. 1981. Lowering plasma cholesterol by raising LDL receptors. *New England Journal of Medicine* 305(9):515-517.

Cardon LR, Bell JI. 2001. Association study designs for complex diseases. *Nature Reviews Genetics* 2(2):91-99.

Cheverud JM, Routman EJ. 1995. Epistasis and its contribution to genetic variance components. *Genetics* 139(3):1455-1461.

Clifford RJ, Edmonson MN, Nguyen C, Buetow KH. 2004. Large-scale analysis of non-synonymous coding region single nucleotide polymorphisms. *Bioinformatics* 20(7):1006-1014.

Cobleigh MA, Vogel CL, Tripathy D, Robert NJ, Scholl S, Fehrenbacher L, Wolter JM, Paton V, Shak S, Lieberman G, Slamon DJ. 1999. Multinational study of the efficacy and safety of humanized anti-HER2 monoclonal antibody in women who have HER2-overexpressing metastatic breast cancer that has progressed after chemotherapy for metastatic disease. *Journal of Clinical Oncology* 17(9):2639-2648.

Deniz Naranjo MC, Munoz Fernandez C, Alemany Rodriguez MJ, Perez Vieitez MC, Irurita Latasa J, Suarez Armas R, Suarez Valentin MP, Sanchez Garcia F. 2004. Gender has a strong modulating effect on the risk of Alzheimer's disease conferred by the apolipoprotein E gene in the population of the Canary Islands, Spain. *Revista de Neurologia* 38(7) 615-618.

Evans WE, McLeod HL. 2003. Pharmacogenomics—drug disposition, drug targets, and side effects. *New England Journal of Medicine* 348(6):538-549.

Gunderson KL, Steemers FJ, Lee G, Mendoza LG, Chee MS. 2005. A genome-wide scalable SNP genotyping assay using microarray technology. *Nature Genetics* 37(5):549-554.

Haines JL, Pericak-Vance MA. 1998. *Approaches to Gene Mapping in Complex Human Diseases*. New York: Wiley-Liss.

Hall JM, Lee MK, Newman B, Morrow JE, Anderson LA, Huey B, King MC. 1990. Linkage of early-onset familial breast cancer to chromosome 17q21. *Science* 250(4988):1684-1689.

Hinds DA, Stuve LL, Nilsen GB, Halperin E, Eskin E, Ballinger DG, Frazer KA, Cox DR. 2005. Whole-genome patterns of common DNA variation in three human populations. *Science* 307(5712):1072-1079.

IOM (Institute of Medicine). 2005. *Implications of Genomics for Public Health*. Washington, DC: The National Academies Press.

Kalow W, Tang BK, Endrenyi L. 1998. Hypothesis: Comparisons of inter- and intra-individual variations can substitute for twin studies in drug research. *Pharmacogenetics* 8(4): 283-289.

Kardia SL, Modell SM, Peyser PA. 2003. Family-centered approaches to understanding and preventing coronary heart disease. *American Journal of Preventive Medicine* 24(2): 143-151.

Klein RJ, Zeiss C, Chew EY, Tsai JY, Sackler RS, Haynes C, Henning AK, SanGiovanni JP, Mane SM, Mayne ST, Bracken MB, Ferris FL, Ott J, Barnstable C, Hoh J. 2005. Complement factor H polymorphism in age-related macular degeneration. *Science* 308(5720): 385-389.

Lander ES, Schork NJ. 1994. Genetic dissection of complex traits. *Science* 265(5181): 2037-2048.

Livingston RJ, von Niederhausern A, Jegga AG, Crawford DC, Carlson CS, Rieder MJ, Gowrisankar S, Aronow BJ, Weiss RB, Nickerson DA. 2004. Pattern of sequence variation across 213 environmental response genes. *Genome Research* 14(10A):1821-1831.

Mancama D, Kerwin RW. 2003. Role of pharmacogenomics in individualising treatment with SSRIs. *CNS Drugs* 17(3):143-151.

Mathew C. 2001. Science medicine and the future—postgenomic technologies: Hunting the genes for common disorders. *British Medical Journal* 322(7293):1031-1034.

McAdams HH, Arkin A. 1997. Stochastic mechanisms in gene expression. *Proceedings of the National Academy of Sciences of the United States of America* 94(3):814-819.

Miller DP, Liu G, De Vivo I, Lynch TJ, Wain JC, Su L, Christiani DC. 2002. Combinations of the variant genotypes of GSTP1, GSTM1, and p53 are associated with an increased lung cancer risk. *Cancer Research* 62(10):2819-2823.

Moorjani S, Roy M, Gagne C, Davignon J, Brun D, Toussaint M, Lambert M, Campeau L, Blaichman S, Lupien P. 1989. Homozygous familial hypercholesterolemia among French Canadians in Quebec Province. *Arteriosclerosis* 9(2):211-216.

Obach RS, Walsky RL, Venkatakrishnan K, Gaman EA, Houston JB, Tremaine LM. 2006. The utility of in vitro cytochrome P450 inhibition data in the prediction of drug-drug interactions. *Journal of Pharmacology and Experimental Therapeutics* 316(1):336-348.

Pirmohamed M, Park BK. 2001. Genetic susceptibility to adverse drug reactions. *Trends in Pharmacological Sciences* 22(6):298-305.

Pritchard JK, Rosenberg NA. 1999. Use of unlinked genetic markers to detect population stratification in association studies. *American Journal of Human Genetics* 65(1):220-228.

Quinn CT, Miller ST. 2004. Risk factors and prediction of outcomes in children and adolescents who have sickle cell anemia. *Hematology/Oncology Clinics of North America* 18(6 SPEC.ISS.):1339-1354.

Rebbeck TR, Walker AH, Phelan CM, Godwin AK, Buetow KH, Garber JE, Narod SA, Weber BL. 1997. Defining etiologic heterogeneity in breast cancer using genetic biomarkers. *Progress in Clinical and Biological Research* 396:53-61.

Rimoin DL, Connor JM, Pyeritz RE, Korf BR, editors. 2002. *Emery and Rimoin's Principles and Practice of Medical Genetics Vol. 2.* 4th edition. New York: Churchill Livingstone.

Risch NJ. 2000. Searching for genetic determinants in the new millennium. *Nature* 405(6788): 847-856.

Satten GA, Flanders WD, Yang Q. 2001. Accounting for unmeasured population substructure in case-control studies of genetic association using a novel latent-class model. *American Journal of Human Genetics* 68(2):466-477.

Sing CF, Stengard JH, Kardia SLR. 2003. Genes, environment, and cardiovascular disease. *Arteriosclerosis, Thrombosis, and Vascular Biology* 23:1190-1196.

Smith. 2005. *The Genomics Age: How DNA Technology Is Transforming the Way We Live and Who We Are.* New York: AMACOM.

Steinberg MH. 2005. Predicting clinical severity in sickle cell anaemia. *British Journal of Haematology* 129(4):465-481.

Stuart MJ, Nagel RL. 2004. Sickle-cell disease. *Lancet* 364(9442):1343-1360.

Syvanen AC. 2005. Toward genome-wide SNP genotyping. *Nature Genetics* (37 Suppl):S5-S10.

Thompson MW, McInnes RR, Willard, editors. 1991. *Thompson & Thompson Genetics in Medicine.* 5th edition. Philadelphia, PA: W.B. Saunders Company.

Wacholder S, Rothman N, Caporaso N. 2000. Population stratification in epidemiologic studies of common genetic variants and cancer: Quantification of bias. *Journal of the National Cancer Institute* 92(14):1151-1158.

Wang X, Tomso DJ, Liu X, Bell DA. 2005. Single nucleotide polymorphism in transcriptional regulatory regions and expression of environmentally responsive genes. *Toxicology and Applied Pharmacology* 207(2 Suppl):84-90.

Wang Z, Fan H, Yang HH, Hu Y, Buetow KH, Lee MP. 2004. Comparative sequence analysis of imprinted genes between human and mouse to reveal imprinting signatures. *Genomics* 83(3):395-401.

Weinshilboum R. 2003. Inheritance and drug response. *New England Journal of Medicine* 348(6):529-537.

Wu JH, Lo SK, Wen MS, Kao JT. 2002. Characterization of apolipoprotein E genetic variations in Taiwanese association with coronary heart disease and plasma lipid levels. *Human Biology* 74(1)25-31.

Zhou W, Liu G, Miller DP, Thurston SW, Xu LL, Wain JC, Lynch TJ, Su L, Christiani DC. 2003. Polymorphisms in the DNA repair genes XRCC1 and ERCC2, smoking, and lung cancer risk. *Cancer Epidemiology, Biomarkers and Prevention* 12(4):359-365.

4

Genetic, Environmental, and Personality Determinants of Health Risk Behaviors

INTRODUCTION AND OVERVIEW

Tobacco use, obesity, and physical inactivity are the greatest preventable causes of morbidity and mortality in the United States (Mokdad et al., 2004). These behaviors involve motivational and reward systems within the individual that develop through gene interactions with the social environment. Therefore, a better understanding of the genetic, social environmental, and individual determinants of risk behaviors, such as tobacco use, unhealthy eating behaviors, and physical inactivity could contribute to improved strategies for primary, secondary, and tertiary disease prevention.

Models of gene, environment, and behavior interactions in disease have been proposed, one of which has been adapted here to illustrate the central role of health risk behaviors (Rebbeck, 2002). Risk behaviors such as tobacco use, unhealthy eating behaviors, and physical inactivity play an important role in models of genetic and environmental interactions in health outcomes. As illustrated in Figure 4-1, gene-environment interactions contribute to the initiation and maintenance of these risk behaviors, which in turn increase risk for poor health outcomes (pathway a). In addition, gene-environment interactions can modify the effects of these risk behaviors on disease states and health outcomes (pathway b) and also can have direct effects on health outcomes (pathway c) (see also models of gene-environment interactions in Chapter 8 and Appendix E).

The goal of this chapter is three-fold: (1) to provide a brief overview of the epidemiology of tobacco use, unhealthy diet/obesity, and physical inactivity in relation to health outcomes; (2) to describe the genetic and environ-

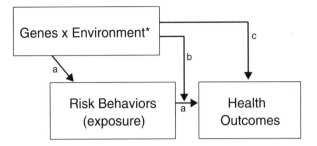

*Refers to main effects and interactions; see also Chapter 8.

FIGURE 4-1 Role of genes, environment, and risk behaviors in health (adapted from Rebbeck, 2002).

mental determinants of these risk behaviors and their underlying motivational systems; and (3) to discuss how the measurement of intermediate phenotypes (recently termed *endophenotypes*), such as personality and temperament, can advance our knowledge of the role of gene-environment interactions in risk behaviors and health.

DEFINITIONS OF HEALTH RISK BEHAVIORS

Although definitions of health risk behaviors vary across studies, there are some generally accepted definitions that will be presented for the purposes of this chapter. With regard to tobacco use, the behavioral definition of smoking used in most prevalence studies includes having smoked more than 100 cigarettes in one's lifetime and smoking every day or most days (CDC, 2005). Increasingly, studies of the determinants of tobacco use, including genetic studies, are using more refined behavioral definitions to characterize trajectories of smoking initiation and progression, as well as phenotypes related to nicotine addiction and smoking persistence (Audrain-McGovern et al., 2004b).

The definition of obesity is more straightforward. The World Health Organization (WHO) defines overweight as having a body mass index (BMI) from 25 to 30, and obesity as a BMI greater than 30 (WHO, 1998). Broadly speaking, physical activity includes any bodily muscular movements that produce energy expenditure (Caspersen et al., 1985; Pate et al., 1995). To reduce health risks, it is recommended that healthy adults engage in at least 150 minutes of moderate intensity physical activity per week (Pate et al., 1995), which can include brisk walking and some forms of aerobic exercise such as running and bicycle riding.

The importance of phenotype definition for investigations of genetic risk factors and gene-environment interaction cannot be overestimated.

Increasingly, studies are focusing on intermediate phenotypes, the intermediate measures of these health behaviors that are considered more proximal
to the biological determinants. For example, in studies of tobacco use,
laboratory-based intermediate phenotypes have included individual differences in the rewarding value of nicotine, the psychophysiological and cognitive effects of nicotine, as well as the effects of nicotine tolerance and
deprivation (Munafo et al., 2005b). In obesity studies, psychological intermediate markers have included the reinforcing value of food, food preferences, food intake, and satiety (see Appendix C for additional discussion).
As discussed in more detail below, these intermediate phenotypes also may
include the dimensions of personality and temperament that are partly
biologically based and that may increase the likelihood that an individual
will engage in health risk behaviors.

TOBACCO USE

Epidemiology and Health Consequences of Tobacco Use

Although the prevalence of tobacco use in adults has declined significantly since the Surgeon General's report in 1965, 23 percent of the American population continues to smoke (NCHS, 2003; CDC, 2004a). Smoking
rates remain higher in persons who have less than a high school education,
compared to college graduates. Furthermore, 18 percent of 13-year-olds
and 58 percent of high school students report having smoked a whole
cigarette (CDC, 2004b).

Tobacco use is the leading cause of preventable mortality in the United
States, accounting for one in five cancer deaths (CDC, 2002; Mokdad et al.,
2004). Furthermore, continued smoking following a diagnosis of cancer
increases the risk of recurrence and reduces the likelihood of survival
(Browman et al., 1993; Kawahara et al., 1998; Khuri et al., 2001; McBride
and Ostroff, 2003). The nicotine in cigarettes is known to have significant
adverse effects on cardiovascular function (Benowitz and Gourlay, 1997),
and smoking cessation following an acute myocardial infarction can reduce
mortality rates (Kinjo et al., 2005). Nicotine, thiocyanate, and other toxins
in cigarette smoke also can impair thyroid, pituitary, and renal function
and contribute to insulin resistance (Kapoor and Jones, 2005). Evidence
from rodent models suggests that nicotine also may alter antibody formation and T-cell function (Friedman and Eisenstein, 2004).

Genetic and Environmental Determinants of Tobacco Use

Motivation to begin smoking is strongly influenced by the social environment, although genetic factors also play a role (Audrain-McGovern et

al., 2004a). Risk factors for smoking initiation in youth include peer and family smoking, family conflict, and exposure to tobacco industry promotional campaigns (Pierce et al., 1998; Choi et al., 2002). In contrast, physical activity has protective effects on youth smoking (Audrain-McGovern et al., 2003a). The importance of the social environment also is supported by evidence for the efficacy of some anti-tobacco media campaigns, smoke-free environment policies, and cigarette taxes (Holm, 1979; Chaloupka et al., 2002).

Once tobacco use has been initiated, smoking cessation can be difficult because of the development of an addiction to nicotine. There is abundant evidence from animal and human studies for an inherited susceptibility to the rewarding effects of nicotine and to nicotine addiction. In fact, data from twin studies indicate that as much as 70 percent of the variance in nicotine addiction is attributable to genetic factors (Sullivan and Kendler, 1999). Investigations of the specific genetic mechanisms that underlie nicotine addiction have focused on candidate genes in neurobiological pathways that play a role in nicotine's reinforcing and addictive effects, including the dopamine, serotonin, and opioid pathways, as well as genetic variation in nicotine metabolic pathways and neuronal nicotinic receptors (Lerman and Berrettini, 2003). While several genetic associations have been reported in the literature, heterogeneity in ascertainment, population stratification, and limitations in phenotype definition have contributed to nonreplication (Lerman and Swan, 2002; Munafo and Flint, 2004; Redden et al., 2005). Given the importance of smoking persistence to health outcomes, efforts are increasing to elucidate the role of inherited genetic variation in response to pharmacotherapies for nicotine dependence (Lerman et al., 2005).

Clearly, tobacco use and nicotine addiction are complex traits arising from the interactions among social-environmental, psychological, and genetic factors (Swan et al., 2003). For example, evidence from twin studies suggests that the importance of genetic factors in cigarette smoking depends, in part, on family functioning (Kendler et al., 2004). Specifically, the heritability estimates for cigarette smoking were lower in families with reports of higher levels of family dysfunction. This finding highlights both the importance of gene-environment interactions in risk behaviors, as well as the potential for identifying and quantifying such interactions through careful research. Furthermore, the genetic effects on the progression to regular smoking among adolescents are greatest among those with higher levels of depressive symptoms (Audrain-McGovern et al., 2004a). Despite awareness of the importance of gene-environment interactions in tobacco use, few molecular genetic studies have incorporated social environmental effects, and few studies of social environment have considered whether such influences are moderated by genetic factors.

UNHEALTHY EATING BEHAVIORS AND OBESITY

Epidemiology and the Health Consequences of Obesity

The WHO defines overweight as having a BMI from 25 to 30, and obesity as a BMI greater than 30 (WHO, 1998). Based on this definition, approximately 57 percent of adult Americans are classified as being over-weight or obese (Flegal et al., 2005), and rates of obesity have increased in recent decades (Allison et al., 1999; CDC, 2000; Flegal et al., 2005). The rising prevalence of obesity in the United States has been linked to increased health risks (Harris, 1998).

Like tobacco use, obesity is a major cause of mortality in the United States, with approximately 325,000 deaths attributable to obesity among nonsmokers (Allison et al., 1999). Obesity is a major risk factor for the development of diabetes, cardiovascular disease (CVD), osteoarthritis, and many forms of cancer (Allison et al., 1999; Bianchini et al., 2002). Poor diet and obesity can also increase treatment complications and reduce the like-lihood of survival following a cancer diagnosis (Pinto et al., 2000; Rock and Demark-Wahnefried, 2002). Although the mechanisms linking obesity to these disease outcomes remains the subject of intense investigation, the adverse health outcomes result in part from alterations in the metabolism of steroid hormones, metabolic alterations including lipid and glucose levels, and increases in the turnover of free fatty acids that lead to insulin resis-tance syndrome (Seidell et al., 1994; Turcato et al., 2000; Rose et al., 2002; Eckel et al., 2002). In addition, excess adiposity has been linked to impaired immune function and increased cortisol secretion (Stallone, 1994), possibly influencing the adverse pathophysiological effects of environmental and psychological stress.

Genetic and Environmental Determinants of Unhealthy Eating Behaviors and Obesity

The development and maintenance of obesity, like tobacco use and nicotine addiction, result from a complex interplay of social, motivational, emotional, and genetic factors (Kopelman, 2000). Increases in obesity preva-lence may be largely attributable to changes in the social environment that support a sedentary lifestyle (e.g., television and video games), the promo-tion of high-calorie fast foods and "supersize" portions, and increased access to vending machines with high-calorie foods in schools and commu-nity settings (Hill and Peters, 1998). Although these environmental factors clearly increase the likelihood of feeding behaviors that lead to obesity in the population as a whole, genetic factors are thought to influence an individual's susceptibility to unhealthy feeding behaviors given a particular social environment, and to his/her likelihood of becoming obese given a particular level of energy intake and expenditure (Costanzo and Schiffman, 1989; Hill and Peters, 1998). There is abundant evidence from animal and human models for genetic contributions to obesity, with 40 to 70 percent of the variability in susceptibility to human obesity attributable to heritable factors (Comuzzie and Allison, 1998). There are single gene disorders that include obesity as part of the syndrome, such as Prader-Willi and Bardet-Biedel; however, such major genetic effects are rare. Mutations studied in rodent models of obesity that are associated with leptin abnormalities also are rare in humans (Kopelman, 2000). For example, a single gene mutation in the melanocortin 4 receptor (MC4R) is thought to account for less than 5 percent of morbid obesity (Vaisse et al., 1998). Molecular genetic studies have identified a very large number of susceptibility genes for multiple obesity phenotypes, including BMI, feeding behavior, and satiety (Comuzzie and Allison, 1998); however, the attributable risks associated with these variants remain unclear. Candidate genes identified in these studies include those coding for agouti signaling proteins, leptin and leptin receptors, and cholecystokinin A receptor (reviewed in Comuzzie and Allison, 1998). Genetic variation in the dopamine transporter and dopamine 2 receptor also has been associated with obesity in some studies (Noble et al., 1994; Epstein et al., 2002). Despite the known complex etiology of obesity, studies of genetic modulation of social environmental exposures are rare. However, there is evidence that fetal nutrition may affect gene expression, possibly altering susceptibility to diet and environmental stressors that promote obesity in later life (Barker et al., 1989; Barker, 1995).

PHYSICAL INACTIVITY

Epidemiology and the Health Consequences of Physical Inactivity

It is recommended that, to reduce health risks, healthy adults engage in at least 150 minutes of moderate intensity physical activity per week (Pate et al., 1995). Despite the positive effects of regular physical activity on breast and colon cancer (McTiernan et al., 1998) and on CVD risk factors (U.S. DHHS, 1996), approximately one-half of adult Americans do not engage in moderate physical activity for at least 30 minutes at least 3 times a week (Sullivan et al., 2005). Engaging in regular physical activity also has important benefits following a cancer diagnosis. A meta-analysis of randomized controlled trials of physical activity interventions concluded that such interventions have significant benefits for cardiovascular respiratory fitness and can reduce cancer treatment side effects (Schmitz et al., 2005b). Physical activity interventions for cancer patients also have been shown to reduce body fat and plasma levels of insulin-like growth factor (Schmitz et al., 2005a).

Genetic and Environmental Determinants of Physical Activity

Levels of physical activity are determined by a complex set of factors. Yet, these determinants are less well studied than those for tobacco use and obesity. Investigations of locomotor activity in inbred mouse strains provide evidence for significant genetic influences on activity when confined to a running wheel (Mhyre et al., 2005). However, it is yet to be determined which behavioral systems underlie this effect (e.g., reward, exploration, or motor drive) or how activity levels would be affected when the animal's environment provided the opportunity for a variety of behaviors requiring different levels of activity for different types of rewards (McClintock, 1981; Hermes et al., 2005). Environmental factors, such as food shortage, can enhance or attenuate mouse strain differences in locomotor activity in response to stimulants, underscoring the importance of gene-environment interactions (Cabib et al., 2000). In humans, approximately 30 to 60 percent of the variance in physical activity and sports participation is due to heritable factors (Perusse et al., 1989; Beunen and Thomis, 1999). A polymorphism in the MC4R gene has been implicated in physical activity levels in nonobese humans and in the general population. This association has been attributed to the role of this receptor in metabolic rate and energy expenditure, however, the precise mechanism is not yet clear (Loos et al., 2005). Features of the social environment that reinforce a sedentary lifestyle (e.g., television, video games, computers) as well as the built environment (large shopping malls located outside of the city, zoning laws prohibiting building businesses within walking distance of homes) contribute to physical inactivity and may modify the effects of genetic predisposition to inactivity.

Specific genetic factors should be examined in conjunction with known social environmental determinants (e.g., media exposure, family and peer influences). For initiation studies, there is a need to focus on critical development periods; for example, early to late adolescence for tobacco use and early childhood through adolescence for obesity and physical inactivity. For studies of persistence and behavior change, there is a need to include investigations of critical periods in adulthood when environmental transitions occur, such as young adulthood (ages 18 to 25).

USING INTERMEDIATE PHENOTYPES TO INVESTIGATE THE EFFECTS OF GENE-ENVIRONMENT INTERACTIONS

Intermediate phenotypes are traits or outcome measures that mediate the effects of gene-environment influences on risk behaviors (see Figure

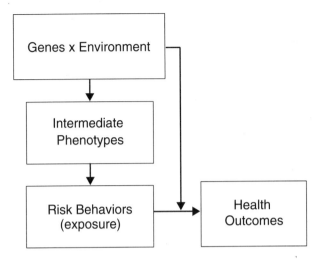

FIGURE 4-2 Intermediate phenotypes of gene-environment effects on risk behaviors and health.

4-2). Such measures tend to be more proximal to the biological determinants than are the risk behaviors themselves, and therefore, they can be assessed with greater experimental control in human models. For example, in studies of tobacco use, laboratory-based intermediate phenotypes have included individual differences in the rewarding value and tolerance of nicotine, its cognitive and autonomic effects, and the effects of nicotine deprivation (Munafo et al., 2005b). Intermediate phenotypes in obesity studies have included the reinforcing value of food, food preferences, food intake, and satiety (see also the commissioned paper on obesity in Appendix C). As discussed in more detail below, these measures also may include dimensions of personality or temperament that are partly biologically based and that may increase the likelihood that an individual will engage in health risk behaviors. In fact, some of the most convincing evidence for gene-environment interactions has been provided by research in these areas. However, while intermediate phenotypes are likely to provide useful research tools, they are quite complex and, therefore, caution should be used when extrapolating the clinical application of such research.

Measuring biological and genetic modifiers of risk also is essential, particularly for predicting whether engaging in a health risk behavior actually results in disease. For example, some people who consume large quantities of animal fat do not necessarily have proportionately high low-density lipoprotein (LDL) levels, which are associated with increased risk for CVD. The Inuit of Greenland, who eat a traditional diet of orsoq, seal, and whale fat, do not have the expected high LDL cholesterol levels or the resultant high rates of CVD (Bjerregaard et al., 1997). Likewise, polymorphisms in apolipoprotein E, a carrier protein important to liver metabolism of LDL, result in different levels of LDL and cardiovascular risk in people who eat similar diets (Miltiadous et al., 2005). A more detailed discussion of both biological and behavioral traits is provided in the following section.

Beyond Risk Behaviors

The inclusion of measurable intermediate phenotypes will assist investigators in the exploration of the relationship among gene-environment interactions, risk behaviors, and health. This may involve incorporating more extensive assessments of biologically based dimensions of personality and temperament and/or incorporating laboratory-based measures of risk behavior propensity (e.g., the rewarding value of nicotine or high-fat foods). Animal and human laboratory models can be performed in parallel to test the effects of genetic factors and environmental influences on intermediate phenotype measures (Blendy et al., 2005). Using genetic animal models and human genetic association studies to stratify populations, the genetic effects on risk behaviors can be measured in the presence and absence of key social environmental cues and stressors.

However, it is not only through risk behaviors like smoking, poor eating habits and obesity, or low exercise levels that gene-environment interactions influence health. Another key pathway that is just as important involves effects of gene-environment interactions on biological characteristics involving neuroendocrine, autonomic, cardiovascular, metabolic, inflammatory, and hemostatic functions. There are several examples in the recent literature that illustrate these gene-environment interaction effects on biomarkers.

The Lys198Asn polymorphism of the Endothelin-1 gene moderates the impact of both obesity and socioeconomic status on systolic blood pressure reactivity to an acute environmental stressor in African American and Caucasian young adults (Treiber et al., 2003). The G308A polymorphism of the TNFα gene moderates the impact of chronic environmental stress, as measured by vital exhaustion levels, on plasma levels of C-reactive protein, a potent risk factor for CVD (Jeanmonod et al., 2004). The extensively studied promoter polymorphism of the serotonin transporter gene (5HTTLPR) moderates the impact of acute mental stress on blood pressure (Williams et al., 2001), an effect that has been cited as one potential mechanism that could be mediating the reported association between the 5HTTLPR long allele and increased risk of myocardial infarction (Fumeron et al., 2002).

Personality and Temperament as Intermediate Phenotypes in Investigations of Risk Behaviors and Health

After many years of distrust and disuse, the concept of a personality trait is once more proving useful in many types of studies. New tools of analysis have made it possible to define and refine the idea of what personality traits actually are, and to demonstrate the universality of certain kinds of individual differences. The term *personality* captures the collective and dynamic organization of all the psychophysical systems that determine the adjustment of the person to his/her environment (Svrakic and Cloninger, 2005). *Temperament* is defined more restrictively as the body's biases as it modulates behavioral responses to and styles of coping with prescriptive physical stimuli, such as danger, stressors, or various types of reward. Personality and temperament are of importance to health professionals because they can underlie certain psychiatric illness (Hirschfeld, 1999). In addition, certain aspects of personality have been associated with increased risk for coronary artery disease and the contraction of human immunodeficiency virus (HIV), psoriasis, ulcerative colitis, and many other diseases that have been described as psychosomatic (McCown, 1993; Tyrer, 1995). Dimensions of temperament may also predispose people to health risk behaviors such as tobacco use.

Personality

A principal reason for the scientific re-birth of personality traits is the use of factor analysis to define and validate them. The "Big Five" model, the one whose use is most widespread and accepted, is based on factor analyses of self-reported descriptions of social and emotional behavior. The five personality domains are: Neuroticism (N, negative affectivity), Extraversion (E versus Introversion), Openness to experience (O), Agreeableness, and Conscientiousness. This model is based on a robust factor structure that has been validated in a variety of populations and cultures using the NEO-Personality Inventory (NEO-PI), a personality test designed to assess normal adult personality (McCrae and Costa, 2002). The population samples were drawn from the United States, Germany, Portugal, Israel, China, Korea, and Japan, and included people from ages 18 to 105 (McCrae et al., 1999; Labouvie-Vief et al., 2000).

Personality traits are consistent and are associated with behavioral trends, coping strategies, and health behaviors. This makes it possible to use them to predict health and life outcomes (Whitbourne, 1987; Bosworth et al., 1999; Caspi and Roberts, 1999), depending on the strength of certain traits. To some extent, Alzheimer's disease (Siegler et al., 1994) and CVD

(Hemingway and Marmot, 1999; Williams et al., 2000) can be predicted from certain personality traits.

Personality, in turn, is influenced by both genes and gene-environment interactions. There is an important body of literature, beginning with a seminal paper by Lesch et al. (1996), that reports associations between genotypes of the promoter polymorphism of the serotonin transporter gene (5HTTLPR) and the personality domains of neuroticism (including facets of anxiety, angry hostility, depression, and impulsiveness) and agreeableness. Lesch and colleagues reported a positive correlation of the 5HTTLPR short allele not only with Harm Avoidance but also with the NEO domain of Neuroticism. There was a negative correlation with Agreeableness, thus relating candidate genes that regulate function of the key neurotransmitter serotonin to personality or temperament.

There have now been three meta-analyses published evaluating studies on the association between the 5HTTLPR polymorphism and anxiety-related traits. Two of these (Schinka et al., 2004; Sen et al., 2004) found that there are reliable associations between 5HTTLLPR and Neuroticism as measured by the NEO-PI, but not Harm Avoidance. A third meta-analysis (Munafo et al., 2005a) found the opposite pattern. Whatever the ultimate outcome of this issue, the weight of the evidence suggests that the five-factor model as assessed by the NEO-PI is reliably associated with variation in one highly studied candidate gene, the serotonin transporter.

There is extensive research showing that psychological factors like depressed affect (as opposed to the illness of major depressive disorder—see below), hostility and anger, and anxiety are associated with increased risk of CVD and the biological and behavioral factors that likely mediate that increased risk (see, for example, Williams et al., 2003a). The critical importance of psychosocial stressors on disease risk has been strongly confirmed in the INTERHEART Study (Rosengren et al., 2004) that examined over 24,000 heart attack patients and controls in countries around the world. The study found that social-environmental factors (such as stress at home or work) and psychological factors (such as depression) were associated with as large an increase in heart attack risk as that associated with biological risk factors (e.g., high blood pressure or high lipids) and with behavioral risk factors (e.g., smoking).

In fact, the psychological risk factor hostility is associated in both prospective and cross-sectional studies with increases in several health risk behaviors, including smoking, overeating/obesity, higher lipid levels, and increased alcohol consumption (Scherwitz et al., 1992; Siegler et al., 1992). Thus it would appear that it is through negative affect and accompanying biological and behavioral characteristics that the social environment influences disease processes in ways that are moderated by genetic factors. There is, moreover, some evidence that the opposite of negative affect (optimism) is associated with more positive outcomes both subjectively and objectively with respect to feelings of well-being and recovery from ill health (Smith and Spiro, 2002), although the genetic associations have not yet been studied.

Temperament

Of the many methods proposed to assess temperament, perhaps the most widely used is the Temperament and Character Inventory (Cloninger et al., 1998). Four major temperament traits have been identified through factor analysis and investigated in many experiments: harm avoidance, novelty seeking, reward dependence, and persistence (Cloninger et al., 1998). The study of temperament underscores the notion that genetic risk factors for a disease may not be mediated directly by gene function related to the diseased system, but rather by genetic risk factors for psychological traits that change neuroendocrine function and regulate gene expression involved in disease.

Harm avoidance is a measure of behavioral inhibition and fearfulness (Cloninger et al., 1998). Some studies suggest that individuals high in harm avoidance may be more susceptible to tobacco use (Etter et al., 2003). Adrenal axis function may mediate this association since corticotropin releasing factor-like proteins also are associated with prolonged nicotine withdrawal and higher rates of relapse (Bruijnzeel and Gold, 2005).

Novelty seeking is a measure of behavioral activation and excitement seeking that includes subscales measuring exploratory excitability, impulsiveness, extravagance, and a tendency to disorder (Cloninger et al., 1998) that may increase susceptibility to tobacco use by increasing the likelihood that an adolescent will be exposed to environments in which tobacco is more available (Tercyak and Audrain-McGovern, 2003). Novelty seekers also have been shown to be more susceptible to effects of tobacco advertising (Audrain-McGovern et al., 2003b). Animal models indicate that both of these associations also may be mediated by individual differences in adrenal axis function (Piazza et al., 1993; Spina et al., 2005).

Two aspects of temperament (reward dependence and persistence) have been linked with craving during abstinence from gambling and alcohol. Reward dependence is a measure of social attachment (Cloninger et al., 1998) and persistence is a measure of perseverance (Svrakic and Cloninger, 2005). Reward dependence is negatively correlated with craving for gambling, while persistence is negatively associated with craving for alcohol (Tavares et al., 2005). Thus, if nicotine addiction shares mechanisms with other addictions, future research may uncover a role for these other aspects of temperament and their underlying neuroendocrine and pharmacological systems (Svrakic and Cloninger, 2005).

Temperament has also been investigated regarding its association with specific disease states, such as Attention Deficit Hyperactivity Disorder (ADHD), which is a risk factor for the initiation and persistence of tobacco use (Lerman et al., 2001; Tercyak et al., 2002). Lynn and colleagues (2005) hypothesized that the dopamine D4 receptor mediated the association between novelty seeking and ADHD. However, they found that the DRD4 gene variant independently predicted ADHD, but not novelty seeking. This finding highlights the complexity of the associations between temperament, genetics, neural intermediate phenotypes, and disease states.

Sex/Gender, Race/Ethnicity, and Personality

Both sex/gender and race/ethnicity have been reported to moderate the effects of genotype on personality dimensions. Gelernter et al. (1998) found that the 5HTTLPR short allele was associated with higher Harm Avoidance scores in males but with lower scores in females; and also that the short allele was associated with higher Neuroticism scores in European Americans but with lower scores in African Americans. Interestingly, these effects parallel moderation of 5HTTLPR effects on a measure of central nervous system serotonin function, cerebrospinal fluid levels of the serotonin major metabolite 5HIAA, by both race/ethnicity (short allele → high 5HIAA in African Americans, low 5HIAA in European Americans) and sex/gender (short allele → high 5HIAA in women, low 5HIAA in men) (Williams et al., 2003b). The mechanisms responsible for these differential effects of 5HTTLPR genotype on personality and brain serotonin levels are not clear at present, but could involve differential patterns of linkage disequilibrium between the 5HTTLPR polymorphism and other sites on the serotonin transporter gene in different population groups, as reported by Gelenter et al. (1999).

Depression, Genes, the Environment, and Health

Emotional or motivational states also can be a critical intermediate phenotype between gene-environment interaction and health risk behaviors, depression being, perhaps, the prototype. Diagnosed depressive disorders (Schulz et al., 2002) such as major depression as well as depressive symptoms (Blazer et al., 2001) have been associated with adverse health outcomes. Depression has been demonstrated to be a risk for a variety of disorders, including diabetes and certain types of cancer, but especially for CVD (Schulz et al., 2000). The factors by which depression leads to poorer health outcomes may include some that are indirect, such as a reduced likelihood seeking health care and of complying with the recommendations of health care professionals. They also may include direct links (Schulz et al., 2000). For example Schulz et al. (2000) suggested that motivational depletion may directly contribute to compromised cardiac function and increased risk for myocardial infarction. A listing of potential mediators of the effects of both depression and hostility/anger on CDV risk would include decreased parasympathetic tone, increased hypothalamic-pituitary-adrenal axis and sympathetic nervous system activation, and increased inflammatory cytokines, and increased platelet activation.

In addition, depression is clearly determined by an interaction between genetic and environmental factors. It has been recognized for many years that depression is in part an inherited trait, and the influence of that heritability persists into later life (Gatz et al., 1992; Kendler, 1996). In studies of twins reared apart, both genetic and environmental factors have been shown repeatedly to contribute to depressive symptoms (Kendler, 1996). In addition, population studies have demonstrated the interaction of genetic polymorphisms and environmental stressors (Caspi et al., 2003; Kendler et al., 2005).

For example, in a prospective, longitudinal study of a representative birth cohort, Caspi et al. (2003) tested the observation that stressful experiences lead to depression in some people, but not in others. A functional polymorphism in the promoter region of the serotonin transporter (5-HT T) gene was found to moderate the influence of stressful life events on depression. Individuals with one or two copies of the short allele of the 5-HT T promoter polymorphism exhibited more depressive symptoms, diagnosable depression, and suicidality in relation to stressful life events than did individuals homozygous for the long allele. This epidemiological study thus provides remarkable evidence for a gene-environment interaction in which an individual's response to environmental stressors is moderated by his/her genetic makeup.

Not only does this study have important implications for the investigation of how gene-environment interactions affect health, the rapid and extensive replication of its gene association results has been unusually significant. The 2003 study by Caspi and colleagues was replicated and extended through findings by Eley et al. (2004), Kaufman et al. (2004), Grabe et al. (2005), and Kendler et al. (2005). The immediate replication of gene association findings has been the exception rather than the rule for such studies. The replication of this gene-environment finding with respect not only to incidence of major depression, but also to depressive symptom levels suggests the likelihood that genetic effects on various endophenotypes are far larger when varying levels of critical environmental exposures are taken into account (Moffitt et al., 2005).

In the Caspi et al. (2003) study, for example, there was no effect of the 5HTTLPR genotype on incidence of major depression in persons with no stressful life events over the preceding five years. In marked contrast, among those with four or more stressful life events, there was no increased incidence of major depression among those with the 5HTTLPRL/L genotype, a 21 percent increase in those with the L/S genotype, and a 33 percent increase among those with the S/S genotype.

It would be hard to overstate the implications of this replicated demonstration of a very large genetic effect on the health of persons only with certain social-environmental exposures. It means that if the appropriate environmental exposures are taken into account, it will be far easier to detect and replicate the effects of genes on disease-relevant endophenotypes than it was when the search for genes was conducted in heterogeneous samples. There is reason to believe that this principle will operate not only with respect to chronic levels of stress over time, but also with respect to single, major life stresses. For example, myocardial infarction is a major life stress in which the presence of the 5HTTLPR short allele predicted increased levels of depression over the ensuing months (Nakatani et al., 2005). The same is suggested by a study on stroke that used a very small sample size (Ramasubbu et al., 2006).

CONCLUSION

A better understanding of risk behaviors is critical to improving the public's health. To date, most efforts have been directed toward modifying risk behaviors, such as programs to increase physical activity or to decrease smoking. Biological augmentation of behavioral modification has been partially successful, such as the use of the nicotine patch for smoking cessation. Health psychologists are increasingly calling attention to the critical role of sociocultural context, a necessary factor to consider if efforts to modify risk behaviors are to be effective. In other words, a risk-prevention program that is effective in one culture may be much less effective in another. Only recently have the genetic contributions of risk behaviors and the environments that lead to the expression of intermediate phenotypes been brought into focus. The recognition that behaviors that increase risk for disease may be driven by genetic factors and modified by social factors presents a rich yet complex paradigm for designing and testing intervention strategies for the future.

REFERENCES

Allison DB, Fontaine KR, Manson JE, Stevens J, VanItallie TB. 1999. Annual deaths attributable to obesity in the United States. *Journal of the American Medical Association* 282(16):1530-1538.

Audrain-McGovern J, Rodriguez D, Moss HB. 2003a. Smoking progression and physical activity. *Cancer Epidemiology, Biomarkers and Prevention* 12(11 Pt 1):1121-1129.

Audrain-McGovern J, Tercyak KP, Shields AE, Bush A, Espinel CF, Lerman C. 2003b. Which adolescents are most receptive to tobacco industry marketing? Implications for counter-advertising campaigns. *Health Communication* 15(4):499-513.

Audrain-McGovern J, Lerman C, Wileyto EP, Rodriguez D, Shields PG. 2004a. Interacting effects of genetic predisposition and depression on adolescent smoking progression. *American Journal of Psychiatry* 161(7):1224-1230.

Audrain-McGovern J, Rodriguez D, Tercyak KP, Cuevas J, Rodgers K, Patterson F. 2004b. Identifying and characterizing adolescent smoking trajectories. *Cancer Epidemiology, Biomarkers and Prevention* 13(12):2023-2034.

Barker DJ. 1995. Fetal origins of coronary heart disease. *British Medical Journal* 311(6998):171-174.

Barker DJ, Osmond C, Golding J, Kuh D, Wadsworth ME. 1989. Growth in utero, blood pressure in childhood and adult life, and mortality from cardiovascular disease. *British Medical Journal* 298(6673):564-567.

Benowitz NL, Gourlay SG. 1997. Cardiovascular toxicity of nicotine: Implications for nicotine replacement therapy. *Journal of the American College of Cardiology* 29(7):1422-1431.

Beunen G, Thomis M. 1999. Genetic determinants of sports participation and daily physical activity. *International Journal of Obesity and Related Metabolic Disorders* 23(Suppl 3):S55-S63.

Bianchini F, Kaaks R, Vainio H. 2002. Overweight, obesity, and cancer risk. *Lancet Oncology* 3(9):565-574.

Bjerregaard P, Mulvad G, Pedersen HS. 1997. Cardiovascular risk factors in Inuit of Greenland. *International Journal of Epidemiology* 26(6):1182-1190.

Blazer DG, Hybels CF, Pieper CF. 2001. The association of depression and mortality in elderly persons: A case for multiple, independent pathways. *Journals of Gerontology. Series A, Biological Sciences and Medical Sciences* 56(8):M505-509.

Blendy JA, Strasser A, Walters CL, Perkins KA, Patterson F, Berkowitz R, Lerman C. 2005. Reduced nicotine reward in obesity: Cross-comparison in human and mouse. *Psychopharmacology* 180(2):306-315.

Bosworth HB, Siegler IC, Brummett BH, Barefoot JC, Williams RB, Vitaliano PP, Clapp-Channing N, Lytle BL, Mark DB. 1999. The relationship between self-rated health and health status among coronary artery patients. *Journal of Aging and Health* 11(4):565-584.

Browman GP, Wong G, Hodson I, Sathya J, Russell R, McAlpine L, Skingley P, Levine MN. 1993. Influence of cigarette smoking on the efficacy of radiation therapy in head and neck cancer. *New England Journal of Medicine* 328(3):159-163.

Bruijnzeel AW, Gold MS. 2005. The role of corticotropin-releasing factor-like peptides in cannabis, nicotine, and alcohol dependence. *Brain Research Reviews* 49(3):505-528.

Cabib S, Orsini C, Le Moal M, Piazza PV. 2000. Abolition and reversal of strain differences in behavioral responses to drugs of abuse after a brief experience. *Science* 289(5478):463-465.

Caspersen CJ, Powell KE, Christenson GM. 1985. Physical activity, exercise, and physical fitness: Definitions and distinctions for health-related research. *Public Health Reports* 100(2):126-131.

Caspi A, Roberts BW. 1999. Personality continuity and change across the life course. In: Pervin LA, John OP, editors. *Handbook of Personality*. 2nd edition. New York: Guilford Press. Pp. 300-326.

Caspi A, Sugden K, Moffitt TE, Taylor A, Craig IW, Harrington H, McClay J, Mill J, Martin J, Braithwaite A, Poulton R. 2003. Influence of life stress on depression: Moderation by a polymorphism in the 5-HTT gene. *Science* 301(5631):386-389.

CDC (Centers for Disease Control and Prevention). 2000. *CDC Factbook 2000/2001: Profile of the Nation's Healh.* Atlanta, GA: CDC.

CDC. 2002. Annual smoking-attributable mortality, years of potential life lost, and economic costs—United States, 1995-1999. *Morbidity and Mortality Weekly Report* 51(14): 300-303.

CDC. 2004a. Cigarette smoking among adults—United States, 2002. *Morbidity and Mortality Weekly Report* 53(19):427-431.

CDC. 2004b. Youth risk behavior surveillance—United States, 2003. *Morbidity and Mortality Weekly Report* 53(SS02):1-96.

CDC. 2005. Cigarette smoking among adults—United States, 2003. *Morbidity and Mortality Weekly Report* 54(20):509-513.

Chaloupka FJ, Cummings KM, Morley C, Horan J. 2002. Tax, price and cigarette smoking: Evidence from the tobacco documents and implications for tobacco company marketing strategies. *Tobacco Control* 11(90001):62-72.

Choi WS, Ahluwalia JS, Harris KJ, Okuyemi K. 2002. Progression to established smoking: The influence of tobacco marketing. *American Journal of Preventive Medicine* 22(4): 228-233.

Cloninger CR, Bayon C, Svrakic DM. 1998. Measurement of temperament and character in mood disorders: A model of fundamental states as personality types. *Journal of Affective Disorders* 51(1):21-32.

Comuzzie AG, Allison DB. 1998. The search for human obesity genes. *Science* 280(5368): 1374-1377.

Costanzo PR, Schiffman SS. 1989. Thinness—not obesity—has a genetic component. *Neuroscience and Biobehavioral Reviews* 13(1):55-58.

Eckel RH, Barouch WW, Ershow AG. 2002. Report of the National Heart, Lung, and Blood Institute-National Institute of Diabetes and Digestive and Kidney Diseases Working Group on the pathophysiology of obesity-associated cardiovascular disease. *Circulation* 105(24):2923-2928.

Eley TC, Sugden K, Corsico A, Gregory AM, Sham P, McGuffin P, Plomin R, Craig IW. 2004. Gene-environment interaction analysis of serotonin system markers with adolescent depression. *Molecular Psychiatry* 9(10):908-915.

Epstein LH, Jaroni JL, Paluch RA, Leddy JJ, Vahue HE, Hawk L, Wileyto EP, Shields PG, Lerman C. 2002. Dopamine transporter genotype as a risk factor for obesity in African-American smokers. *Obesity Research* 10(12):1232-1240.

Etter JF, Pelissolo A, Pomerleau C, De Saint-Hilaire Z. 2003. Associations between smoking and heritable temperament traits. *Nicotine & Tobacco Research* 5(3):401-409.

Flegal KM, Graubard BI, Williamson DF, Gail MH. 2005. Excess deaths associated with underweight, overweight, and obesity. *Journal of the American Medical Association* 293(15):1861-1867.

Friedman H, Eisenstein TK. 2004. Neurological basis of drug dependence and its effects on the immune system. *Journal of Neuroimmunology* 147(1-2):106-108.

Fumeron F, Betoulle D, Nicaud V, Evans A, Kee F, Ruidavets JB, Arveiler D, Luc G, Cambien F. 2002. Serotonin transporter gene polymorphism and myocardial infarction: Etude Cas-Temoins de l'Infarctus du Myocarde (ECTIM). *Circulation* 105(25):2943-2945.

Gatz M, Pedersen NL, Plomin R, Nesselroade JR, McClearn GE. 1992. Importance of shared genes and shared environments for symptoms of depression in older adults. *Journal of Abnormal Psychology* 101(4):701-708.

Gelernter J, Kranzler H, Coccaro EF, Siever LJ, New AS. 1998. Serotonin transporter protein gene polymorphism and personality measures in African American and European American subjects. *American Journal of Psychiatry* 155(10):1332-1338.

Gelernter J, Cubells JF, Kidd JR, Pakstis AJ, Kidd KK. 1999. Population studies of polymorphisms of the serotonin transporter protein gene. *American Journal of Medical Genetics* 88(1):61-66.

Grabe HJ, Lange M, Wolff B, Volzke H, Lucht M, Freyberger HJ, John U, Cascorbi I. 2005. Mental and physical distress is modulated by a polymorphism in the 5-HT transporter gene interacting with social stressors and chronic disease burden. *Molecular Psychiatry* 10(2):220-224.

Harris MI. 1998. Diabetes in America: Epidemiology and scope of the problem. *Diabetes Care* 21(Suppl 3):C11-C14.

Hemingway H, Marmot M. 1999. Psychosocial factors in the etiology and prognosis of coronary heart disease: Systematic review of prospective cohort studies. *British Medical Journal* 318:1460-1467.

Hermes GL, Jacobs LF, McClintock MK. 2005. The sectored foraging field: A novel design to quantify spatial strategies, learning, memory, and emotion. *Neurobiology of Learning and Memory* 84(1):69-73.

Hill JO, Peters JC. 1998. Environmental contributions to the obesity epidemic. *Science* 280(5368):1371-1374.

Hirschfeld RM. 1999. Personality disorders and depression: Comorbidity. *Depression and Anxiety* 10(4):142-146.

Holm S. 1979. A simple sequentially rejective multiple test procedure. *Scandinavian Journal of Statistics* 6:65-70.

Jeanmonod P, von Kanel R, Maly FE, Fischer JE. 2004. Elevated Plasma C-reactive protein in chronically distressed subjects who carry the A allele of the TNF-alpha-308 G/A polymorphism. *Psychosomatic Medicine* 66(4):501-506.

Kapoor D, Jones TH. 2005. Smoking and hormones in health and endocrine disorders. *European Journal of Endocrinology / European Federation of Endocrine Societies* 152(4): 491-499.

Kaufman J, Yang BZ, Douglas-Palumberi H, Houshyar S, Lipschitz D, Krystal JH, Gelernter J. 2004. Social supports and serotonin transporter gene moderate depression in maltreated children. *Proceedings of the National Academy of Sciences of the United States of America* 101(49):17316-17321.

Kawahara M, Ushijima S, Kamimori T, Kodama N, Ogawara M, Matsui K, Masuda N, Takada M, Sobue T, Furuse K. 1998. Second primary tumours in more than 2-year disease-free survivors of small-cell lung cancer in Japan: The role of smoking cessation. *British Journal of Cancer* 78(3):409-412.

Kendler KS. 1996. Major depression and generalised anxiety disorder. Same genes, (partly) different environments—revisited. *British Journal of Psychiatry* (30) Suppl.:68-75.

Kendler KS, Aggen SH, Prescott CA, Jacobson KC, Neale MC. 2004. Level of family dysfunction and genetic influences on smoking in women. *Psychological Medicine* 34(7):1263-1269.

Kendler KS, Kuhn JW, Vittum J, Prescott CA, Riley B. 2005. The interaction of stressful life events and a serotonin transporter polymorphism in the prediction of episodes of major depression: A replication. *Archives of General Psychiatry* 62(5):529-535.

Khuri FR, Kim ES, Lee JJ, Winn RJ, Benner SE, Lippman SM, Fu KK, Cooper JS, Vokes EE, Chamberlain RM, Williams B, Pajak TF, Goepfert H, Hong WK. 2001. The impact of smoking status, disease stage, and index tumor site on second primary tumor incidence and tumor recurrence in the head and neck retinoid chemoprevention trial. *Cancer Epidemiology, Biomarkers and Prevention* 10(8):823-829.

Kinjo K, Sato H, Sakata Y, Nakatani D, Mizuno H, Shimizu M, Sasaki T, Kijima Y, Nishino M, Uematsu M, Tanouchi J, Nanto S, Otsu K, Hori M. 2005. Impact of smoking status on long-term mortality in patients with acute myocardial infarction. *Circulation Journal* 69(1):7-12.

Kopelman PG. 2000. Obesity as a medical problem. *Nature* 404(6778):635-643.

Labouvie-Vief G, Diehl M, Tarnowski A, Shen J. 2000. Age differences in adult personality: Findings from the United States and China. *Journals of Gerotology: Series B: Psychological Sciences & Social Sciences,* 55B(1):4-17.

Lerman C, Berrettini W. 2003. Elucidating the role of genetic factors in smoking behavior and nicotine dependence. *American Journal of Medical Genetics Part B: Neuropsychiatric Genetics* 118(1):48-54.

Lerman C, Swan GE. 2002. Non-replication of genetic association studies: Is DAT all, folks? *Nicotine & Tobacco Research* 4(3):247-249.

Lerman C, Audrain J, Tercyak K, Hawk LW Jr, Bush A, Crystal-Mansour S, Rose C, Niaura R, Epstein LH. 2001. Attention-Deficit Hyperactivity Disorder (ADHD) symptoms and smoking patterns among participants in a smoking-cessation program. *Nicotine & Tobacco Research* 3(4):353-359.

Lerman C, Patterson F, Berrettini W. 2005. Treating tobacco dependence: State of the science and new directions. *Journal of Clinical Oncology* 23(2):311-323.

Lesch KP, Bengel D, Heils A, Sabol SZ, Greenberg BD, Petri S, Benjamin J, Muller CR, Hamer DH, Murphy DL. 1996. Association of anxiety-related traits with a polymorphism in the serotonin transporter gene regulatory region. *Science* 274(5292):1527-1531.

Loos RJ, Rankinen T, Tremblay A, Perusse L, Chagnon Y, Bouchard C. 2005. Melanocortin-4 receptor gene and physical activity in the Quebec Family Study. *International Journal of Obesity* 29(4):420-428.

Lynn DE, Lubke G, Yang M, McCracken JT, McGough JJ, Ishii J, Loo SK, Nelson SF, Smalley SL. 2005. Temperament and character profiles and the dopamine D4 receptor gene in ADHD. *American Journal of Psychiatry* 162(5):906-913.

McBride CM, Ostroff JS. 2003. Teachable moments for promoting smoking cessation: The context of cancer care and survivorship. *Cancer Control* 10(4):325-333.

McClintock M. 1981. Simplicity from complexity: A naturalistic approach to behavior and neuroendocrine function. *New Directions for Methodology of Social and Behavioral Science* 8:1-19.

McCown W. 1993. Personality factors predicting failure to practice safer sex by HIV positive males. *Personality and Individual Differences* 14(4):613-615.

McCrae RR, Costa PT Jr. 2002. *Personality in Adulthood: A Five-Factor Theory Perspective.* 2nd edition. New York: Guilford Press.

McCrae RR, Costa PT Jr, Pedroso de Lima M, Simoes A, Ostendorf F, Angleitner A, Marusic I, Bratko D, Caprara GV, Barbaranelli C, Chae JH, Piedmont RL. 1999. Age differences in personality across the adult life span: Parallels in five cultures. *Developmental Psychology* 35(2):466-477.

McTiernan A, Ulrich C, Slate S, Potter J. 1998. Physical activity and cancer etiology: Associations and mechanisms. *Cancer Causes and Control* 9(5):487-509.

Mhyre TR, Chesler EJ, Thiruchelvam M, Lungu C, Cory-Slechta DA, Fry JD, Richfield EK. 2005. Heritability, correlations and in silico mapping of locomotor behavior and neurochemistry in inbred strains of mice. *Genes, Brain, and Behavior* 4(4):209-228.

Miltiadous G, Xenophontos S, Bairaktari E, Ganotakis M, Cariolou M, Elisaf M. 2005. Genetic and environmental factors affecting the response to statin therapy in patients with molecularly defined familial hypercholesterolaemia. *Pharmacogenetics and Genomics* 15(4):219-225.

Moffitt TE, Caspi A, Rutter M. 2005. Strategy for investigating interactions between measured genes and measured environments. *Archives of General Psychiatry* 62(5):473-481.

Mokdad AH, Marks JS, Stroup DF, Gerberding JL. 2004. Actual causes of death in the United States, 2000. *Journal of the American Medical Association* 291(10):1238-1245.

Munafo MR, Flint J. 2004. Meta-analysis of genetic association studies. *Trends in Genetics* 20(9):439-444.

Munafo MR, Clark T, Flint J. 2005a. Does measurement instrument moderate the association between the serotonin transporter gene and anxiety-related personality traits? A meta-analysis. *Molecular Psychiatry* 10(4):415-419.

Munafo MR, Shields AE, Berrettini WH, Patterson F, Lerman C. 2005b. Pharmacogenetics and nicotine addiction treatment. *Pharmacogenomics* 6(3):211-223.

Nakatani D, Sato H, Sakata Y, Shiotani I, Kinjo K, Mizuno H, Shimizu M, Ito H, Koretsune Y, Hirayama A, Hori M. 2005. Influence of serotonin transporter gene polymorphism on depressive symptoms and new cardiac events after acute myocardial infarction. *American Heart Journal* 150(4):652-658.

NCHS (National Center for Health Statistics). 2003. *Health, United States, 2003 with Chartbook on Trends in the Health of Americans.* Hyattsville, MD: NCHS.

Noble EP, Noble RE, Ritchie T, Syndulko K, Bohlman MC, Noble LA, Zhang Y, Sparkes RS, Grandy DK. 1994. D2 dopamine receptor gene and obesity. *International Journal of Eating Disorders* 15(3):205-217.

Pate RR, Pratt M, Blair SN, Haskell WL, Macera CA, Bouchard C, Buchner D, Ettinger W, Heath GW, King AC, et al. 1995. Physical activity and public health. A recommendation from the Centers for Disease Control and Prevention and the American College of Sports Medicine. *Journal of the American Medical Association* 273(5):402-407.

Perusse L, Tremblay A, Leblanc C, Bouchard C. 1989. Genetic and environmental influences on level of habitual physical activity and exercise participation. *American Journal of Epidemiology* 129(5):1012-1022.

Piazza PV, Deroche V, Deminiere JM, Maccari S, Le Moal M, Simon H. 1993. Corticosterone in the range of stress-induced levels possesses reinforcing properties: Implications for sensation-seeking behaviors. *Proceedings of the National Academy of Sciences of the United States of America* 90(24):11738-11742.

Pierce JP, Choi WS, Gilpin EA, Farkas AJ, Berry CC. 1998. Tobacco industry promotion of cigarettes and adolescent smoking. *Journal of the American Medical Association* 279(7): 511-515.

Pinto BM, Eakin E, Maruyama NC. 2000. Health behavior changes after a cancer diagnosis: What do we know and where do we go from here? *Annals of Behavioral Medicine* 22(1):38-52.

Ramasubbu R, Tobias R, Buchan AM, Bech-Hansen NT. 2006. Serotonin transporter gene promoter region polymorphism associated with poststroke major depression. *Journal of Neuropsychiatry and Clinical Neurosciences* 18(1):96-99.

Rebbeck TR. 2002. The contribution of inherited genotype to breast cancer. *Breast Cancer Research* 4(3):85-89.

Redden DT, Shields PG, Epstein L, Wileyto EP, Zakharkin SO, Allison DB, Lerman C. 2005. Catechol-O-methyl-transferase functional polymorphism and nicotine dependence: An evaluation of nonreplicated results. *Cancer Epidemiology, Biomarkers and Prevention* 14(6):1384-1389.

Rock CL, Demark-Wahnefried W. 2002. Nutrition and survival after the diagnosis of breast cancer: A review of the evidence. *Journal of Clinical Oncology* 20(15):3302-3316.

Rose DP, Gilhooly EM, Nixon DW. 2002. Adverse effects of obesity on breast cancer prognosis, and the biological actions of leptin (review). *International Journal of Oncology* 21(6):1285-1292.

Rosengren A, Hawken S, Ounpuu S, Sliwa PK, Zubaid M, Almahmeed WA, Ngu Blackett K, Sitthi-Amorn C, Sato H, Yusuf PS. 2004. Association of psychosocial risk factors with risk of acute myocardial infarction in 11,119 cases and 13, 648 controls from 52 countries (the INTERHEART study): Case control study. *Lancet* 364(0438):953-962.

Scherwitz LW, Perkins LL, Chesney MA, Hughes GH, Sidney S, Manolio TA. 1992. Hostility and health behaviors in young adults: The CARDIA Study. Coronary Artery Risk Development in Young Adults Study. *American Journal of Epidemiology* 136(2):136-145.

Schinka JA, Busch RM, Robichaux-Keene N. 2004. A meta-analysis of the association between the serotonin transporter gene polymorphism (5-HTTLPR) and trait anxiety. *Molecular Psychiatry* 9(2):197-202.

Schmitz KH, Ahmed RL, Hannan PJ, Yee D. 2005a. Safety and efficacy of weight training in recent breast cancer survivors to alter body composition, insulin, and insulin-like growth factor axis proteins. *Cancer Epidemiology, Biomarkers and Prevention* 14(7):1672-1680.

Schmitz KH, Holtzman J, Courneya KS, Masse LC, Duval S, Kane R. 2005b. Controlled physical activity trials in cancer survivors: A systematic review and meta-analysis. *Cancer Epidemiology, Biomarkers and Prevention* 14(7):1588-1595.

Schulz R, Beach SR, Ives DG, Martire LM, Ariyo AA, Kop WJ. 2000. Association between depression and mortality in older adults: The Cardiovascular Health Study. *Archives of Internal Medicine* 160(12):1761-1768.

Schulz R, Drayer RA, Rollman BL. 2002. Depression as a risk factor for non-suicide mortality in the elderly. *Biological Psychiatry* 52(3):205-225.

Seidell JC, Andres R, Sorkin JD, Muller DC. 1994. The sagittal waist diameter and mortality in men: The Baltimore Longitudinal Study on Aging. *International Journal of Obesity and Related Metabolic Disorders* 18(1):61-67.

Sen S, Burmeister M, Ghosh D. 2004. Meta-analysis of the association between a serotonin transporter promoter polymorphism (5-HTTLPR) and anxiety-related personality traits. *American Journal of Medical Genetics. Part B, Neuropsychiatric Genetics* 127(1):85-89.

Siegler IC, Peterson BL, Barefoot JC, Williams RB. 1992. Hostility during late adolescence predicts coronary risk factors at mid-life. *American Journal of Epidemiology* 136(2): 146-154.

Siegler IC, Dawson DV, Welsh KA. 1994. Caregiver ratings of personality change in Alzheimer's disease patients: A replication. *Psychology and Aging* 9:464-466.

Smith TW, Spiro A. 2002. Personality, health and aging: Prolegomenon for the next generation. *Journal of Research in Personality* 36:363-394.

Spina L, Fenu S, Longoni R, Rivas E, Di Chiara G. 2005. Nicotine-conditioned single-trial place preference: Selective role of nucleus accumbens shell dopamine D(1) receptors in acquisition. *Psychopharmacology* 184(3-4):447-455.

Stallone DD. 1994. The influence of obesity and its treatment on the immune system. *Nutrition Reviews* 52(2 Pt 1):37-50.

Sullivan PF, Kendler KS. 1999. The genetic epidemiology of smoking. *Nicotine & Tobacco Research* 1(Suppl 2): S51-S57; discussion S69-S70.

Sullivan PW, Morrato EH, Ghushchyan V, Wyatt HR, Hill JO. 2005. Obesity, inactivity, and the prevalence of diabetes and diabetes-related cardiovascular comorbidities in the U.S., 2000-2002. *Diabetes Care* 28(7):1599-1603.

Svrakic D, Cloninger C. 2005. Personality Disorders. In: Kaplan B, Kaplan V, editors. *Kaplan & Sadock's Comprehensive Textbook of Psychiatry*. Philadelphia, PA: Lippincott Williams and Wilkins. Pp. 3593-3603.

Swan GE, Hudmon KS, Jack LM, Hemberger K, Carmelli D, Khroyan TV, Ring HZ, Hops H, Andrews JA, Tildesley E, McBride D, Benowitz N, Webster C, Wilhelmsen KC, Feiler HS, Koenig B, Caron L, Illes J, Cheng LS. 2003. Environmental and genetic determinants of tobacco use: Methodology for a multidisciplinary, longitudinal family-based investigation. *Cancer Epidemiology, Biomarkers and Prevention* 12(10):994-1005.

Tavares H, Zilberman ML, Hodgins DC, el-Guebaly N. 2005. Comparison of craving between pathological gamblers and alcoholics. *Alcoholism, Clinical and Experimental Research* 29(8):1427-1431.

Tercyak KP, Audrain-McGovern J. 2003. Personality differences associated with smoking experimentation among adolescents with and without comorbid symptoms of ADHD. *Substance Use and Misuse* 38(14):1953-1970.

Tercyak KP, Lerman C, Audrain J. 2002. Association of attention-deficit/hyperactivity disorder symptoms with levels of cigarette smoking in a community sample of adolescents. *Journal of the American Academy of Child and Adolescent Psychiatry* 41(7):799-805.

Treiber FA, Barbeau P, Harshfield G, Kang HS, Pollock DM, Pollock JS, Snieder H. 2003. Endothelin-1 gene Lys198Asn polymorphism and blood pressure reactivity. *Hypertension* 42(4):494-499.

Turcato E, Bosello O, Di Francesco V, Harris TB, Zoico E, Bissoli L, Fracassi E, Zamboni M. 2000. Waist circumference and abdominal sagittal diameter as surrogates of body fat distribution in the elderly: Their relation with cardiovascular risk factors. *International Journal of Obesity and Related Metabolic Disorders* 24(8):1005-1010.

Tyrer P. 1995. Somatoform and personality disorders: Personality and the soma. *Journal of Psychosomatic Research* 39(4):395-397.

U.S. DHHS (U.S. Department of Health and Human Services). 1996. *Physical Activity and Health: A Report of the Surgeon General.* Atlanta, GA: CDC, National Center for Chronic Disease Prevention and Health Promotion.

Vaisse C, Clement K, Guy-Grand B, Froguel P. 1998. A frameshift mutation in human MC4R is associated with a dominant form of obesity. *Nature Genetics* 20(2):113-114.

Whitbourne SK. 1987. Personality development in adulthood and old age: Relationships among identity style, health and well-being. In: Schaie KW, editor. *Annual Review of Gerontology and Geriatrics Vol. 7.* New York: Springer. Pp. 189-216.

WHO (World Health Organization). 1998. *Obesity: Preventing and Managing the Global Epidemic: Report of the WHO Consultation of Obesity.* Geneva: WHO. WHO Technical Report Series No. 894.

Williams JE, Paton CC, Siegler IC, Eigenbrodt M, Nieto J, Tyroler HA. 2000. Anger proneness predicts CHD risk: A prospective analysis from the Atherosclerosis Risk in Communities Study. *Circulation* 101:2034-2039.

Williams RB, Marchuk DA, Gadde KM, Barefoot JC, Grichnik K, Helms MJ, Kuhn CM, Lewis JG, Schanberg SM, Stafford-Smith M, Suarez EC, Clary GL, Svenson IK, Siegler IC. 2001. Central nervous system serotonin function and cardiovascular responses to stress. *Psychosomatic Medicine* 63(2):300-305.

Williams RB, Barefoot JC, Schneiderman N. 2003a. Psychosocial risk factors for cardiovascular disease: More than one culprit at work. *Journal of the American Medical Association* 290(16):2190-2192.

Williams RB, Marchuk DA, Gadde KM, Barefoot JC, Grichnik K, Helms MJ, Kuhn CM, Lewis JG, Schanberg SM, Stafford-Smith M, Suarez EC, Clary GL, Svenson IK, Siegler IC. 2003b. Serotonin-related gene polymorphisms and central nervous system serotonin function. *Neuropsychopharmacology* 28(3):533-541.

5

Sex/Gender, Race/Ethnicity, and Health

In the search for a better understanding of genetic and environmental interactions as determinants of health, certain fundamental aspects of human identity pose both a challenge and an opportunity for clarification. Sex/gender and race/ethnicity are complex traits that are particularly useful and important because each includes the social dimensions necessary for understanding its impact on health and each has genetic underpinnings, to varying degrees.

Although there have been numerous genetic studies of sex and gender—and more recently race and ethnicity—over the past several decades, detailed information about the extent of our genetic similarities and differences did not reach the public's attention until the completion of the Human Genome Project. With base pair comparisons possible across the individuals sequenced, the estimate that any two humans are 99.9 percent the same has raised our awareness that all humans are incredibly similar at the genetic level. Paradoxically, the evidence of vast numbers of DNA base pairs at which humans differ also became known at this time. It is estimated currently that any two people will differ at approximately 3 million positions along their genomes. Although there is some evidence that information about an individual's sex or ancestry would provide information about the likelihood that he/she carries one allele versus another, it is typically a matter of probability—not a discrete or absolute determinant (even for the Y chromosome). While there is growing evidence of a number of significant differences between males and females in terms of health and health outcomes (IOM, 2001), "considerable controversy remains about the existence and importance of racial differences in genetic effects, particularly for complex diseases" (Ioannidis et al., 2004).

Previous chapters have discussed the contributions of the social environment, behavior, psychological factors, physiological mechanisms, and genetic variation to health. This chapter highlights the fact that the contributions of these variables are not monolithic and that fundamental individual traits, such as sex/gender and race/ethnicity, can change their meaning and health impact in different contexts. These complex traits are multifaceted, and the goal is to tease apart the facets at different levels of organization in order to identify which of them directly modulate health. This is a reciprocal process, because these various domains in turn inform our understanding of sex/gender and race/ethnicity. Failing to distinguish these different facets, both in the aggregate and within each level of analysis, will compromise the ability to obtain a more fine-grained understanding of how the different aspects of these fundamental individual traits interact to influence health.

SEX/GENDER

Although the terms *sex* and *gender* are often used interchangeably, they, in fact, have distinct meanings. *Sex* is a classification based on biological differences—for example, differences between males and females rooted in their anatomy or physiology. By contrast, *gender* is a classification based on the social construction (and maintenance) of cultural distinctions between males and females. Gender refers to "a social construct regarding culture-bound conventions, roles, and behaviors for, as well as relations between and among, women and men, boys and girls" (Krieger, 2003).

Differences in the health of males and females often reflect the simultaneous influence of both sex and gender. Not only can gender relations influence the expression of biological traits, but also sex-associated biological characteristics can contribute to amplify gender differentials in health (Krieger, 2003). The relative contributions of gender relations and sex-linked biology to health differences between males and females depend on the specific health outcome under consideration. In some instances, sex-linked biology is the sole determinant of a health outcome—for example gonadal digenesis among women with Turner's syndrome (due to X-monosomy). In other instances, gender relations account substantially for observed gender differentials for a given health outcome—for example the higher prevalence of needle-stick injuries among female compared to male health care workers, which is in turn attributed to the gender segregation of the health care workforce. The prevalence of HIV infection through needle-stick injury is higher among female health care workers because the majority of doctors are men, the majority of nurses and phlebotomists are women, and drawing blood is relegated to nurses and phlebotomists (who are mostly women) (Ippolito et al., 1999).

In yet other instances, gender relations can act synergistically with sex-linked biology to produce a health outcome. For example, the risk of hypospadias is higher among male infants born to women exposed to potential endocrine-disrupting agents at work. In this example, maternal exposure to the endocrine-disrupting agent (e.g., phthalates) arises because of gender segregation in the labor market (e.g., exposure among hair-dressers who are mainly women). Once exposure occurs, the risk of the outcome is predicated on sex-linked biology and is different for women and men, as well as for female and male fetuses, because only women can be pregnant, and exposure can lead to the outcome (hypospadias) only among male fetuses (all examples cited in Krieger, 2003).

Finally, in some instances, sex-linked biology can be obscured by the influence of gender relations in producing health differentials between women and men. For example, women's lower risk of coronary heart disease (CHD) prior to menopause often has been ascribed to the cardioprotective effects of endogenous estrogens (a sex difference), but at the same time, the male/female differential in heart disease also may reflect a diagnostic artifact; that is, the underdetection of heart disease among women caused by an unconscious bias among physicians to ascribe the symptoms of a real heart attack among premenopausal women to some other disorder (a gender difference) (McKinlay, 1996). Arber and colleagues (2006) demonstrated the presence of such bias in a randomized experimental study involving video-vignettes of a scripted consultation in which patients presented with standardized symptoms of CHD. The videotaped consultations were identical in terms of symptoms, but the patients' gender, age (55 versus 75), class, and race varied. A probability sample of 256 primary care doctors from the United States and the United Kingdom viewed these video-vignettes and the results demonstrated that the diagnosis and patient management decisions were significantly affected by the patient's gender. Women were asked fewer questions and received fewer diagnostic tests compared to men. The authors found evidence of "gendered ageism," in which middle-aged women presenting with classic symptoms of CHD were asked the least amount of questions and prescribed the fewest CHD-related medications (Arber et al., 2006).

Besides the behavior of health care providers, a number of other social processes are recognized as contributing to gender inequalities in health. At the macro (or societal) level, these include the gender segregation of the labor force (alluded to above) and gender discrimination. Gender segregation of the workforce and gender discrimination together contribute to the persistence of the gender wage gap—that is the fact that women earn less than men in paid employment (Reskin and Padavic, 1994). The gender wage gap in turn contributes to the feminization of poverty. Women—

particularly female heads of households—are over-represented among poor households in virtually every society. The adverse health effects of poverty (see Chapter 2 of this report) therefore fall disproportionately on women and their children. At the societal level, indicators of women's economic autonomy or lack thereof (e.g., rates of poverty among women, the size of the gender wage gap, and the proportion of women in managerial and technical professions) have been shown to closely mirror women's health status (mortality and rates of disability) (Kawachi et al., 1999).

Within households, gender relations also are characterized by the unequal division of labor (e.g., care giving roles are more often assumed by women), as well as by the unequal exercise of authority and power. Women with paid work are more likely than men to engage in the "second shift" (Hochschild, 1989), taking on responsibilities for childcare, housework, and care giving. The stresses associated with care giving, particularly providing care for ill spouses, have been linked to adverse health outcomes, such as cardiovascular disease (Lee et al., 2003).

Men and women differ biologically because their primary reproductive hormones are different. Less well recognized are the sex differences in certain aspects of immune function that stem from the fact that women and men face different immune challenges. In women, but not in men, successful reproduction requires the support of "foreign bodies"—sperm and a developing fetus. Moreover, as is the case for many other mammalian species, other aspects of male and female biology also may differ because they have different roles in caring for offspring or function in different ecological niches, thus reducing parental competition. For example, a brief stressor mimicking a burrow collapse results in a more pronounced long-term innate inflammatory response in female rats than in male rats exposed to the same stressor (Hermes et al., 2006). Given that females become aggressive during lactation and may likely suffer from wounding, selection would favor those who can mount an inflammatory response that is effective enough to enable them to survive at least long enough to wean their nursing pups. Given that males do not behave paternally in this species, a selection pressure at this juncture of the reproductive lifespan would not be as strong.

The central point is that sex differences in health and risk for disease are not simply minor correlates of differences in reproductive hormones. They also result from deeply embedded highly coordinated physiological systems that have evolved to serve sex-specific functions. For example, women must have sufficient energy reserves to sustain the huge metabolic demands of pregnancy and lactation. Thus, it is not surprising to see sex differences in energy metabolism. In men, insulin functions as a negative feedback signal in the regulation of fat metabolism, reducing body fat, but this does not occur in women, where it serves to conserve women's fat stores (Hallschmid et al., 2004). Sex hormones have both genomic and nongenomic effects on the accumulation, distribution, and metabolism of adipose tissue, including the regulation of leptin (Mayes and Watson, 2004). Leptin has long-term effects on the regulation of body weight, mediated through appetite, energy expenditure and body temperature. Marked sex differences can be seen in levels of leptin, which in men (but not women) are associated with hypertension (Sheu et al., 1999). Moreover, leptin stimulates cellular components of innate immunity, stimulating T-cells, macrophages, and neutrophils, as well as preventing the programmed cell death of neutrophils (apoptosis) (Bruno et al., 2005). Indeed, leptin is increased during infections. Thus, fat metabolism and immune functions are differentially controlled in men and women, and the implications for disease risk and treatment are only now beginning to be explored.

In recent years, there has been an increased focus on understanding the differences and similarities between females and males at the societal level (i.e., behaviors, lifestyles, environment), at the level of the whole organism, and at the cellular and molecular levels (IOM, 2001) (see Table 5-1). There is, of course, huge variation in the degree of overlap in the physical traits of men and women. Sexual dimorphism is typically reserved for traits for which the difference is relatively large, such as height (population overlap of one standard deviation—10 percent of men are smaller than the average woman), while smaller differences are typically termed as sexually differentiated, such as hand shape (Williams et al., 2000).

A significant number of studies have documented the differences between sexes across the lifespan. Genetic and physiological make up, in addition to an individual's personal experiences and interactions with the environment, can play a large part in observed sex differences such as varying incidence and severity of disease. This may be the result of differences in exposure to the risk factors, the routes of exposure and processing of a foreign agent, and cellular responses to the body. Differences cannot simply be attributed to hormones. Sex affects behavior, perception, and health in multiple complex ways. Differences in the sex chromosomes are but one factor, although a significant one for a small number of diseases influenced by gene dosage (i.e., specific to the X chromosome), or for genes found only on the Y chromosome (IOM, 2001).

In order to understand the impact of sex/gender on health, it will be necessary to deeply appreciate that it is not a simple categorical variable, ultimately definable by the presence or absence of the Y chromosome. Rather, it is a multifaceted variable, biologically, psychologically and socially, with each facet having different effects on health and risk for disease. Each facet is oriented along dimensions that typically covary so strongly that many assume that they are inseparable (see the typical phenotypes of sex/gender in Table 5-1). However, there can be variance, if not sex reversals, along a given dimen-

TABLE 5-1   The Independent Dimensions of Sex/Gender in Humans

7 Dimensions of Sex Differences	Typical Phenotypes		Variants
	Female	Male	
Genetic	XX	XY	XO (Turner Syndrome), XYY (Kleinfelter Syndrome), XY′ (Y deletions), genetic mosaics
Gonadal	Ovary	Testis	Streak gonads, ovatestis
Hormone Profiles	Estrogens > androgens	Androgens > estrogens	Androgen receptors with low binding affinity, adrenal androgens
Reproductive Tract	Uterus, fallopian tubes	Vas deferens, prostate	True hermaphrodite (hemiuterus), fused mullarian ducts
External Genitalia	Labia, clitoris	Scrotum, penis	Hypospadia, microphallus, vaginal agenesis and hypoplasia
Secondary Sex Characteristics	Breasts	Beard	Gynecomastia, hirsute
Anatomy and Metabolism	Wide pelvis inlet and outlet, abdominal fat, delayed neutrophil apoptosis	Tall, fast-twich muscles, low cardiac levels of heat shock protein 27	Such variables typically are continuous, as is height; thus, variants typically have values in the range typical of the opposite sex
Gender Identity	I am a woman	I am a man	Turnim man, third gender, guevodoces, intersex
Sexual Orientation	Men erotic	Women erotic	Homosexuality, lesbian, bisexuality
Sex Role	Primary caregiver homemaker	Construction worker, firefighter	Such variables are typically continuous, as is parental care, and so variants are typically having values in the range typical of the opposite sex
Psychological Processes: Cognition, Emotion, Social Styles	Verbal fluency, emotional intelligence, social aggression	Visuospatial reasoning, physical aggression	Such psychological processes typically are continuous, as is verbal fluency; thus, variants typically have values in the range typical of the opposite sex

sion without comparable variation in the others. This disassociation clearly demonstrates their independence. Thus, future research on the impact of interactions among social, behavioral, and genetic factors on health must determine which of these facets and dimensions contribute directly to sex differences in health and which are merely correlates.

An example helps to illustrate human variation. There are XY individuals with a genetic variant of the androgen receptor who are unambiguously heterosexual women and who are engaged in feminine social roles ranging from actresses to Olympic athletes. They have testes and hormone levels higher than those of pubertal boys. But, because their androgen receptors do not bind androgen, their genitalia, secondary sex characteristics, and musculature are fully differentiated as women. Until the Olympic committee changed its definition of sex from genetic to hormonal sex, such women had to compete as men. These women share the health risk of gonadal cancer, and typically their testes—their source of estrogens—are removed. However, their social roles—as actresses or Olympic athletes, for example—are better predictors of cardiovascular health and risk for muscle injury.

Moreover, sex/gender differences in health represent another arena that demonstrates powerfully that taking only a statistical approach to the problem of gene-environment interactions, and simply dividing variance in health into main effects and interactions, blinds researchers to the multitude of inseparable gene-environment interactions that have co-evolved to enable survival and successful reproduction. An excellent model for conducting research on development in dynamic terms was put forth in the National Research Council/Institute of Medicine (NRC/IOM) report entitled *From Neurons to Neighborhoods: The Science of Early Childhood Development* (2000).

The constructs of race and ethnicity, which have similar limitations and complexity as sex and gender, are explored in the following section.

RACE/ETHNICITY

Unlike sex, race is not firmly biologically based but rather is a "construct of human variability based on perceived differences in biology, physical appearance, and behavior" (IOM, 1999). According to Shields and colleagues (2005),

> *with the exception of the health disparities context, in which self-identified race remains a socially important metric, race should be avoided or used with caution and clarification, as its meaning encompasses both ancestry . . . and ethnicity . . .*

Both race and ethnicity can be potent predictors for disease risk; however, it is important to emphasize the distinction between correlation and causation and to explore interactions among factors, while rejecting a unidirectional model that moves from genotype to phenotype.

With the increased attention being given to racial disparities in health, the definition of race has come under increased scientific scrutiny. Race continues to be one of the most politically charged subjects in American life, because its associated sociocultural component often has led to categorizations that have been misleading and inappropriately used (Kittles and Weiss, 2003). Definitions of race involve descriptions that are embedded in cultural as well as biological factors, and a careful distinction must be made between race as a statistical risk factor and as causal genetic variables (Kittles and Weiss, 2003). Thus, genetics cannot provide a single all-purpose human classification scheme that will be adequate for addressing all of the multifaceted dimensions of health differentials. It may be found that some alleles associated with destructive or protective factors related to disease and health are created, modified, or triggered by cultural and contextual factors.

Race also is notoriously difficult to define and is inconsistently reported in the literature and in self-reports. Self-report has been the classic measure for race and is still reliable in some cases given certain caveats. The usefulness of the data derived from self-reports of race in health research, however, has been the subject of much debate (Risch et al., 2002; Cooper et al., 2003; Burchard et al., 2003). In 2003, Burchard and colleagues wrote the following:

> Excessive focus on racial or ethnic differences runs the risk of undervaluing the great diversity that exists among persons within groups. However, this risk needs to be weighed against the fact that in epidemiologic and clinical research, racial and ethnic categories are useful for generating and exploring hypotheses about environmental and genetic risk factors, as well as interactions between risk factors, for important medical outcomes. Erecting barriers to the collection of information such as race and ethnic background may provide protection against the aforementioned risks; however, it will simultaneously retard progress in biomedical research and limit the effectiveness of clinical decision-making.

Although there are requirements for reporting race in specific categories in federally sponsored research, the Office of Management and Budget directive that set out this requirement notes that these are not scientific categories. The National Institutes of Health (NIH) has reiterated that researchers should collect any additional data that would be more useful or appropriate for their specific projects. Researchers would advance our understanding of race and ethnicity by addressing factors that are related to race such as geographic area of ancestry or by providing greater detail about ancestors. In the 2000 Census, less than 3 percent (6.8 million) of the total population reported being of mixed race, and 7 percent of these 6.8 million people reported a heritage that included 3 or more races (Grieco and Cassidy, 2001). However, even those who report one race may have very complex backgrounds in terms of geography. For example, a black American could have origins in East Africa, West Africa, North Africa, or the Caribbean.

NIH has prescribed that all research projects will involve a good faith effort to include minorities when appropriate. By requiring funded research to make appropriate accommodations for minority subject recruitment, NIH has encouraged scientists to begin to consider issues of race, ethnicity, and culture in research as never before. Some of the emphasis on learning more about minority populations arises from the acknowledgement of the stark disparities in health when comparisons are made across racial groups.

Health Disparities and Race

Disadvantages in health exist for many groups such as Pacific Islanders, Hispanics, and Native Americans, when compared to Caucasians. Asians on many accounts are found to have more positive health profiles but are not without disadvantages in comparison with Caucasians (Whitfield et al., 2002). Literature on health disparities has documented African American/ Caucasian differences in major causes of death such as hypertension, diabetes, fatal stroke, and heart disease. The gap in health seems to be greatest between the ages of 51 and 63 (Hayward et al., 2000). Despite the 30-year trend toward convergence, the age-adjusted mortality rate from all causes of death for African Americans remains 1.3 times greater than that of Caucasians. This differential produces a life expectancy gap between African Americans and Caucasians of 5.3 years for men and 4.4 years for women (Hoyert et al., 2006). Furthermore, it also appears that African Americans are less likely to survive to middle age, and if they do, they are more likely to have health problems (Hayward et al., 2000).

Health disparities are a major public health concern and are a major emphasis of research across the country and across many disciplines. Genetic, social, and behavioral studies have shown that there are a large number of correlated differences across ethnic groups at the genetic, cultural, and environmental levels. From a methodological point of view, any comparison across ethnic groups from a single disciplinary vantage point will have a tremendous confounding issue. It is only by studying the multiple levels and risk factors simultaneously within subgroups (defined by ethnicity, geography, genetic backgrounds, and exposures to the environment) that we will begin to understand how specific combinations of environmental factors combine with specific combinations of genetic factors to give rise to health differences.

Race and Genetic Variation

Geographic origin, patterns of migration, selection, and historic events can lead to development of populations with very different genetic allele frequencies. Historically, to the extent that barriers such as large deserts or bodies of water, high mountains, or major cultural factors impeded communication and interaction of people, mating was restricted within group, producing genetic marker differences and thus, differences in the presence of specific disease-related alleles (see Box 5-1) (Kittles and Weiss, 2003). In line with this, Burchard and colleagues (2003) found that population genetic research of the last 20 years shows that the largest genetic differences occur between groups separated by continents. However, an analysis of 134 meta-analyses of genetic association studies by Ioannidis et al. (2004) found "at least 85% of genetic variation is accounted for by within-population interindividual differences, not by differences between groups."

Claims about correlations among genetic variation and race vary widely. Self-identified race/ethnicity corresponds highly to genetic cluster categories according to Tang and colleagues (2005); of the 3,636 individuals studied, less than 1 percent exhibited differences between their self-identified race/ethnicity and genetic cluster membership. However Bamshad (2005) in his review of the literature suggests that while genetic ancestry and geographic ancestry are correlated, race and genetic ancestry is only modestly related.

Research into differences among population groups often uses single nucleotide polymorphism (SNP) markers to identify phenotypic variation. SNPs may affect a given phenotype at multiple levels so that a given protein is altered in its sequence, in its proper place in the organism, and in its proper development time. A codon may be altered that leads to protein with an altered amino acid sequence which results in either an inactive or a hyperactive form of the protein in every cell where the protein is expressed. A part of the promoter may be altered such that a protein is absent in some of its normal tissues but not in others or is present in the wrong tissue or at the wrong time. An mRNA splice site may be altered such that protein isoforms are inappropriately expressed in a given tissue. A target sequence may be altered leading to aberrant targeting of the protein to cellular compartments. An untranslated sequence in the 3′-end of the gene may be altered to give a longer or shorter period of existence for a given mRNA. Finally, an epigenetic mechanism may be altered leading to changes in developmental timing of a particular protein.

Due to evolutionary history, sequence is more highly conserved in cod-

BOX 5-1
The Importance of Ancestral Origin

Despite the complexities and care that must be taken in attributing phenotypic differences to genetic differences among races, much may be gained by focusing on disorders that occur more frequently within a well-defined population. Ethnic groups or groups that share common ancestry have the same total frequency of genetic disorders (Clayton-Smith and Donnai, 2002; Rimoin et al., 2002). However, each differs in the frequency of different specific genetic disorders, which may have occurred through a variety of mechanisms, such as founder effects and bottlenecks. One of these mechanisms is natural selection, when heterozygotes have a selective advantage. In the homozygous state, however, these alleles lead to deleterious disorders. For example, sickle cell disease is thought to occur with high frequency in populations originating in Africa where malaria is common, because the heterozygotes are relatively resistant to malaria.

By studying these diseases within the populations in which they are most common, it has been possible to identify the genes responsible for some of these disorders, knowledge that can then be used to alter the incidence of disease. Exciting outgrowths of these investigations would be the development of genotype-specific prevention strategies and the eventual development of disease-specific treatments to benefit affected individuals—although the possibility of incomplete penetrance always must be considered, especially for complex diseases.

For example, in addition to having a shared religion, a cultural heritage, an oral tradition, and a written language, Jews also share a common gene pool, dating back to their common origins almost 4,000 years ago. Although the frequency of genetic diseases in general is no greater in Jews than in any other ethnic group, this shared genetic background has resulted in certain hereditary diseases occurring at a higher frequency in individuals of Jewish ancestry (Abel, 2001). Because of the historical migrations of Jews out of Israel over the millennia and the subsequent centuries of long geographic separation of segments of the Jewish community, there are disorders that are more common among certain subgroups within the Jewish community; such as Ashkenazi, Sephardi, and Persian.

ing regions when compared to noncoding regions. This feature creates the following situation in the genetic research of traits of great importance for public health: the interactions of SNPs with environment will be subtle and so will require large studies comprised of large cohorts carefully phenotyped for large numbers of environmental factors and genotyped for thousands of SNPs. Yet another challenge facing investigation using SNPs is that the bulk of SNPs found are not located in the conserved coding regions. Coordination of researchers involved in studies of humans, of other mammalian systems, of protein biochemistry and site-directed mutagenesis, and of cellular biology will be required to understand the interaction of genes and

- Diseases, such as Tay-Sachs disease and Gaucher disease, are relatively common among Ashkenazi Jews. Many other less common disorders (Bloom syndrome, Familial Dysautonomia, Niemann Pick disease and Canavan's syndrome) also occur at higher incidence in individuals with a European Jewish heritage (Kaback, 2001; Brady, 2006).
- More common among other branches of Jewish people is Familial Mediterranean Fever and Beta-thalassemia in Sephardic Jews, while Persian Jews experience Inclusion Body Myopathy more often (Shohat et al., 1992; Zeharia et al., 2005).
- Of more far-reaching clinical impact is the recognition that several of the most common diseases of mankind are seen at a high frequency within Jewish families, and within Jewish families there are characteristic molecular genetic markers for these diseases. Among these are coronary artery disease, inflammatory bowel disease (ulcerative colitis and Crohn's disease), diabetes, and certain forms of cancer, such as breast and colon cancer (McClain et al., 2005).

Screening for carriers of Tay-Sachs disease has virtually eliminated this once common and devastating disorder among Jews. Research into the biochemical basis of Gaucher disease has led to enzyme replacement therapy, which is of enormous benefit to affected individuals. Testing the relatives of individuals with genetic disorders (such as cystic fibrosis or Canavan's disease) can help prevent the recurrence of these disorders in the family. Screening for mutations associated with breast cancer has relieved the anxiety of many women who have seen their female relatives develop cancer and has allowed for more careful follow-up of those who are at higher risk. Additional research will undoubtedly lead to more effective screening and treatment programs for other disorders that affect Jewish families (McGinness and Kaback, 2002).

Other subpopulations have higher frequencies of certain diseases that have strong genetic contributions. Cystic fibrosis is more common in the Scots and Irish, while thalassemias are more prevalent in Mediterranean populations. Hemochromatosis is associated with a mutant allele (C282Y) that is found in all European groups and at especially high frequency (8 to 10 percent) in northern Europeans, but it is virtually absent in non-Caucasian groups (Merryweather-Clarke et al., 2000).

environment required to make an impact on public health in the United States.[1]

The use of SNPs also may aid in understanding variations in health outcomes among racial/ethnic groups. Using a sample that included a small number (less than 50 each) of African Americans, Hispanics, Asians, and Europeans, Smith et al. (2001) found that distribution of genetic variants

---

[1]The committee would like to thank Kent Taylor, Ph.D., Associate Director, Genotyping Laboratory, Medical Genetics Institute at Cedars Sinai Medical Center for his explication of SNP variation.

showed a median difference of 15 to 20 percent at both the microsatellite and SNP markers. Additionally, 10 percent of all markers showed a difference of 40 percent or more. To the extent that findings from this study reflect the larger population, one would hypothesize that an allele with 20 percent or greater frequency in one racial group would also be found in another racial group, while those with a frequency below 20 percent would most likely be race-specific.

According to Burchard (2003), "race-specificity of variants is particularly common among Africans, who display greater genetic variability than other racial groups and have a larger number of low-frequency alleles." Burchard concludes that variation among racial groups in the occurrence of variant alleles underlying disease or normal phenotypes may lead to differences in occurrence of the phenotypes themselves. For example, in some studies of hypertension, variation of SNPs at different allelic frequencies from one population to another suggest that higher rates of hypertension found in African Americans may be related to the alternations in DNA that vary by group (Cui et al., 2003; Erlich et al., 2003). Prior to drawing conclusions, however, one must consider alternative explanations that include gene-environment interactions as possible contributors to observed disparities (Whitfield and McClearn, 2005).

Arguments that genetic factors cannot be a major cause of health disparities arise out of a paradigm of genetic research that focuses on independent effects of genetics. Research on health disparities is an important opportunity to integrate biological knowledge with social and behavioral knowledge in order to better understand the determinants of disease. Social factors are certainly key contributors, but there is evidence that those factors do not account for all health differences (Braun, 2002). Conversely, solely focusing on molecular genetics ignores the dynamic nature of populations of DNA and the complex relationships among genes, organisms, and environment.

Considerable literature exists concerning how environmental processes, events, and circumstances contribute to development and behavior in ways that influence health as well. Some of these environmental factors are negative and are found to be more prevalent in the development of minorities. Some research suggests that African Americans may experience events and circumstances that have sociocultural origins that significantly influence development over the life course (Levine, 1982; Spencer et al., 1985; McLoyd and Randolph, 1985; Jackson, 1985; Jackson and Chatters, 1986). These sociocultural influences contribute to differences between racial groups as well as to differences between individuals within groups (Krauss, 1980; Levine, 1982; Jackson and Chatters, 1986). Sources of individual differences in health and behavior in African Americans have implications for the quality of late life as well as quantity of late life (years of life remaining). The multiple jeopardy hypothesis, for example, holds that negative environmental, social, and economic conditions during the early years of life for African Americans detrimentally affect social, psychological, and biological conditions in late life (Jackson, 1989). Although this hypothesis attempts to explain health differentials experienced by African Americans relative to Caucasians, it is critical to remember that there is considerable individual variability in these conditions *within* the African American population and within other minority populations.

In the search for the environmental origins of health differentials among ethnic groups, much of the earlier research focused on behaviors and social structures (NRC, 2001). The complexity of variables within racial groups presents challenges to identifying single, simple causes for poor health among racial/ethnic minorities. For example, environmental and behavioral variability among Hispanics evinces similarities and differences among its subgroups. This racial/ethnic (Hispanic) category consists of people from more than 20 different origins, but the people share a common language. Conversely, the groups within the Hispanic category significantly differ in their regional concentrations in the United States (e.g., Mexicans in the Southwest, Puerto Ricans in the Northeast, and Cubans in the Southeast) (NRC, 2001). In the United States, a significant relationship between race/ethnicity and foreign birth status also is found (NRC, 2001). Contrasts between immigrants and their U.S.-born peers suggest an advantage in health status to those who are foreign born (Singh and Yu, 1996; Hummer et al., 1999), at least until they become oriented to American culture. Then the advantage decreases (Vega and Amaro, 1994).

Perhaps the most studied social variable in the search for environmental origins of health differentials is socioeconomic status (SES) (see Chapter 2). For example, substantial differences exist between African Americans and Caucasian Americans with regard to their socioeconomic position. Thus, according to the U.S. Census Bureau's Current Population Survey (DeNavas et al., 2005), the median income for African American households was $30,134 in 2004 (the latest year for which data are available), compared to $48,977 among non-Hispanic Caucasian Americans. Poverty rates among African American households are nearly three times as high (24.7 percent in 2004), compared to Caucasian households (8.6 percent). Comparing households reporting similar levels of income, African American households report substantially lower levels of net wealth compared to Caucasian Americans (Conley, 1999). These differences in income and wealth are partly attributable to differences in average educational attainment when comparing African Americans (17.6 percent of whom reported having bachelor's degree or higher in 2004) to Caucasian Americans (30.6 percent of whom had a bachelor's degree or higher) (U.S. Census Bureau, 2005). Racial differences in intergenerational transfers of wealth, the growth of home equity over time, and access to federal programs that facilitated home ownership after World War II have played an even larger role in racial disparities in wealth over time (Oliver and Shapiro, 1997). African Americans also report higher levels of uninsurance (19.7 percent in 2004) compared to Caucasian Americans (11.3 percent) (DeNavas et al., 2005).

Research reveals that these socioeconomic differences between races account for a substantial portion of the racial disparity in health outcomes (IOM, 2000). At the same time, adjusting for socioeconomic differences does not completely eliminate racial disparities for all health outcomes (e.g., infant mortality). In other words, there is an independent contribution of racial/ethnic status to disparities in specific health outcomes. These residual health differences may result from the adverse health consequences of perceived discrimination for African Americans (IOM, 2000), from potential differences in biological susceptibility to disease, and/or from gene-environment interactions.

A universal finding is that people with higher indices of SES (education, income, and occupational grade) have lower mortality rates and lower rates of most diseases. However, more research is needed on how particular markers of SES show linear or nonlinear effects on health status (NRC, 2001). These gradients will be critical to understand in examining how genetic influences vary in social environments.

One of the future and formidable challenges to using the information ascertained from adding genetic information to examinations of health differentials is to gain an understanding of the underlying effect genes have on health within these complex environments. It may be found that the polymorphisms that occur in genotypes are destructive or protective factors related to disease and health that are created, modified, or triggered by cultural and contextual factors (Whitfield, 2005; Whitfield and McClearn, 2005).

CONCLUSION

Sex-linked biology and gender relations, as well as the concepts of race and ethnicity, require conceptual clarity in order to determine the interactive influences of each in giving rise to health differentials. To narrowly focus on such concepts impedes an appreciation of the rich variety among humans, however attention must be given to these and other categories in order to conduct meaningful research assessing the impact on health of interactions among social, behavioral, and genetic factors. For example, although a consistent genetic effect across racial groups can result in genetic variants with a common biological effect, that effect can be modified by both environmental exposures and the overall admixture of the population. The challenge is to parse out how health outcomes are influenced by genetic variations, behavioral and cultural practices, and social environments independently and as they interact with each others, while recognizing that sex, gender, race, and ethnicity may play important roles in their own right and because of their social meanings.

REFERENCES

Abel EL. 2001. *Jewish Genetic Disorders: A Layman's Guide.* Jefferson, NC: McFarland & Company.

Arber S, McKinlay J, Adams A, Marceau L, Link C, O'Donnell A. 2006. Patient characteristics and inequalities in doctors' diagnostic and management strategies relating to CHD: A video-simulation experiment. *Social Science and Medicine* 62(1):103-115.

Bamshad M. 2005. Genetic influences on health: Does race matter? *Journal of the American Medical Association* 294(8):937-946.

Brady RO. 2006. Enzyme replacement for lysosomal diseases. *Annual Review of Medicine* 57:283-296.

Braun L. 2002. Race, ethnicity, and health: Can genetics explain disparities? *Perspectives in Biology and Medicine* 45(2):159-174.

Bruno A, Conus S, Schmid I, Simon HU. 2005. Apoptotic pathways are inhibited by leptin receptor activation in neutrophils. *Journal of Immunology* 174(12):8090-8096.

Burchard EG, Ziv E, Coyle N, Gomez SL, Tang H, Karter AJ, Mountain JL, Perez-Stable EJ, Sheppard D, Risch N. 2003. The importance of race and ethnic background in biomedical research and clinical practice. *New England Journal of Medicine* 348(12):1170-1175.

Clayton-Smith J, Donnai D. 2002. Human malformations. In: Rimoin D, Connor J, Pyeritz R, Korf B, editors. *Emery and Rimoin's Principles and Practice of Medican Genetics.* 4th edition. New York: Churchill Livingstone. Pp. 488-500.

Conley D. 1999. *Being Black, Living in the Red: Race, Wealth, and Social Policy in America.* Los Angeles: University of California Press.

Cooper RS, Kaufman JS, Ward R. 2003. Race and genomics. *New England Journal of Medicine* 348(12):1166-1170.

Cui J, Zhou X, Chazaro I, DeStefano AL, Manolis AJ, Baldwin CT, Gavras H. 2003. Association of polymorphisms in the promoter region of the PNMT gene with essential hypertension in African Americans but not in whites. *American Journal of Hypertension* 16(10): 859-863.

DeNavas-Walt C, Proctor BD, Hill Lee C (U.S. Census Bureau). 2005. *Income, Poverty, and Health Insurance Coverage in the United States: 2004.* Washington, DC: U.S. Government Printing Office.

Erlich PM, Cui J, Chazaro I, Farrer LA, Baldwin CT, Gavras H, DeStefano AL. 2003. Genetic variants of WNK4 in whites and African Americans with hypertension. *Hypertension* 41(6):1191-1195.

Grieco E, Cassidy R. 2001. *Overview of Race and Hispanic Origin: Census 2000 Brief.* Washington, DC: U.S. Census Bureau.

Hallschmid M, Benedict C, Schultes B, Fehm HL, Born J, Kern W. 2004. Intranasal insulin reduces body fat in men but not in women. *Diabetes* 53(11):3024-3029.

Hayward MD, Crimmins EM, Miles TP, Yu Y. 2000. The significance of socioeconomic status in explaining the racial gap in chronic health conditions. *American Sociological Review* 65(6):910-930.

Hermes GL, Rosenthal L, Montag A, McClintock MK. 2006. Social isolation and the inflammatory response: Sex differences in the enduring effects of a prior stressor. *American Journal of Physiology. Regulatory, Integrative and Comparative Physiology* 290(2): R273-282.

Hochschild A. 1989. *The Second Shift*. New York: Viking.

Hoyert DL, Heron M, Murphy SL, Kung HC (National Center for Health Statistics). 2006. *Deaths: Final Data for 2003*. [Online]. Available: www.cdc.gov/nchs/products/pubs/pubd/hestats/finaldeaths03/finaldeaths03.htm [accessed June 7, 2006].

Hummer RA, Rogers RG, Nam CB, LeClere FB. 1999. Race/ethnicity, nativity, and U.S. adult mortality. *Social Science Quarterly (University of Texas Press)* 80(1):136-153.

Ioannidis JPA, Ntzani EE, Trikalinos TA. 2004. 'Racial' differences in genetic effects for complex diseases. *Nature Genetics* 36(12):1312-1318.

IOM (Institute of Medicine). 1999. *The Unequal Burden of Cancer: An Assessment of NIH Research and Programs for Ethnic Minorities and the Medically Underserved*. Washington DC: National Academy Press.

IOM. 2000. *Promoting Health: Intervention Strategies from Social and Behavioral Research*. Washington, DC: National Academy Press.

IOM. 2001. *Exploring the Biological Contributions to Human Health: Does Sex Matter?* Washington, DC: National Academy Press.

Ippolito G, Puro V, Heptonstall J, Jagger J, De Carli G, Petrosillo N. 1999. Occupational human immunodeficiency virus infection in health care workers: Worldwide cases through September 1997. *Clinical Infectious Diseases* 28(2):365-383.

Jackson J. 1985. Race, national origin, ethnicity, and aging. In: Binstock R, Shanas E, editors. *Handbook of Aging and Social Sciences*. New York: Van Nostrand-Reinhold. Pp. 264-303.

Jackson J, Chatters LNH. 1986. The subjective life quality of Black Americans. In: Andrews F, editor. *Research on the Quality of Life*. Ann Arbor, MI: Institute for Social Research.

Jackson JPC. 1989. Physical health conditions of middle and aged blacks. In: Markides K, editor. *Aging and Health: Perspectives on Gender, Race, Grace, Ethnicity, and Class*. Newbury Park, CA: Sage. Pp. 111-176.

Kaback MM. 2001. Screening and prevention in Tay-Sachs disease: Origins, update, and impact. *Advances in Genetics* 44:253-265.

Kawachi I, Kennedy BP, Gupta V, Prothrow-Stith D. 1999. Women's status and the health of women and men: A view from the States. *Social Science and Medicine* 48(1):21-32.

Kittles RA, Weiss KM. 2003. Race, ancestry, and genes: Implications for defining disease risk. *Annual Review of Genomics and Human Genetics* 4(1):33-67.

Krauss I. 1980. Between and within group comparisons in aging research. In: Poon L, editor. *Aging in the 1980's: Psychological Issues*. Washington, DC: American Psychological Association. Pp. 542-551.

Krieger N. 2003. Genders, sexes, and health: What are the connections—and why does it matter? *International Journal of Epidemiology* 32(4):652-657.

Lee S, Colditz GA, Berkman LF, Kawachi I. 2003. Caregiving and risk of coronary heart disease in U.S. women: A prospective study. *American Journal of Preventive Medicine* 24(2):113-119.

Levine E. 1982. Old people are not alike: Social class, ethnicity/race, and sex are bases for important differences. In: Sieber J, editor. *The Ethics of Social Research*. New York: Springer-Verlag. Pp. 127-144.

Mayes JS, Watson GH. 2004. Direct effects of sex steroid hormones on adipose tissues and obesity. *Obesity Review* 5(4):197-216.

McClain MR, Nathanson KL, Palomaki GE, Haddow JE. 2005. An evaluation of BRCA1 and BRCA2 founder mutations penetrance estimates for breast cancer among Ashkenazi Jewish women. *Genetics in Medicine: Official Journal of the American College of Medical Genetics* 7(1):34-39.

McGinness M, Kaback M. 2002. Heterozygote testing and carrier screening. In: Rimoin D, Connor J, Pyeritz R, Korf B, editors. *Emery and Rimoin's Principles and Practice of Medical Genetics*. 4th edition. New York: Churchill Livingstone. Pp. 752-762.

McKinlay JB. 1996. Some contributions from the social system to gender inequalities in heart disease. *Journal of Health and Social Behavior* 37(1):1-26.

McLoyd VC, Randolph SM. 1985. Secular trends in the study of Afro-American children: A review of child development, 1936-1980. *Monographs of the Society for Research in Child Development* 50(4-5):78-92.

Merryweather-Clarke AT, Pointon JJ, Jouanolle AM, Rochette J, Robson KJ. 2000. Geography of HFE C282Y and H63D mutations. *Genetic Testing* 4(2):183-198.

NRC (National Research Council). 2001. *New Horizons in Health: An Integrative Approach.* Washington, DC: National Academy Press.

NRC/IOM. 2000. *From Neurons to Neighborhoods: The Science of Early Childhood Development.* Washington, DC: National Academy Press.

Oliver ML, Shapiro TM. 1997. *Black Wealth, White Wealth: A New Perspective on Racial Inequality.* New York: Routledge Press.

Reskin B, Padavic I. 1994. *Women and Men at Work.* Thousand Oaks, CA: Pine Forge Press.

Rimoin D, Connor M, Pyeritz R, Korf B. 2002. Nature and frequency of genetic disease. In: Rimoin D, Connor J, Pyeritz R, Korf B, editors. *Emery and Rimoin's Principles and Practice of Medical Genetics.* 4th edition. New York: Churchill Livingstone. Pp. 55-59.

Risch N, Burchard E, Ziv E, Tang H. 2002. Categorization of humans in biomedical research: Genes, race and disease. *Genome Biology* 3(7):comment 2007.1-2007.12.

Sheu WH, Lee WJ, Chen YT. 1999. High plasma leptin concentrations in hypertensive men but not in hypertensive women. *Journal of Hypertension* 17(9):1289-1295.

Shields A, Fortun M, Hammonds EM, King PA, Lerman C, Rapp R, Sullivan PF. 2005. The use of race variables in genetic studies of complex traits and the goal of reducing health disparities: A transdisciplinary perspective. *American Psychologist* 60(1):77-103.

Shohat M, Bu X, Shohat T, Fischel-Ghodsian N, Magal N, Nakamura Y, Schwabe AD, Schlezinger M, Danon Y, Rotter JI. 1992. The gene for familial Mediterranean fever in both Armenians and non-Ashkenazi Jews is linked to the alpha-globin complex on 16p: Evidence for locus homogeneity. *American Journal of Human Genetics* 51(6):1349-1354.

Singh GK, Yu SM. 1996. Adverse pregnancy outcomes: Differences between U.S.- and foreign-born women in major U.S. racial and ethnic groups. *American Journal of Public Health* 86(6):837-843.

Smith MW, Lautenberger JA, Shin HD, Chretien JP, Shrestha S, Gilbert DA, O'Brien SJ. 2001. Markers for mapping by admixture linkage disequilibrium in African American and Hispanic populations. *American Journal of Human Genetics* 69(5):1080-1094.

Spencer MB, Brookens GR, Allen WR. 1985. *Beginnings: The Social and Affective Development of Black Children.* Hillsdale, NJ: Lawrence Erlbaum Associates.

Tang H, Quertermous T, Rodriguez B, Kardia SL, Zhu X, Brown A, Pankow JS, Province MA, Hunt SC, Boerwinkle E, Schork NJ, Risch NJ. 2005. Genetic structure, self-identified race/ethnicity, and confounding in case-control association studies. *American Journal of Human Genetics* 76(2):268-275.

U.S. Census Bureau. 2005. *Table A: Summary Measures of the Educational Attainment of the Population, Ages 25 and Over: 2004.* [Online]. Available: www.census.gov/Press-Release/www/releases/archives/04eductableA.xls.

Vega WA, Amaro H. 1994. Latino outlook: Good health, uncertain prognosis. *Annual Review of Public Health* 15:39-67.

Whitfield KE. 2005. Studying biobehavioral aspects of health disparities among older adult minorities. *Journal of Urban Health* 82(2 Suppl 3):iii103-110.

Whitfield KE, McClearn G. 2005. Genes, environment, and race: Quantitative genetic approaches. *American Psychologist* 60(1):104-114.

Whitfield KE, Brandon DT, Wiggins SA. 2002. Sociocultural influences in genetic designs of aging: Unexplored perspectives. *Experimental Aging Research* 28(4):391-405.

Williams TJ, Pepitone ME, Christensen SE, Cooke BM, Huberman AD, Breedlove NJ, Breedlove TJ, Jordan CL, Breedlove SM. 2000. Finger-length ratios and sexual orientation. *Nature* 404(6777):455-456.

Zeharia A, Fischel-Ghodsian N, Casas K, Bykhovskaya Y, Tamari H, Lev D, Mimouni M, Lerman-Sagie T. 2005. Mitochondrial myopathy, sideroblastic anemia, and lactic acidosis: An autosomal recessive syndrome in Persian Jews caused by a mutation in the PUS1 gene. *Journal of Child Neurology* 20(5):449-452.

6

Embedded Relationships Among Social, Behavioral, and Genetic Factors

Over the past several decades, there has been an exponential increase in our understanding of the social, behavioral, and genetic components of health and disease. Accompanying that understanding is a need to more fully connect and integrate knowledge across all levels of these determinants of health. Such integration will provide a better understanding of how social factors are translated into physiological effects on cellular responses, including changes in gene expression. Likewise, the genomics revolution, catalyzed by the Human Genome Project, has stimulated widespread interest in how genetic variations may influence human behavior and response to social factors.

The previous chapters have implicitly used a linear, if not hierarchical, model to describe the strengths of and lacunae in our current understanding of reciprocal interactions among the various levels of organization: social factors, individual behavior and experience, physiological systems, and gene function. In this chapter we explore how future work must recognize that such a linear approach does not fully reflect the integrated nature of the social and physical environment and gene function that is the salient feature of biological systems. Instead, we must use a variety of models in order to address the fact that rarely is there a one-to-one relationship between genes and a trait.

Indeed, with only ~30,000 genes in the human genome, most genes are likely to serve different functions at different times and in different environments (McClintock et al., 2005). Moreover, the selection of our genome occurred when our ancestors migrated, through the interaction with different social and physical environments. This affected not only their life-span trajectories, fertility, health, and disease and survival rates, but also those of their children and grandchildren (see Chapter 5 for additional discussion). Thus information in the genome is inextricably linked with the cellular, physiological, psychological, social, and physical environments in which it functions over a lifetime, and many of these nongenetic factors are passed on to subsequent generations.

One of the limitations of a purely hierarchical perspective to integrating knowledge across levels is that, in reality, the effects of variation at any one level (e.g., gene, gene transcript, protein, metabolite, or tissue) are actually embedded in another level and are not simply "underneath" or "above" the other level. A well-established hierarchy is illustrated by the ways in which DNA is transcribed into messenger RNA, which is then translated into protein, which in turn is appropriately folded and chemically modified in order to perform a specific function in protein complexes. Conversely, an example of the complex, nonhierarchical, and embedded nature of biological information is the fact that some DNA variations affect transcription but are not found in the messenger RNA; other variants are transcribed and affect translation but are not found in the translated protein; and still others are transcribed, translated, and ultimately affect protein function. The following subsections further illustrate this concept and its implications for assessing the impact of associations and interactions among social, behavioral, and genetic factors on health.

THINKING FROM THE BOTTOM UP: GENOMIC INFORMATION INFLUENCING GENE EXPRESSION

The Human Genome Project and many other international efforts have been focused on understanding the nature of the genome and its variations. Millions of single nucleotide polymorphisms (SNPs) have been identified (e.g., see dbSNP from the National Center for Biotechnology Information[1]), and investigators around the world are engaged in performing genetic association studies in order to better understand the influences of these variations on measures of health and disease. It is well known that genetic variations within a gene can alter its expression both quantitatively and qualitatively. For example, mutations within the promoter region of a gene can influence when, where, and how much a particular gene is expressed (i.e., transcribed in messenger RNA). Currently, most gene expression studies ignore individual-level variation in gene expression due to genetic variation. However, over the past few years several landmark stud-

---

[1]See www.bioinfo.org.cn/relative/dbSNP%20Home%20Page.htm.

ies in humans have shown that genetic variation within a gene can have profound effects on gene expression. In a study by Lo et al. (2003), allele-specific expression of 602 transcribed SNPs was examined, and 54 percent of genes showed preferential expression of one allele over another. At least 25 percent of the 602 transcribed SNPs showed more than a four-fold difference in expression between the two alleles. Cheung et al. (2005) have demonstrated that the expression level of genes is highly heritable in humans and map onto different regions of the genome. In a small study of 14 pedigrees, variation in more than 1,000 genes expressed in human lymphocyte cell lines (out of 3,554 genes examined) was significantly heritable and linked to regions of the genome. Further, they found that only 374 of these 1,000 genes with heritable expression patterns showed evidence of possible mutations in their own gene region that directly affected transcription levels. Using a genome-wide association approach with >770,000 SNPs, Cheung et al. (2005) found 27 genes with the greatest evidence of inherited expression patterns could be divided into 2 approximately equal subsets—those with SNP associations in their genomic region (cis-effects) and those with SNP associations on different chromosomes (trans-effects). Functional analysis using allele-specific binding assays (HaploChip assay) were then used to confirm the results from the SNP association study. By utilizing transcriptomic and genomic data simultaneously, new insights into the causes of variability in gene expression are being discovered. This type of research (discussed in the following sections) could be very beneficial to understanding why some people in a population have adverse responses to environmental exposures while others do not.

Transcriptomics Technologies

Transcriptomics is a term used to describe the genome-wide measurement of mRNA transcripts in a particular tissue or cell line. The two main technologies used for genome-wide measurements of gene expression (mRNA expression) are DNA microarrays and serial analysis of gene expression (SAGE). In DNA microarray technology, thousands of known DNA sequences are bound systematically to a solid platform, and mRNA that has been extracted from a particular sample (and fluorescently labeled) is then hybridized to the DNA sequences. In contrast, SAGE is a high-throughput technology based on the sequencing of short sequence tags within each mRNA transcript found within a particular tissue. It provides a method of directly sampling the population of mRNAs in a cell rather than being restricted to preselected gene transcripts that have been placed on a chip. A number of important issues have been identified involving the reproducibility and standardization associated with these technologies and the massive datasets they produce. Progress in data quality and data sharing has been facilitated in large part by the creation of the Minimum Information About a Microarray Experiment guidelines by the Microarray Gene Expression Data Society (Brazma et al., 2001; Ball et al., 2002a; Ball et al., 2002b; Ball et al., 2004a; Ball et al., 2004b). The tremendous emphasis on data sharing of transcriptomic studies has been a major asset to the scientific community, both as a source of independent data that can be used as a means of validating results in diverse sample populations and in cross-species comparisons. It also has stimulated the development of new knowledge about global patterns of gene expression that are associated with particular cellular systems (Malek et al., 2002; Stuart et al., 2003).

The field of transcriptomics also has catalyzed the development of many novel statistical and pattern recognition methods, as researchers initially struggled to analyze massive amounts of data to identify genes whose expression profiles were found to be altered, co-regulated, or representative of key pathways thought to be activated by environmental exposures. Cluster analysis has been one commonly used tool for multidimensional visualization and the discernment of underlying subgroups of individuals with similar expression profiles (reviewed by Brun et al., 2004). Network models and supervised machine learning algorithms also have been important in generating new insights about key pathways in disease development or even predicting disease outcomes using these high-dimensional data. Through advances in bioinformatics it is now possible to merge gene expression data with additional data sources in order to aid the investigative process. For example, the hundreds of genes found to be associated with a disease or environmental exposure in a transcriptomic study can easily be linked to PubMed abstracts or the associated Medical Subject Heading terms (Jenssen et al., 2001; Fink et al., 2003; Doniger et al., 2003; Djebbari et al., 2005). Likewise, merging gene expression results with SNP databases, genetic linkage databases, epigenetic information on imprinting, comparative genomic hybridization arrays, proteomic databases, and metabolic pathway databases provides an unparalleled opportunity for integration across the levels of the molecular universe that characterizes our human biology. For example, the Gene Ontology Project (www.geneontology.org) attempts to classify gene products, assigning proteins to groups specifying their molecular function, the biological process to which they contribute, and their cellular component (Ashburner et al., 2000). Similarly, using Enzyme Commission numbers, genes can be mapped to metabolic and signaling pathway databases such as the Kyoto Encyclopedia of Genes and Genomes (KEGG) (www.genome.ad.jp/kegg) (Kanehisa, 2002). In general, microarray technology is an incredibly powerful tool used to investigate complex gene expression relationships on a genome-wide scale, and it likely will be invaluable in assessing the relationships among social, behavioral, and genetic factors as they relate to health and disease.

Epigenetic Phenomenon

Epigenesis originated as a term to describe the processes in embryonic development that transforms the undifferentiated cells in the newly fertilized egg into a complex, multitissue organism. Today, it is used in a much broader sense to represent everything from the general concept of the forces that shape how an individual's genotype gives rise to a particular phenotype (Waddington, 1957; Petronis, 2003) to the specific molecular mechanisms by which cells differentiate, age, change metabolic functions, or even transform into cancerous cells (Jablonka and Lamb, 2002). The most well-known mechanism for the epigenetic regulation of cell phenotypes is DNA methylation, which turns off a gene or gene region (i.e., keeps it from being expressed) by changing the chemical structure of the DNA (Jaenisch and Bird, 2003). Many different factors can affect the methylation pattern of genes and thus affect their expression. For example, as a normal part of human development genes are turned on and off using methylation processes stimulated by other gene products in the embryo, fetus, newly born infant, child, adolescent, and aging adult. Environmental factors such as infection and diet are also known to affect gene methylation. For example, the work of Waterland and Jirtle (2004) suggests that prenatal and postnatal nutrition can have long-lasting epigenetic effects on an adult's predisposition to obesity, cardiovascular disease (CVD), type 2 diabetes, and cancer. Rett syndrome is an example of a clinical syndrome typified by mental retardation and autistic-like behaviors that arises through the failure of these methylation processes (Shahbazian and Zoghbi, 2002).

In addition to the growing body of research on the environmental and developmental factors that affect epigenesis, there also is evidence that epigenetic patterns of gene expression may be inherited and can affect genetically inherited diseases. For example, through a process known as genetic imprinting, the methylation pattern in a parent is passed onto offspring through the germline and in some cases this has been associated with differential disease patterns.

In general, epigenetic phenomena are thought to govern a very wide array of biological processes that determine how genotypes interact with environmental factors in a complex, dynamic fashion to give rise to phenotypic variability both across individuals with the same genotype or same set of environmental exposures as well as across a person's lifetime. The ways in which these epigenetic processes impart a kind of cellular memory of activity and experience that is passed to daughter cells indicates that the timing of particular environmental exposures may be key to the development of particular diseases for individuals with particular genotypes. It also has been suggested that this cellular memory may lead individuals to select particular environments, thus creating a correlation between genotypes and environments (Carey, 2003; Gottesman and Hanson, 2005). From the standpoint of assessing the associations and interactions among social, behavioral, and genetic factors, epigenetic processes are likely to play a major role in determining how these seemingly disparate factors operate together to give rise to the distribution of disease in a population. These processes also are likely to explain differences in research results across studies or populations when only simple single biomarkers or social indicators are examined.

An increasing number of studies are starting to relate changes in DNA methylation patterns to altered patterns of gene expression that are associated with disease risk (reviewed by Jones, 2005). These observations have led to the development of technologies that are capable of scanning the genome for altered patterns of DNA methylation (e.g., Kaminsky et al., 2005). Nickel, cadmium, and xenobiotics (such as diethylstilbesterol or DES) all have been shown to affect gene methylation (Sutherland and Costa, 2003; Bombail et al., 2004). Methylation, as a means of inhibiting gene expression semi-permanently, means that some toxicological agents could have permanent effects on the genomic capacity of the individual to adapt to changing environments, including other toxic agents in their environment. CpG array-based technology is quickly advancing and now allows for the simultaneous detection of altered DNA methylation, histone acetylation, and gene expression (Shi et al., 2003). As this field progresses, it will be important to integrate epigenetic and genetic approaches in order to better model the risk of disease caused by environmental toxicants. Models of how to merge epigenotype and genotype information are now starting to emerge (Bjornsson et al., 2004), and more theoretical, as well as applied, work is needed in this area of toxicogenomics.

THINKING FROM THE BOTTOM UP: GENOMIC INFORMATION EMBEDDED IN BIOCHEMICAL SYSTEMS

At the molecular level, SNPs are simple DNA substitutions of one A, T, G, or C base in a DNA sequence for another. By knowing which portions of the DNA sequence actually code for the protein sequence, it is possible to predict whether a DNA sequence change (i.e., an SNP) will change the sequence of the protein. If it does, then it is quite possible that the activity of the protein will be altered and thus affect its metabolic or biochemical functionality. Currently, there are 30,000 SNPs identified that alter the DNA sequence of a gene in a way that alters the protein sequence it encodes. Approximately 60 percent of known genes have at least one SNP with a frequency of 1 percent or greater that changes its protein sequence. Moving our perspective to the level of biochemical and physiological systems, it can be seen that these variations in protein sequence now constitute a source of variation in the metabolic functionality of cellular and physiological systems.

It may be helpful to consider this from the perspective of the human population. With more than 6 million SNPs identified and each SNP giving rise to 3 possible genotypes in the population, there are $>3^{6 \text{ million}}$ possible genome types. Analogously, if 30,000 of these are translated into protein sequence differences, there are $>3^{30,000}$ unique proteomes possible in the population. This variation in proteome types will impact how social and behavioral factors are translated into variation in health and disease. In other words, many genomic variations are embedded in protein variations that are embedded in variability in cellular and physiological systems. It also should be noted that not all SNPs have a functional effect. Determining whether a particular SNP is associated with a disease, that is, actually having a biological effect, rather than being a correlate of the functional polymorphism, currently is consuming much time and effort.

Proteomics Technologies

Proteomics is the study of the full collection of proteins that make up our cellular and metabolic machinery. Because proteins are dynamically created and turned over as a part of normal cellular processes, proteins change in both quantity and activity depending on diet, stress, physical activity, and other environmental exposures. Each protein may be present in multiple chemically modified forms, and these protein modifications may be more critical to its metabolic or biochemical function than the amount of protein that is found in the cell (Mann and Jensen, 2003). Two major approaches used to measure the large collection of proteins in cells are gel-based proteomics and "shotgun" proteomics. In the gel-based approach, proteins are first separated by electrophoresis and then further resolved by another separation method (e.g., pH). Shotgun proteomic analysis involves relatively random digestion of complex protein mixtures followed by mass spectrometry analysis (Yates, 1998; Washburn et al., 2002). Another type of proteome analysis that has attracted widespread attention is proteomic profiling—a spectral profile of the proteins in a tissue or biofluid (e.g., serum)—using matrix-assisted laser desorption ionization-time of flight mass (Chaurand et al., 1999; Petricoin et al., 2002; Villanueva et al., 2004). The signals in these spectra contain hundreds or thousands of signals and represent intact proteins, as well as protein fragments, that collectively reflect the cellular protein machinery. This approach has been used for discovering novel biomarkers of diseases, especially cancers, that could be used for early detection (Conrads et al., 2004; Baggerly et al., 2005).

Metabonomic Technologies

Metabonomics (also known as metabolomics) is the analysis of small molecular products of biochemical and physiological processes. Since metabolism is a highly complex, dynamic, and adaptive set of systems, measurement of the metabonome, as well as proteomes and transcriptomes, is expected to change in response to diet, stress, physical environment, circadian rhythms, physical activity, developmental changes, and aging, as well as during disease development. The range of metabolic molecules is quite large, spanning from electrolytes to short-chain proteins to large lipid molecules or exogenous compounds (e.g., diet and drugs) that represent both anabolic and catabolic processes from multiple tissues and organ systems. The two main technologies for measuring the metabonome are nuclear magnetic resonance (NMR) spectroscopy and mass spectrometry. Although NMR has been used more extensively, mass spectrometry-based methods have much greater sensitivity and can detect molecules at up to 10,000-fold lower levels than NMR (Wilson et al., 2005; Brown et al., 2005). In some cases, this level of sensitivity is necessary; however, in most cases it is probably not needed. NMR spectra of urine contain thousands of signals representing thousands of metabolites (Nicholson et al., 2002). Using pattern recognition approaches, NMR spectra can be compared across samples to identify distinguishing patterns that reflect differences in environmental exposures. Given the current sophisticated algorithms for data processing and analysis, it is possible to chemically identify most of the peaks in a complex metabonomic spectra (Beckwith-Hall et al., 1998; Holmes et al., 1998) and in some cases tissue-specific injury or disease (Azmi et al., 2002; Griffin et al., 2004). By quantifying metabolite levels and mapping them onto known metabolic pathways new inferences can be drawn about the biochemical and cellular consequences of certain diseases (Griffin et al., 2004). Interestingly, metabomonic studies also are raising awareness of the important role that gut flora (estimated 1.5 kg/person) play in augmenting normal metabolism and how they may be a significant source of metabolic variability across individuals (Nicholson et al., 2005).

THINKING FROM THE TOP DOWN: SOCIAL FACTORS INFLUENCING CELLS, TISSUES, AND PHYSIOLOGY

In contrast to the "bottom up" approach, in a "top down" approach, external and human behavioral factors are mapped onto an individual's psychological response, which can then alter proteins, metabolites, and physiological processes. In some cases, these factors can influence signal transduction, which is a key pathway for modulating gene expression in response to environmental signals. As discussed earlier, variation in the target gene may affect how the signal is translated into a change in gene expression. This is a molecular example of gene-environment interaction.

One well-documented example of the top down approach of thinking is the study of the pathways involved in the physiological effects of stress. The direct connection between stress stimuli and the response of the neuroendocrine system was demonstrated by the work of Walter Cannon in the 1920s (Cannon, 1932). The expression "fight or flight" was first used by Cannon to illustrate the body's primitive physiological responses to perceived threats and other external stressors such as exposure to heat or cold.

The Effects of Stress

A vast body of research has been devoted to the study of the effects of stress on many biological processes throughout the life course, including CVD, immune function, and child development. Because psychologists, physiologists, and the general public use the word *stress* in many varying ways (Engle, 1985), there is no one agreed upon definition for the term. Individual perceptions of stress and the resulting response to the stressor depend on genetics, events that occur during early development, prior experiences with the stressor, and behavior, such as lifestyle choices (McEwen and Seeman, 1999). When using stress in relation to animals the term typically is used to describe the body and brain's various responses to the presence of a threat that could compromise the physical or psychological well-being of the animal (Selye, 1973; Selye, 1975). The complex physiologic response to stress alters the natural priorities set by the body and can result in substantial effects on normal health maintenance and development (Johnson et al., 1992).

Brain structures that mediate stress response (e.g., hypothalamus and brainstem) are also responsible for regulating vital body functions such as heart rate, respiration, digestion, reproduction, growth and development, sleep-wake cycles, and the establishment of energy stores in the absence of stress. When presented with a threat that surpasses the limits of the body's available resources and capabilities, the brain initiates the intricate pathways and feedback loops of the hypothalamic-pituitary-adrenal system (HPA). The HPA stimulates the production and release of steroid hormones, such as glucocorticoids, and neurotransmitters, such as catecholamines. The release of cortisol, a glucocorticoid, and epinephrine, a catecholamine that is also referred to as adrenaline, results in a multitude of effects that allow for a quick protective response against the threat in the short term, but have the potential for adverse effects if continued for an extended period of time.

The presence of cortisol and epinephrine activates and potentiates some biological processes of the body, while deactivating and dampening others.

Important sympathetic nervous system responses elicited by increased levels of cortisol and epinephrine include increased heart rate and respiration, increased blood flow to muscles, mobilization of white blood cells in anticipation of injury, and the degradation of energy stores, thus increasing levels of blood sugar. Increased levels of cortisol and epinephrine also suppress blood flow to the digestive system, dampen immune responses involved with fighting infection, and inhibit growth and reproductive hormones. Neurological effects of these important neurochemicals include sharpening vigilance and attention, while suppressing unnecessary short-term memory and learning functions (IOM, 2000).

It has been postulated that exposure to stress at early life stages may have effects on the stress response system that persist throughout the entire life course. Meaney et al. (1996) used animal studies to demonstrate that infantile rats exposed to short-term stress, such as handling, had decreased HPA activity, thus depressing responses to stressors throughout the life course. Conversely, rats exposed to prolonged stressors, such as maternal separation, physical trauma, and administration of endotoxins, had increased HPA activity, thus exacerbating response to stressors throughout the life course. In addition to these lasting HPA effects, increased levels of mRNA for corticotropin-releasing hormone (CRH) and arginine vasopressin (AVP), initiators of the stress response, were observed in the hypothalamus. Evidence from this study further indicated that exposure to stress early in life also affects the gene expression of glucocorticoid receptors, explaining the high levels of CRH and AVP, which are typically regulated through a negative-feedback loop involving the glucocorticoid receptors. These findings indicate that exposure to stress early in life can have monumental effects on the development of the HPA system and future responsivity to stressors that are presented throughout life (Meaney et al., 1996).

Chronic stress is implicated in many negative health outcomes that include diminished immune response, arthrosclerosis, resistance to glucocorticoids, and reproductive dysfunctions (Cavigelli and McClintock, 2003). Individuals exposed to chronic stress can suffer from *allostatic load*, which is the accumulation of negative physiologic effects such as those listed above. It is associated with persistent high levels of catecholamines and glucocorticoids, as well as the continued struggle to achieve allostasis during times of chronic stress exposure. Genes, early development, and behaviors such as diet and exercise, and tobacco and alcohol use (see Chapter 4 for further discussions of these behaviors) all contribute to an individual's allostatic load (McEwen and Seeman, 1999). In addition to allostatic load, Cavigelli and McClintock (2003) found that individuals with naturally high levels of glucocorticoid produced in response to stress also have decreased longevity.

Stress and CVD

The deleterious effects of stress on CVD were clearly defined in the 1950s by Selye (1956). Since that time, substantial evidence has amassed that supports the role of psychological factors in the etiology and progression of CVD. For example, Manuck et al. (1988) used animal studies to illustrate an increased rate of atherosclerotic plaque buildup in individuals with chronically high blood pressure and elevated levels of catecholamines as a result of persistent socially stressful situations. The buildup of atherosclerotic plaque is a factor in the development or complication of CVD, such as heart attack or stroke.

Another factor implicated in the risk of CVD is cholesterol. High blood concentration of cholesterol and other lipids due to prolonged exposure to stress can increase the risk of developing arthrosclerosis and the risks of additional heart disease complications. Stoney et al. (1999a; 1999b) found that levels of cholesterol in the blood varied according to the degree of perceived stress, and operated independently of modifications to health behaviors that are traditionally associated with cholesterol levels such as diet and exercise. Studies of more mild exposure to stress for shorter durations have also revealed elevated levels of cholesterol, specifically low-density lipoproteins, triglycerides, and other molecules associated with negative health outcomes and cardiovascular disease (Stoney et al., 1999a; Stoney et al., 1999b). Traditionally it has been assumed that levels of cholesterol in the blood are indirectly linked to chronic stress through the direct effects stress has on health behavior (i.e., diet choices and physical activity). However, Stoney has proposed a model of direct effect between stress and lipid concentration. This new model hypothesizes that exposure to short-term stressors that activate the sympathetic nervous system also reduces lipase activity, the enzymes responsible for lipid metabolism and storage, thus increasing the blood lipid level in times of stress (Stoney et al., 1999a; Stoney et al., 1999b).

Stress and Immune Function

A considerable amount of evidence has established a relationship between stress and the suppression of certain aspects of the immune system. It has also been determined that immune function during times of stress can be mediated by different factors in humans. After performing a meta-analysis of available literature, Herbert and Cohen (1993) determined for example that duration of exposure to stress played an important role in the level of the immune response. As previously mentioned, acute stress has a protective immune response. This is exhibited by increased levels of suppressor/cytotoxic T-cells. However, the presence of prolonged exposure diminishes levels of these important immune responders (Herbert and Cohen, 1993). This study also evaluated the differences in immune responses between objective stress and self-reported subjective stress. When compared to self-reported stress, objective stress clearly had larger alterations of immune system response in terms of measured levels of natural killer (NK) cell activity and Immuglobulin A levels in saliva (Herbert and Cohen, 1993).

Finally, the study analyzed whether stress originating from interpersonal/social situations had more of an impact on immune response than stress that was the result of nonsocial factors. Despite the inconclusive findings about which stress resulted in a greater immune response, it appears that social and nonsocial stressors induce different types of immune responses. Stress related to social experiences resulted in changes in the helper-to-suppressor ratio as well as the percent of suppressor/cytotoxic T in the blood, while nonsocial stressors elicited changes in the number B-cells and T-cells present and the percentage of helper T-cells in the blood (Herbert and Cohen, 1993).

A specific example of stress altering immune responses comes from animal studies conducted by Stefanski et al. (2005) that examined levels of serum immune cells and corticosterones in response to pregnancy and social stress. Under normal circumstances and in the absence of stress, levels of corticosterones gradually increased throughout the pregnancy, while levels of immune cells such as specific T-cells—CD4 CD4 T, CD8 T— B-cells, and lymphocyte proliferation, continually decreased. When confronted with daily social stressors throughout the course of pregnancy, levels of corticosterones drastically increased, while levels of immune cells such as NK cells, B-cells, and lymphocyte proliferation were all substantially reduced (Stefanski et al., 2005) (see Chapter 7 for a more extensive discussion of animal models in relation to stress response).

In addition to the negative health outcomes associated with stress and the immune response, evidence shows that both positive and negative emotional styles can also affect one's susceptibility to viral infection. Cohen et al. (2003) found a dose-response relationship between exhibiting a positive emotional style (i.e., feelings of happiness and being relaxed) and the risk of developing a cold after a systematic laboratory exposure to the virus.

Child Development

By the time young children reach one year of age, stress-mediating brain structures such as the amygdala are fully matured and allow children to experience fear, anxiety, and stress. Late infancy is marked by the natural stress response of fear when confronted by unfamiliar people (Bronson, 1971; Waters et al., 1975) and the response of anxiety when removed from the presence of recognizable caregivers (Ainsworth and Bell, 1970; Sroufe, 1979). These likely responses to short-term stressors play an important part in the emotional development of children and are not expected to have long-term adverse effects. However, as previously mentioned, stress responses inhibit normal growth and developmental processes that are an essential part of a healthy childhood, thus long-term or repeated exposures to stressors is likely to have negative effects on normal development (IOM, 2000).

Animal studies show that infants are particularly susceptible to stressful events, such as neglect, that have the potential to permanently alter the HPA system, resulting in hyperactive stress responses (Meaney et al., 1996; Denenberg, 1999). Decreased maternal attention such as licking and grooming have also been implicated in the development of more stress-reactive animals (Liu et al., 1997). Introducing an infant that is genetically predisposed to be more stress-reactive into the care of an adoptive mother that is genetically predisposed to be less stress-reactive causes the infant to develop with a higher than expected stress tolerance, implying a role of nurture in addition to the genetic predisposition that determines the characteristics of stress response.

Primate studies demonstrate the importance of maternal presence during early life stages. Monkeys that are separated from their biological mothers at a very young age and reared with a cloth surrogate, but provided with daily peer interactions, are less socially inept than monkeys reared in complete isolation. However, the monkeys reared with the cloth surrogate still produce a number of physiological indicators that point toward anxiety and fear (Suomi, 1991). When faced with stress, these animals produce higher levels of stress response neurochemicals such as glucocorticoids and catecholamines. Other studies indicate that monkeys reared without a cloth surrogate and only in the presence of infant peers exhibit parallel hyperactive stress responses to those reared with the surrogate (Champoux et al., 1989; Champoux et al., 1992).

As discussed, early life inputs, such as maternal presence and attention, can be crucial to the normal development of the stress response system as children grow into adults. These key inputs can keep stress response activity in check and result in the maturation of a response system that is capable of rapidly shutting down responses when the stressor has been removed. However, lack of this positive input can create a system that is hyperactive and unable to modulate responses to stimuli (NRC/IOM, 2000).

The 2000 National Research Council/Institute of Medicine report, *From Neurons to Neighborhoods: The Science of Early Childood Development,* highlights the importance and difficulty of crossing between disciplines to understand the multiple factors that influence early childhood development. The report recommends pursuing integrative science that includes:

a) understanding how experience is incorporated into the developing nervous system and how the boundaries are determined that differentiate deprivation from sufficiency and sufficiency from enrichment; b) understanding how biological processes, including neurochemical and neuroendocrine factors, interact with environmental influences to affect the development of complex behaviors, including self-regulatory capacities, prosocial or anti-social tendencies, planning and sustained attention, and adaptive responses to stress; c) describing the dynamics of geneenvironmental interactions that underlie the development of behavior and contribute to differential susceptibility to risk and capacity for resilience; and d) elucidating the mechanism that underlie nonoptimal birth outcomes and developmental disabilities (NRC/IOM, 2000).

MOLECULAR MECHANISMS OF GENE-ENVIRONMENT INTERACTION

From a molecular perspective, gene-environment interaction can mean different things to different researchers. For example, gene-environment interaction could refer to the regulation of gene transcription (i.e., gene expression) by signals from the environment binding to appropriate cell surface receptors and stimulating a signal transduction pathway that carries the molecular signal into the nucleus and eventually binding to the DNA in the promoter region of the gene to stimulate or inhibit its expression. In this case, environmental variation will increase variation in gene expression. From another perspective, gene-environment interaction could refer to how a DNA mutation in the gene alters its expression in response to the environment. In this case, genetic variation is contributing to variation in gene expression even in the absence of environmental variation. From a third perspective, gene-environment interactions occur when a DNA mutation changes the protein sequence encoded by a gene and the altered protein has a different activity than the nonmutant and acts differently when performing its role in a system that is processing an environmental factor. In this case, the molecularly embedded genetic information in the protein isoforms carried by the individual is translated into metabolic features that represent a gene-environment interaction.

Our understanding of the molecular mechanisms of gene-environment interactions are likely to continue to expand as the "omic" technologies deliver more insight into the high-dimensional microcosms that selforganize into the macro properties of human biology that have been fine tuned to adapt to social, behavioral, and physical environments.

THE NEED FOR SYSTEMS APPROACHES

The preceding sections of this chapter described many different levels and agents of influence on health ranging from the social to the genomic to the chemical. One of the most important contributions of the research of the past few decades, climaxing for geneticists with the Human Genome Project, is that it is pushing scientists toward a more holistic view of human biology. As scientists try to put the pieces of the puzzle together, the natural step beyond examining single agents of health and disease is to move toward a systems view. Recently, there has been a resurgence in the amount of attention that has been given to systems biology because of the vast amount of data that can now be collected at the genomic, transcriptomic, proteomic, and metabolomic levels. However, systems theories and methods have a long tradition in science. The development of path analysis by Sewall Wright in the 1940s—a correlational approach—was one of the first attempts at studying states and relationships among many variables in order to understand the whole. This work more recently has evolved to use the sophisticated statistical method called Structural Equation Modeling (Hoyle, 1995; Maruyama, 1997), which has been used successfully in the behavioral and social sciences. The development of a general systems theory approach by Bertalanffy (1968) to describe dynamical systems catalyzed the development of new methods of analysis such as Biochemical System Theory (Savageau, 1976) and Metabolic Control Theory (Kacser and Burns, 1973). Arthur Guyton's work using control theory to model the regulation of physiological systems (Guyton, 1976) is another important example of the use of systems concepts to model wholes from parts.

In attempting to build bridges between social, behavioral, and genetic information about health and disease, investing in new systems approaches is likely to yield many new insights in areas of investigations such as how small nonlinear effects result in significant health outcomes. One of the most difficult aspects of integrating this knowledge into a systems approach is that the information is organized somewhat but not exactly hierarchically. For example, a traditional hierarchical view of biology looks something like this: DNA → mRNA → protein → protein interactions → metabolic pathway → metabolic networks → cells → tissues → organs → organisms → populations → ecologies. However, there also is feedback from the ecology to the organism to metabolic pathways to the DNA, which does not strictly follow the same pathways. Biological information has several important features: it operates on multiple hierarchical levels of organization at the same time and thus is indeed embedded. It is processed in complex networks. These information networks are typically robust, such that many single perturbations will not greatly affect them. There are key nodes in the network where perturbations may have profound effects;

these offer powerful targets for the understanding and manipulation of the system (Ideker et al., 2001). The central task of a systems approach is to (a) comprehensively gather information from each of the distinct levels, (b) examine relationships among the agents of the system, (c) hypothesize system topologies, (d) integrate data into predictive mathematical models of the system, (e) test predictions, and (f) identify key regulatory signals and relationships where intervention could stimulate new outcomes.

There are a growing number of publicly available molecular databases and systems analysis software programs that could be used for initiating systems modeling of social, behavioral, and genetic interactions. For instance, the Database of Interacting Proteins (Xenarios et al., 2001), the Biomolecular Interaction Network Database (Bader et al., 2001), and the Munich Information Center for Protein Sequences of the German National Center for Environment and Health (Mewes et al., 1999) contain searchable catalogs of known protein-protein interactions; the Transcription Factors Database (Wingender et al., 2000) and The Promoter Database of *Saccharomyces cerevisiae* (Zhu and Zhang, 1999) catalog interactions between proteins and DNA (i.e., transcription factor interactions), and databases of metabolic pathways also recently have been established (e.g., EcoCyc [Karp et al., 2000], KEGG [Ogata et al., 1999], and What Is There [Selkov et al., 1998]). A growing number of databases are also under development for storing the now sizeable number of mRNA-expression datasets (Ermolaeva et al., 1998; Stoeckert et al., 1999; Hawkins et al., 1999; Ringwald et al., 2000; Aach et al., 2000); companies, such as Affymetrix, Rosetta, Spotfire, Informax, Incyte, Gene Logic, and Silicon Genetics, market gene-expression databases commercially. Notably lacking from this list, however, are repositories of information on the behavioral and social components of the system. Work toward developing publicly available information on these levels could open up significant possibilities for the computer modeling of health outcomes.

The development and practice of systems approaches to model social, behavioral, and genetic interactions involves a number of requirements that will pose particular challenges for researchers. These include: (a) bridging disciplinary and language barriers encountered by teams of social scientists, behavioral scientists, molecular biologists, geneticists, and computational scientists; (b) the need for high-throughput facilities for molecular technologies, such as DNA sequencing, DNA arrays, genotyping, proteomics, metabonomics, and tissue arrays; (c) a lack of integrated public health, medical, and biological informatics systems; (d) the need to develop novel analytical tools and efficient, powerful computational infrastructures; (e) a lack of integration of discovery-driven and hypothesis-driven science; and (f) the need to develop diverse partnerships among academia, community, industry, and government.

To address these challenges and advance our understanding of the complex contributions to health of social, behavioral, and genetic factors, it becomes imperative to move toward conducting research that assesses the interactions of these variables (see Chapter 8 for a detailed discussion of interactions). Therefore, the committee makes the following recommendations:

Recommendation 1: Conduct Transdisciplinary, Collaborative Research. *The National Institutes of Health (NIH) should develop Requests for Applications (RFAs) to study the impact on health of interactions among social, behavioral, and genetic factors and their interactive pathways (i.e., physiological). Such transdisciplinary research should involve the genuine collaboration of social, behavioral, and genetic scientists. Genuine collaboration is essential for the identification, incorporation, analysis, and interpretation of the multiple variables used.*

Recommendation 2: Measure Key Variables Over the Life Course and Within the Context of Culture. *The NIH should develop RFAs for studies of interactions that incorporate measurement, over the life course and within the context of culture, of key variables in the important domains of social, behavioral, and genetic factors.*

Essential social variables include educational attainment, income and wealth, occupational status, social networks/social support, and the work conditions that have been linked consistently and robustly to health outcomes. Behavioral and psychological variables include tobacco/alcohol/drug use, eating behavior, physical activity, temperament, perceived stress and coping, perceived social support, emotional state, and motivation. Essential genetic factors include the DNA sequence variation, structural chromosomal changes, gene expression, epigenetic modifications, and downstream targets of gene expression. Physiological measures should consider relevant hormones, neurotransmitters, signaling molecules, and cell types that serve as transducing mechanisms between the social world and genetics. Furthermore, candidate physiological measures should be selected that recognize biological and clinical relevance; practical application in the context of large-scale field studies; interactions among multiple physiological systems that are traditionally compartmentalized (e.g., the nervous system, the endocrine system, and the immune system); intracellular pathways that mediate the interaction between gene function and physiological systems; and the role of a given physiological measure in multiple biological systems. Finally, because of the complexity encountered in variables related to sex/gender and race/ethnicity, such variables must be considered and analyzed from a variety of perspectives, including social, cultural, psychological, historical, political, genetic, and geographic/ancestral.

Additionally, as discussed previously and in Chapter 8, the study of interactions will require new modeling strategies, the use of profiling approaches, and the conduct of research in diverse groups and settings. Therefore, the committee proposes the following recommendations:

> **Recommendation 3: Develop and Implement New Modeling Strategies to Build More Comprehensive, Predictive Models of Etiologically Heterogeneous Disease.** *The NIH should emphasize research aimed at developing and implementing such models (e.g., pattern recognition, multivariate statistics, and systems-oriented approaches) for incorporating social, behavioral, and genetic factors and their interactive pathways (i.e., physiological) in testable models within populations, clinical settings, or animal studies.*

> **Recommendation 4: Investigate Biological Signatures.** *Researchers should use genomic, transcriptomic, proteomic, metabonomic, and other high-dimensional molecular approaches to discover new constellations of genetic factors, biomarkers, and mediating systems through which interactions with social environment and behavior influence health.*

> **Recommendation 5: Conduct Research in Diverse Groups and Settings.** *The NIH should encourage research on the impact of interactions among social, behavioral, and genetic factors and their interactive pathways (i.e., physiological) on health that emphasizes diversity in groups and settings. Furthermore, NIH should support efforts to ensure that the findings of such research are validated by replication in independent studies, translated to patient-oriented research, conducted and applied in the context of public health, and used to design preventive and therapeutic approaches.*

Transdisciplinary research assessing the impact on health of interactions among social, behavioral, and genetic factors has the potential to bring to the fore new understanding of disease risk. Such an understanding could lead to the development of more effective interventions and, ultimately, to improved health for individuals and populations. This research provides an exciting opportunity to advance our understanding and our impact on improving health.

REFERENCES

Aach J, Rindone W, Church GM. 2000. Systematic management and analysis of yeast gene expression data. *Genome Research* 10(4):431-445.

Ainsworth MD, Bell SM. 1970. Attachment, exploration, and separation: Illustrated by the behavior of one-year-olds in a strange situation. *Child Development* 41(1):49-67.

Ashburner M, Ball CA, Blake JA, Botstein D, Butler H, Cherry JM, Davis AP, Dolinski K, Dwight SS, Eppig JT, Harris MA, Hill DP, Issel-Tarver L, Kasarskis A, Lewis S, Matese JC, Richardson JE, Ringwald M, Rubin GM, Sherlock G. 2000. Gene ontology: Tool for the unification of biology. The Gene Ontology Consortium. *Nature Genetics* 25(1): 25-29.

Azmi J, Griffin JL, Antti H, Shore RF, Johansson E, Nicholson JK, Holmes E. 2002. Metabolic trajectory characterisation of xenobiotic-induced hepatotoxic lesions using statistical batch processing of NMR data. *Analyst* 127(2):271-276.

Bader GD, Donaldson I, Wolting C, Ouellette BF, Pawson T, Hogue CW. 2001. BIND—The Biomolecular Interaction Network Database. *Nucleic Acids Research* 29(1):242-245.

Baggerly KA, Morris JS, Edmonson SR, Coombes KR. 2005. Signal in noise: Evaluating reported reproducibility of serum proteomic tests for ovarian cancer. *Journal of the National Cancer Institute* 97(4):307-309.

Ball CA, Sherlock G, Parkinson H, Rocca-Sera P, Brooksbank C, Causton HC, Cavalieri D, Gaasterland T, Hingamp P, Holstege F, Ringwald M, Spellman P, Stoeckert CJ Jr, Stewart JE, Taylor R, Brazma A, Quackenbush J. 2002a. Standards for microarray data. *Science* 298(5593):539.

Ball CA, Sherlock G, Parkinson H, Rocca-Sera P, Brooksbank C, Causton HC, Cavalieri D, Gaasterland T, Hingamp P, Holstege F, Ringwald M, Spellman P, Stoeckert CJ Jr, Stewart JE, Taylor R, Brazma A, Quackenbush J. 2002b. The underlying principles of scientific publication. *Bioinformatics* 18(11):1409.

Ball C, Brazma A, Causton H, Chervitz S, Edgar R, Hingamp P, Matese JC, Parkinson H, Quackenbush J, Ringwald M, Sansone SA, Sherlock G, Spellman P, Stoeckert C, Tateno Y, Taylor R, White J, Winegarden N. 2004a. Standards for microarray data: An open letter. *Environmental Health Perspectives* 112(12):A666-A667.

Ball CA, Brazma A, Causton H, Chervitz S, Edgar R, Hingamp P, Matese JC, Parkinson H, Quackenbush J, Ringwald M, Sansone SA, Sherlock G, Spellman P, Stoeckert C, Tateno Y, Taylor R, White J, Winegarden N. 2004b. Submission of microarray data to public repositories. *PLoS Biology* 2(9):E317.

Beckwith-Hall BM, Nicholson JK, Nicholls AW, Foxall PJ, Lindon JC, Connor SC, Abdi M, Connelly J, Holmes E. 1998. Nuclear magnetic resonance spectroscopic and principal components analysis investigations into biochemical effects of three model hepatotoxins. *Chemical Research in Toxicology* 11(4):260-272.

Bertalanffy L. 1968. *General System Theory: Foundations, Development, Applications*. New York: George Brazillier.

Bjornsson HT, Fallin MD, Feinberg AP. 2004. An integrated epigenetic and genetic approach to common human disease. *Trends in Genetics* 20(8):350-358.

Bombail V, Moggs JG, Orphanides G. 2004. Perturbation of epigenetic status by toxicants. *Toxicology Letters* 149(1-3):51-58.

Brazma A, Hingamp P, Quackenbush J, Sherlock G, Spellman P, Stoeckert C, Aach J, Ansorge W, Ball CA, Causton HC, Gaasterland T, Glenisson P, Holstege FC, Kim IF, Markowitz V, Matese JC, Parkinson H, Robinson A, Sarkans U, Schulze-Kremer S, Stewart J, Taylor R, Vilo J, Vingron M. 2001. Minimum information about a microarray experiment (MIAME)-toward standards for microarray data. *Nature Genetics* 29(4):365-371.

Bronson G. 1971. Fear of the unfamiliar in human infants. In: Schaffer H, editor. *The Origin of Human Social Relations*. London: Academic Press.

Brown SC, Kruppa G, Dasseux JL. 2005. Metabolomics applications of FT-ICR mass spectrometry. *Mass Spectrometry Reviews* 24(2):223-231.

Brun C, Herrmann C, Guenoche A. 2004. Clustering proteins from interaction networks for the prediction of cellular functions. *BMC Bioinformatics* 5:95.

Cannon WB. 1932. *The Wisdom of the Body*. New York: Norton.

Carey G. 2003. *Human Genetics for the Social Sciences*. Thousand Oaks, CA: Sage.

Cavigelli SA, McClintock MK. 2003. Fear of novelty in infant rats predicts adult corticosterone dynamics and an early death. *Proceedings of the National Academy of Sciences of the United States of America* 100(26):16131-16136.

Champoux M, Coe S, Schanberg S, Kuhn C, Suomi S. 1989. Hormonal effects of early rearing conditions in the infant Rhesus monkey. *American Journal of Primatology* 19:111-117.

Champoux M, Byrne E, Delizio R, Suomi S. 1992. Motherless mothers revisited: Rhesus maternal behavior and rearing history. *Primates* 33(2):251-255.

Chaurand P, Stoeckli M, Caprioli RM. 1999. Direct profiling of proteins in biological tissue sections by MALDI mass spectrometry. *Analytical Chemistry* 71(23):5263-5270.

Cheung VG, Spielman RS, Ewens KG, Weber TM, Morley M, Burdick JT. 2005. Mapping determinants of human gene expression by regional and genome-wide association. *Nature* 437(7063):1365-1369.

Cohen S, Doyle WJ, Turner RB, Alper CM, Skoner DP. 2003. Emotional style and susceptibility to the common cold. *Psychosomatic Medicine* 65(4):652-657.

Conrads TP, Hood BL, Issaq HJ, Veenstra TD. 2004. Proteomic patterns as a diagnostic tool for early-stage cancer: A review of its progress to a clinically relevant tool. *Molecular Diagnosis* 8(2):77-85.

Denenberg VH. 1999. Commentary: Is maternal stimulation the mediator of the handling effect in infancy? *Developmental Psychobiology* 34(1):1-3.

Djebbari A, Karamycheva S, Howe E, Quackenbush J. 2005. MeSHer: Identifying biological concepts in microarray assays based on PubMed references and MeSH terms. *Bioinformatics* 21(15):3324-3326.

Doniger SW, Salomonis N, Dahlquist KD, Vranizan K, Lawlor SC, Conklin BR. 2003. MAPPFinder: Using Gene Ontology and GenMAPP to create a global gene-expression profile from microarray data. *Genome Biology* 4(1):R7.

Engle B. 1985. Stress is a noun! No, a verb! No, an adjective. In: Field T, McCabe P, Sneiderman N, editors. *Stress and Coping*. Vol. 1. Hillsdale, NJ: Erlbaum. Pp. 3-12.

Ermolaeva O, Rastogi M, Pruitt KD, Schuler GD, Bittner ML, Chen Y, Simon R, Meltzer P, Trent JM, Boguski MS. 1998. Data management and analysis for gene expression arrays. *Nature Genetics* 20(1):19-23.

Fink JL, Drewes S, Patel H, Welsh JB, Masys DR, Corbeil J, Gribskov M. 2003. 2HAPI: A microarray data analysis system. *Bioinformatics* 19(11):1443-1445.

Gottesman II, Hanson DR. 2005. Human development: Biological and genetic processes. *Annual Review of Psychology* 56:263-286.

Griffin JL, Bonney SA, Mann C, Hebbachi AM, Gibbons GF, Nicholson JK, Shoulders CC, Scott J. 2004. An integrated reverse functional genomic and metabolic approach to understanding orotic acid-induced fatty liver. *Physiological Genomics* 17(2):140-149.

Guyton AC. 1976. Interstitial fluid pressure and dynamics of lymph formation. Introduction. *Federation Proceedings* 35(8):1861-1862.

Hawkins V, Doll D, Bumgarner R, Smith T, Abajian C, Hood L, Nelson PS. 1999. PEDB: The Prostate Expression Database. *Nucleic Acids Research* 27(1):204-208.

Herbert TB, Cohen S. 1993. Stress and immunity in humans: A meta-analytic review. *Psychosomatic Medicine* 55(4):364-379.

Holmes E, Nicholls AW, Lindon JC, Ramos S, Spraul M, Neidig P, Connor SC, Connelly J, Damment SJ, Haselden J, Nicholson JK. 1998. Development of a model for classification of toxin-induced lesions using 1H NMR spectroscopy of urine combined with pattern recognition. *NMR in Biomedicine* 11(4-5):235-244.

Hoyle RH. 1995. *Structural Equation Modeling: Concepts, Issues, and Applications.* Thousand Oaks, CA: Sage Publications.

Ideker T, Galitski T, Hood L. 2001. A new approach to decoding life: Systems biology. *Annual Review of Genomics and Human Genetics* 2(1):343-372.

Jablonka E, Lamb MJ. 2002. The changing concept of epigenetics. *Annals of the New York Academy of Sciences* 981:82-96.

Jaenisch R, Bird A. 2003. Epigenetic regulation of gene expression: How the genome integrates intrinsic and environmental signals. *Nature Genetics* 33(Suppl):245-254.

Jenssen TK, Laegreid A, Komorowski J, Hovig E. 2001. A literature network of human genes for high-throughput analysis of gene expression. *Nature Genetics* 28(1):21-28.

Johnson EO, Kamilaris TC, Chrousos GP, Gold PW. 1992. Mechanisms of stress: A dynamic overview of hormonal and behavioral homeostasis. *Neuroscience and Biobehavioral Reviews* 16(2):115-130.

Jones PA. 2005. Overview of cancer epigenetics. *Seminars in Hematology* 42(3 Suppl 2): S3-S8.

Kacser H, Burns JA. 1973. The control of flux. *Symposia of the Society for Experimental Biology* 27:65-104.

Kaminsky ZA, Assadzadeh A, Flanagan J, Petronis A. 2005. Single nucleotide extension technology for quantitative site-specific evaluation of metC/C in GC-rich regions. *Nucleic Acids Research* 33(10):E95.

Kanehisa M. 2002. The KEGG database. *Novartis Foundation Symposium* 247:91-101; discussion 101-103, 119-128, 244-252.

Karp PD, Riley M, Saier M, Paulsen IT, Paley SM, Pellegrini-Toole A. 2000. The EcoCyc and MetaCyc databases. *Nucleic Acids Research* 28(1):56-59.

Liu D, Diorio J, Tannenbaum B, Caldji C, Francis D, Freedman A, Sharma S, Pearson D, Plotsky PM, Meany MJ. 1997. Maternal care, hippocampal glucocorticoid receptors, and hypothalamic-pituitary-adrenal responses to stress. *Science* 277:1659-1662.

Lo HS, Wang Z, Hu Y, Yang HH, Gere S, Buetow KH, Lee MP. 2003. Allelic variation in gene expression is common in the human genome. *Genome Research* 13(8):1855-1862.

Malek RL, Irby RB, Guo QM, Lee K, Wong S, He M, Tsai J, Frank B, Liu ET, Quackenbush J, Jove R, Yeatman TJ, Lee NH. 2002. Identification of Src transformation fingerprint in human colon cancer. *Oncogene* 21(47):7256-7265.

Mann M, Jensen ON. 2003. Proteomic analysis of post-translational modifications. *Nature Biotechnology* 21(3):255-261.

Manuck SB, Kaplan JR, Adams MR, Clarkson TB. 1988. Studies of psychosocial influences on coronary artery atherogenesis in cynomolgus monkeys. *Health Psychology* 7(2): 113-124.

Maruyama GM. 1997. *Basics of Structural Eqaution Modeling.* Thousand Oaks, CA: Sage Publications.

McClintock MK, Conzen SD, Gehlert S, Masi C, Olopade F. 2005. Mammary cancer and social interactions: Identifying multiple environments that regulate gene expression throughout the life span. *Journals of Gerontology: Series B* 60B(Special Issue 1):32-41.

McEwen B, Seeman T. 1999. *Allostatic Load and Allostasis.* [Online]. Available: www.macses.ucsf.edu/Research/Allostatic/notebook/allostatic.html#Allostasis [accessed May 22, 2006].

Meaney MJ, Diorio J, Francis D, Widdowson J, LaPlante P, Caldji C, Sharma S, Seckl JR, Plotsky PM. 1996. Early environmental regulation of forebrain glucocorticoid receptor gene expression: Implications for adrenocortical responses to stress. *Developmental Neuroscience* 18(1-2):49-72.

Mewes HW, Heumann K, Kaps A, Mayer K, Pfeiffer F, Stocker S, Frishman D. 1999. MIPS: A database for genomes and protein sequences. *Nucleic Acids Research* 27(1):44-48.

Nicholson JK, Connelly J, Lindon JC, Holmes E. 2002. Metabonomics: A platform for studying drug toxicity and gene function. *Nature Review Drug Discovery* 1(2):153-161.

Nicholson JK, Holmes E, Wilson ID. 2005. Gut microorganisms, mammalian metabolism and personalized health care. *Nature Review Microbiology* 3(5):431-438.

NRC/IOM (National Research Council/Institute of Medicine). 2000. *From Neurons to Neighborhoods: The Science of Early Childhood Development.* Washington, DC: National Academy Press.

Ogata H, Goto S, Sato K, Fujibuchi W, Bono H, Kanehisa M. 1999. KEGG: Kyoto Encyclopedia of Genes and Genomes. *Nucleic Acids Research* 27(1):29-34.

Petricoin EF, Ardekani AM, Hitt BA, Levine PJ, Fusaro VA, Steinberg SM, Mills GB, Simone C, Fishman DA, Kohn EC, Liotta LA. 2002. Use of proteomic patterns in serum to identify ovarian cancer. *Lancet* 359(9306):572-577.

Petronis A. 2003. Epigenetics and bipolar disorder: New opportunities and challenges. *American Journal of Medical Genetics. Part C, Seminars in Medical Genetics* 123(1):65-75.

Ringwald M, Eppig JT, Kadin JA, Richardson JE. 2000. GXD: A Gene Expression Database for the laboratory mouse: Current status and recent enhancements. The Gene Expresison Database group. *Nucleic Acids Research* 28(1):115-119.

Savageau, MA. 1976. *Biochemical Systems Analysis.* Reading, MA: Addison-Wesley Publishing Co.

Selkov E Jr, Grechkin Y, Mikhailova N, Selkov E. 1998. MPW: The Metabolic Pathways Database. *Nucleic Acids Research* 26(1):43-45.

Selye H. 1956. *The Stress of Life.* New York: McGraw-Hill.

Selye H. 1973. The evolution of the stress concept. *American Scientist* 61(6):692-699.

Selye H. 1975. Confusion and controversy in the stress field. *Journal of Human Stress* 1(2): 37-44.

Shahbazian MD, Zoghbi HY. 2002. Rett syndrome and MeCP2: Linking epigenetics and neuronal function. *American Journal of Human Genetics* 71(6):1259-1272.

Shi H, Wei SH, Leu YW, Rahmatpanah F, Liu JC, Yan PS, Nephew KP, Huang TH. 2003. Triple analysis of the cancer epigenome: An integrated microarray system for assessing gene expression, DNA methylation, and histone acetylation. *Cancer Research* 63(9):2164-2171.

Sroufe L. 1979. Socioemotional development. In: Osofsky J, editor. *Handbook of Infant Development.* New York: John Wiley & Sons. Pp. 462-515.

Stefanski V, Raabe C, Schulte M. 2005. Pregnancy and social stress in female rats: Influences on blood leukocytes and corticosterone concentrations. *Journal of Neuroimmunology* 162(1-2):81-88.

Stoeckert CJ Jr, Salas F, Brunk B, Overton GC. 1999. EpoDB: A prototype database for the analysis of genes expressed during vertebrate erythropoiesis. *Nucleic Acids Research* 27(1):200-203.

Stoney CM, Bausserman L, Niaura R, Marcus B, Flynn M. 1999a. Lipid reactivity to stress: II. Biological and behavioral influences. *Health Psychology* 18(3):251-261.

Stoney CM, Niaura R, Bausserman L, Matacin M. 1999b. Lipid reactivity to stress: I. Comparison of chronic and acute stress responses in middle-aged airline pilots. *Health Psychology* 18(3):241-250.

Stuart JM, Segal E, Koller D, Kim SK. 2003. A gene-coexpression network for global discovery of conserved genetic modules. *Science* 302(5643):249-255.

Suomi S. 1991. Adolescent depression and depressive symptoms: Insights from longitudinal studies with Rhesus monkeys. *Journal of Youth and Adolescence* 20(2):273-287.

Sutherland JE, Costa M. 2003. Epigenetics and the environment. *Annals of the New York Academy of Sciences* 983:151-160.

Villanueva J, Philip J, Entenberg D, Chaparro CA, Tanwar MK, Holland EC, Tempst P. 2004. Serum peptide profiling by magnetic particle-assisted, automated sample processing and MALDI-TOF mass spectrometry. *Analytical Chemistry* 76(6):1560-1570.

Waddington CH. 1957. *The Strategy of the Genes: A Discussion of Some Aspects of Theoretical Biology*. New York: The MacMillan Company.

Washburn MP, Ulaszek R, Deciu C, Schieltz DM, Yates JR 3rd. 2002. Analysis of quantitative proteomic data generated via multidimensional protein identification technology. *Analytical Chemistry* 74(7):1650-1657.

Waterland RA, Jirtle RL. 2004. Early nutrition, epigenetic changes at transposons and imprinted genes, and enhanced susceptibility to adult chronic diseases. *Nutrition* 20(1): 63-68.

Waters E, Matas L, Sroufe LA. 1975. Infants' reactions to an approaching stranger: description, validation, and functional significance of wariness. *Child Development* 46(2): 348-356.

Wilson ID, Plumb R, Granger J, Major H, Williams R, Lenz EM. 2005. HPLC-MS-based methods for the study of metabonomics. *Journal of Chromatography B Analytical Technologies in the Biomedical and Life Sciences* 817(1):67-76.

Wingender E, Chen X, Hehl R, Karas H, Liebich I, Matys V, Meinhardt T, Pruss M, Reuter I, Schacherer F. 2000. TRANSFAC: An integrated system for gene expression regulation. *Nucleic Acids Research* 28(1):316-319.

Xenarios I, Fernandez E, Salwinski L, Duan XJ, Thompson MJ, Marcotte EM, Eisenberg D. 2001. DIP: The Database of Interacting Proteins: 2001 update. *Nucleic Acids Research* 29(1):239-241.

Yates J. 1998. Mass spectrometry and the age of the proteome. *Journal of Mass Spectrometry* 33(1):1-19.

Zhu J, Zhang MQ. 1999. SCPD: A promoter database of the yeast *Saccharomyces cerevisiae*. *Bioinformatics* 15(7-8):607-611.

7

Animal Models

ROLE OF ANIMAL MODELS[1]

Rationale

Most studies identifying gene-environment interactions that are risk factors for disease in humans rely on observational studies of naturally occurring genetic polymorphisms and environmental variability. These correlational research designs, although a rich source of testable hypotheses, cannot provide definitive evidence for the causal effects of genes, environments, or their interaction. Basic research using animal models is a feasible way to establish causal relationships in the reciprocal interactions among social, behavioral, and genetic contributors to health and disease. Thus, animal studies are an important complement to clinical and community-based research.

Specifically, animal models can be used to conduct studies for which different aspects of social, behavioral, and genetic factors can be controlled or standardized to a significantly larger extent than can be done in human studies. Animal models enable the manipulation of single variables or specific groups of variables in a highly controlled context. In some cases, animal models provide opportunities to establish causality through studies examining the temporal sequence of events or studies involving the removal

---

[1]The commissioned paper submitted by Steve W. Cole was used in the preparation of this chapter.

followed by the add-back of hypothesized mediators. Such controlled removal and add-back can be achieved at the genetic, protein, physiological, behavioral, or social-environment level. Animal models also allow for invasive examination of organ, tissue, and region-specific mechanisms at the physiological, cellular, and molecular levels. Also animals with short reproductive cycles and life spans provide an invaluable tool for conducting developmental and life-span studies, and animal models enable the conduct of breeding experiments and genetic manipulation that facilitate the elucidation of inherited traits and genetic effects.

Strategies for Linking Animal and Human Research

Modeling Known Interactions and Diseases in Humans

Animal research can serve as models of gene-environment interactions and diseases identified in humans. In the case of social control of disease processes, the choice of species to be studied depends on the level of social interactions that needs to be examined. For example, rodent models can demonstrate how differences in social status, population density, or early experiences interact with genetic makeup to affect susceptibility to disease (e.g., examine effects of social factors in knockout or knockin animals [or inbred strains] that differ in susceptibility to infection, cancer, autoimmunity). The advantages of rodent models include significant control over genetic, physiological, behavioral, and social factors and relatively short reproductive, developmental, and life cycles. They are amenable to studying a variety of important psychosocial variables, including social isolation, social relationships, attachment, parenting, temperament, and motivational states.

However, nonhuman primate models, which offer limited control over genetic factors and have a longer life span, may be best suited to examine the consequences of more complex social factors, such as those involving cooperation or trust. For example, after bouts of aggression, nonhuman primates demonstrate reconciliatory behavior that is thought to be important for maintaining cooperative social hierarchies (de Waal, 2000). Some aspects of human behavior (e.g., optimism, hope, guilt) may be studied in animals only when the investigator can demonstrate a robust animal model with multiple behavioral paradigms as well as shared neural mechanisms.

In addition, animal models developed for traditional biomedical research are also powerful models for studying the psychosocial modulation of known mechanisms of specific human diseases. There are many animal species, strains, and transgenic models developed through biomedical science, that have been well characterized in terms of the genetic, molecular, and cellular processes underlying human disease. Studying these animals in a variety of psychosocial paradigms, based on variables identified through survey, epidemiological, and human experimental research, can test hypothesized causal relations derived from correlational data in humans.

Fundamental Biology: Nontraditional Laboratory Animals

It is essential to study animals as evolved biological systems in which surviving and reproducing in particular social and physical environments have selected a constellation of interactions between social, behavioral, physiological systems, and gene function. Doing so reveals insights and principles that also underlie human health and disease but that are not salient in the modern world or in a typical biomedical approach. Moreover, ethology and evolutionary biology recognize that individual differences are not necessarily just "noise," but represent different evolved strategies for survival in different contexts. Taking an ethological approach to variation in strategies reveals the range of gene-environment interactions that occur within species as they have evolved in their natural ethological and ecological contexts.

Studies of deer mice (*Peromyscus maniculatus*), who live in highly seasonal environments, reveal that function of the immune system requires significant energy, so much so that during winter an animal trades off entering puberty and becoming reproductive in order to sustain the energetic requirements of fighting infectious disease (Prendergast and Nelson, 2001; Nelson, 2004). It is not the demands of the cold weather itself that signals this trade-off, but rather the shortened days that precede seasonal temperature change, allowing the animal to modulate relative balance of immune function and reproduction in anticipation of the energetic demands of winter.

In house sparrows, immune activity increases energy expenditure, illustrating the energetic costs of immune function that could otherwise be deployed to growth (Martin et al., 2003). Such animal research, set in an ecological context, provides a powerful animal model for such trade-offs in humans. When social structure restricts resources and results in a population living in an environment with a high pathogen load, slower growth can result, as is the case of children in the lowlands of Bolivia. This presumably happens because the allocation of energetic resources to immune function has been diverted from growth (McDade, 2005). This dynamic interaction between social access to energy stores, pathogen interaction, fat deposition, and growth likely involves leptin, a pleiotropic molecule with cytokine properties that is produced by fat cells during an inflammatory response (Faggioni et al., 2001; Fantuzzi, 2005).

Limitations and Power for Generalization to Humans

One danger in using animal models is "overspecifying" what is being measured—that is, interpreting the animal's behavior anthropomorphically, without measuring different facets of the behavior in order to clearly demonstrate what behavioral system is being measured. For example, claims are made about genetic or brain mechanisms in spatial learning and intelligence when mice perform well or show deficits in a Morris water maze. In this task however, the mouse is required to do something it did not evolve to do—swim. Moreover, while swimming to avoid drowning, this non-aquatic species is required to navigate a circular pool to find a submerged platform—again, an improbable scenario. In fact, performance in a Morris water maze can be affected by the rodent's ability to handle stress, degree of thigmotaxis (the tendency to stay close to a solid surface), and the ability to inhibit a fixed-action pattern (Day and Schallert, 1996). Thus, when an enriched environment aids recovery from a stroke, measured by improved performance in a Morris water maze, it is essential to determine which of these behavioral systems is being affected and not assume that it is spatial learning and cognitive performance, which is the most salient aspect of the test to human investigators (Ronnback et al., 2005).

Conversely, it is also a mistake to assume that human psychosocial traits that affect disease are uniquely human and that humans do not have psychological processes in common with animals. This is an error commonly made when human psychological states are measured with verbal accounts of subjective experience—for example, "I do not feel I have people I can turn to for social support" or "I feel overwhelmed." Such verbal reports are certainly unique to humans, but nonetheless they are likely based on psychological processes and behavioral traits that have commonality with animal systems, especially when their underlying neuroendocrine mechanisms are similar. The parallel is readily accepted in nonemotional domains. The study of human hunger utilizes self-reports: "I feel hungry" or "I feel sated." Yet, few question that animals are an excellent model for teasing apart the diverse aspects of hunger and satiety as a motivational state. Indeed, rodent models have been a powerful tool for teasing apart multiple facets of hunger, ranging from taste, chewing, insulin, leptin, and hypothalamic activity to gastrointestinal activity; there are far more independent factors than have been intuitively obvious (White, 1986; Morley, 1990; Hall and Swithers-Mulvey, 1992; Williams et al., 2001; Changizi et al., 2002; Oka et al., 2003). Thus, social animals can be powerful models of psychosocial effects on disease and gene expression, enabling the identification of transduction pathways from the social world to disease as well as the multiple functions of such pathways. Even such seemingly unique human social activities as making business decisions involve neuroendocrine mechanisms conserved across mammals, if not other species (Morse, 2006).

DEFINITIONS FROM ANIMAL RESEARCH

Animal research has clarified concepts that are key to understanding the effects of social environment on health and disease and gene function, extending and moderating the conclusions based on epidemiological studies in humans. These concepts include genetics, immune and neuroendocrine function, causality, pleitropy, and life-span fitness.

Genetics

Genetics requires a broad conception that includes both functional genomics (intra-individual changes in gene expression over time) and the more traditional topic of structural polymorphism (interindividual variations in DNA sequence or epigenetic characteristics). This broad conceptualization is essential because social influences on gene transcription are fairly well studied, while few studies have examined the relationships between social factors and genetic polymorphisms. That such effects exist is likely because structural polymorphisms generally exert their effects in the context of expressed genes.

Physiology: The Missing Link

An essential role of animal research is to test the relationship between presumptive genetic influences (e.g., inferred from studies of heritability) and defined genetic influences (e.g., effects attributable to the expression of specific genes or epigenetic characteristics). The immune system includes classical immune cells (e.g., leukocytes) as well as other cellular contexts relevant to disease pathogenesis or host defense, such as somatic cells responding to pathogens through innate immune responses (e.g., "danger signals" produced by Toll-like receptors, Type I interferon production). The neuroendocrine system also is broadly defined to include not only true neurally driven hormone production (e.g., hypothalamic-pituitary-adrenal [HPA] axis), but also neuroeffector processes that do not necessarily involve systemic hormone distribution (e.g., local effects of neurotransmitter release from autonomic or sensory neurons or neuropeptides such as vasopressin and oxytocin).

Part of the reason so few genetic determinants of immune response currently are presently known may be an overly restrictive focus on "immune system" genes. Polymorphisms in many "nonimmune" genes, which are regulated by the psychosocial environment through physiological systems, may also influence leukocyte function and/or the pathogenesis of diseases involving immune or inflammatory components. For example, catecholamines are known to influence several aspects of leukocyte function (Sanders and Straub, 2002; Kavelaars, 2002), and polymorphisms in genes encoding their alpha—and beta—adrenergic receptors are associated with differential incidence of asthma, parasitic infections, and cardiovascular disease (Ramsay et al., 1999; Ulbrecht et al., 2000; Ukkola et al., 2001; Weiss, 2005; Thakkinstian et al., 2005; Lanfear et al., 2005). Glucocorticoids, another physiological system exquisitely sensitive to the psychosocial environment, play a key role in regulating inflammatory gene expression (Webster et al., 2002), and polymorphisms in the glucocorticoid receptor gene (NR3C1) have been linked to cardiovascular and autoimmune disease (Lin et al., 1999; Ukkola et al., 2001; Jiang et al., 2001; Dobson et al., 2001; van Rossum et al., 2002; Lin et al., 2003).

Causality

Mediating and moderating variables often are inferred in human studies through multivariate statistical analysis (Baron and Kenny, 1986). A moderating variable is one that changes the way an independent variable is related to a dependent variable (e.g., sex differences in the relationship between reported symptoms and risk for cardiac disease). A mediating variable is one that statistically accounts for the association between an independent and dependent variable in a study (e.g., cortisol levels may be a better predictor of disease onset than feelings of stress). However, the disease process may be mediated by autonomic tone, not measured in the study, and not cortisol itself.

In the animal literature, however, these terms have more stringent criteria. Studies demonstrate "mediation" only when a hypothesized intermediate factor has been experimentally manipulated to block the effects of some upstream influence on a downstream outcome within a transitive causal chain. "Moderation" is reserved for cases in which one variable is experimentally manipulated to alter the causal effect of a second manipulated variable on an observed outcome. A statistical interaction is not sufficient. The strongest evidence for genetic moderation comes from studies in which both genes and environment are experimentally manipulated, but few studies meet this criterion. Based on this fact alone, it can be concluded that much remains to be learned about the interaction between genes and the social environment in the context of immune system function and disease.

Context, Pleitropy, and Lifetime Fitness

Behavioral ecologists have elegantly and dramatically revealed that we cannot expect human studies to reveal monolithic or simple linear relationships among social behavior, hormones, immune function, and disease. The reality of surviving and reproducing over a lifetime in a changing environment has selected for genetic and physiological traits that are highly context dependent. This is enabled in part by both genetic and physiological pleitropy, where the same gene or molecule can have very different functions in different physiological systems. For example, genes encoding for the major histocompatibility complex produce a molecule that is involved not only in the presentation of pathogen proteins to a T-cell, but also in selection of mates, choice of communal nesting partners, and guiding neurons in the development of the nervous system (Manning et al., 1992; Jordan and Bruford, 1998; Huh et al., 2000; Jacob et al., 2002; Rock and Shen, 2005). Social isolation in rats accelerates puberty, seemingly enhancing fertility and fitness in the young animal, yet it accelerates reproductive senesce, reducing fitness when considered over the life span (LeFevre and McClintock, 1991; Zehr et al., 2001).

A rich and rigorously tested example is the relationship between social interactions, immune function, fertility, and fitness in side-blotched lizards (Svensson et al., 2001). The females of this species have two genetic morphs—one with yellow throats and the other with orange. In addition, throat color is correlated with steroid hormones that have physiological pleitropic effects on behavior and fertility. In both morphs, high population density, and its attendant aggressive encounters and pathogen exposure, is associated with a decreased antibody production to an antigen.

One might assume, as is often is done in laboratory and human studies, that the lower antibody production is associated with greater mortality and lower fitness. In the field, however, this relationship to fitness (survival after the female's first clutch) is seen only in the yellow morphs; in the orange morphs higher survival actually is associated with lower antibody production. The orange morph is particularly sensitive to the energetic costs of immune function, and at high densities it suppresses immunity as well as disperses. That is, within a species, immune function is density dependent. The orange morph invests in large clutches of eggs, consistent with reduced investment in immunity, and their daughters have reduced antibody production. The yellow morphs produce smaller clutches, and their daughters have high antibody production.

This system has resulted in a strong correlation of traits driven by different loci—that is throat color and antibody production. Because males prefer to mate with females of the rare color morph at any given time, the population of females oscillates between predominantly yellow and orange morphs, each with a different relationship among antibody production, reproductive strategies, and fitness. The relationship of social interactions, genetics, and immunity in humans, with their exquisite adaptability to a wide variety of environments and social structures, cannot be expected to be simpler. As research progresses, the concepts of genetic and physiological pleitropy, context dependence, and taking a life-span perspective on costs and benefits will be essential.

IDENTIFYING GENE-SOCIAL ENVIRONMENT INTERACTIONS AFFECTING HEALTH AND DISEASE

Early Life Experience

Meaney et al. have conducted a comprehensive series of studies showing that early life events, such as maternal separation, handling, or natural variations in maternal care, induce long-term changes in endocrine and behavioral responses to stress that are observed well into adulthood (Meaney, 2001). Using cross-fostering studies, these authors showed that changes in both maternal behavior and stress reactivity can be transmitted through nongenomic mechanisms across generations (Francis et al., 1999). Moreover, these authors also showed that the changes resulting from differences in maternal care are due to "environmental programming" that permanently alters gene expression and has downstream effects on stress-axis responsivity (Meaney and Szyf, 2005). Such epigenetic programming of stress reactivity is mediated by changes in hippocampal glucocorticoid receptor gene expression that are regulated by differences in maternal care and mediated by methylation of the consensus sequence for the transcription factor NGFI-A, which activates glucocorticoid receptor gene expression in the hippocampus (Fish et al., 2004). Increased DNA methylation prevents NGFI-A binding to the promoter for the glucocorticoid receptor gene and hence inhibits transcription, ultimately reducing expression of hippocampal glucocorticoid receptors (Fish et al., 2004). Reduced receptor levels result in reduced sensitivity to corticosterone-mediated negative feedback, which may result in increased and prolonged reactivity of the HPA axis.

These studies illustrate that socially relevant environmental and behavioral factors can induce epigenetic changes in specific brain regions that translate into long-lasting differences in stress reactivity. These experiments provide an excellent example of the advantages that are found in the use of animal models. Aspects of these findings are now being translated to human subjects (Pruessner et al., 2004). In addition, pre- and postnatal exposure to social stressors has been shown to induce significant effects on social and sexual behavior, endocrine responses, and brain sex steroid receptor distribution in adulthood in guinea pigs (Kaiser et al., 2003; Kaiser and Sachser, 2005), and prenatal social stress also has been shown to masculinize female behavior in adulthood (Sachser and Kaiser, 1996).

It may be assumed from these studies that higher stress reactivity may transfer into greater chronic stress burden, which is known to adversely affect immune function and health. Other studies also have shown that early life experiences involving social stressors are related to increased alcohol consumption (Fahlke et al., 2000) and dysregulated immune responses that last well into adulthood (Coe et al., 1989). However, some studies indicate that early life stressors actually enhance certain measures of immune function in adulthood (Coe et al., 1992). Nonhuman primate studies also have shown that exposure to mild early life stressors strengthens emotional and neuroendocrine stress responses in adulthood (Parker et al., 2005). Therefore, animal and human studies are needed to further examine the downstream psychophysiological and health consequences of variations in maternal care and other aspects of early life experience and to determine why factors such as early life stressors show adaptive effects in some studies but maladaptive effects in others.

Temperament

The earliest indications that social factors might affect individual health came from clinical observations of increased vulnerability to cancer and infectious disease among "socially withdrawn" individuals. A surprisingly large number of clinical studies have shown that socially inhibited or introverted individuals are at increased risk for immune-mediated infectious diseases, allergies, and hypersensitivity responses (Kagan et al., 1991; Cole et al., 1997; Cole et al., 1999; Cohen et al., 2003; Cole et al., 2003). Studies by Cavigelli and McClintock have demonstrated the long-term health consequences in rats of differences in temperament, such as increased fear of novelty (neophobia) and stress reactivity (Cavigelli and McClintock, 2003). Neophobia was measured using a modification of the open field arena that was designed to quantify an animal's degree of locomotion and interaction with novel objects. The authors showed that males from the same litter that demonstrate a high degree of neophobia and corticosterone stress responses to novelty during infancy maintain these characteristics as adults. They also showed that the predominant cause of death is the development of tumors in neophobic and neophilic animals, and that high neophobic males die sooner than their low neophobic brothers. The authors suggest that increased neuroendocrine reactivity of the high neophobic animals may be a mechanism that contributes to increased mortality over the life span of the animal. These studies demonstrate the usefulness of using rodent models for conducting life-span studies.

Other studies of social and behavioral development have linked socially inhibited behavior to individual differences in central nervous system information processing, brain neurotransmitter activity, and reactivity of the autonomic nervous system and HPA to social stimuli (Kagan et al., 1988; Kalin et al., 1998; Miller et al., 1999; Byrne and Suomi, 2002; Schwartz et al., 2003; Cavigelli and McClintock, 2003; Kalin and Shelton, 2003). In primate models, socially withdrawn behavior is a prospective risk factor for increased simian immunodeficiency virus (SIV) pathogenesis following a controlled viral challenge (Capitanio et al., 1999). Specific immune parameters mediating differential disease vulnerability have not been well defined in humans. However, selective breeding of mice to enhance socially inhibited behavior has been found to induce correlated reductions in natural killer (NK) cell numbers and cytotoxic activity (Petitto et al., 1993; Petitto et al., 1999), and decreases in T lymphocyte numbers, proliferative potential, and cyto-kine production (Petitto et al., 1994). Conversely, selective breeding for immune responses (e.g., antibody production) can produce correlated changes in social behavior (Vidal and Rama, 1994).

Social Isolation

Observational epidemiologic and clinical studies in humans have repeatedly found increased morbidity and mortality among people with limited social contact (House, 2001; Hawkley and Cacioppo, 2003; Cacioppo and Hawkley, 2003; Cohen, 2004) and those recently bereaved of close social partners (Schaefer et al., 1995; Martikainen and Valkonen, 1996; Li et al., 2003). Experimental evidence from human laboratory studies suggests that social relationships protect health in part by decreasing neuroendocrine responses to exogenous threats (Uchino et al., 1996; Sachser et al., 1998). Other behavioral mechanisms also may contribute to the health-protective effects of social relationships, including economic support (e.g., facilitating health care), reference group support for healthy behavior (e.g., discouraging tobacco or heavy alcohol use), and behavioral assistance with health services utilization (e.g., assistance in accessing treatment, adhering to medical regimens). The relative contributions of behavioral versus neuroendocrine mechanisms to isolation-linked health risks are not well understood in humans. However, experimental manipulation of social contact in animal models can alter long-term neuroendocrine function in ways that increase the risk of organic disease (e.g., isolation enhances hormone production rates to increase breast cancer incidence in social rodent models) (McClintock et al., 2005). In observational human studies, subjective social isolation (loneliness) has been linked to reduced vaccine-induced antibody responses and leukocyte proliferative activity (Glaser et al., 1992; Pressman et al., 2005).

Social isolation, which generally consists of housing animals individually instead of in groups, has been used as a stressor (Angulo et al., 1991; Chida et al., 2005). Isolation may indeed be stressful for animals that live in groups in their natural environments. However, it is important to keep in mind that some effects of isolation "stress" may be due to increased sensitivity or reactivity of the animal to external stimuli (e.g., handling) when the animal is no longer accustomed to being around or near other animals. Therefore, rather than or in addition to being a stressor itself, social isolation may increase stress reactivity or stress responsivity, which may be a potential confounder if it is not the focus of study.

Social Affiliation and Support

Studies using different species of voles have begun to elucidate genetic and hormonal mechanisms mediating complex social behaviors such as those involving monogamy versus polygamy (Young et al., 1998; Young et al., 2001). Male prairie voles show increased partner preference for a female with whom they are paired following stressful conditions that result in elevations of plasma corticosterone or following pharmacologically induced increases in plasma corticosterone, with females showing the opposite effect of exposure to stress (DeVries et al., 1996). Vasopressin-1a receptor (V1aR) gene transfer into the ventral forebrain region of male prairie voles (a monogamous species) increases affiliative behavior and strengthens partner preference (Pitkow et al., 2001). Interestingly, similar gene transfer into the ventral forebrain region of meadow voles significantly increased partner preference formation in this polygamous species (Lim et al., 2004), and transfer of vole V1aR in the rat septum increased social discrimination and social behavior in rats (Landgraf et al., 2003). In contrast, V1aR gene knockout mice show deficits in social recognition and anxiety-related behavior (Bielsky et al., 2004). Moreover, variations in microsatellite segments in the 5' region of the transcription start site for the V1aR gene differs in terms of length and regulatory control of gene expression among different individuals and is associated with individual differences in receptor expression and behavioral characteristics (Hammock and Young, 2005).

These studies suggest that some complex social and behavioral traits may be strongly modulated by changes in gene expression in critical areas of the brain. Such differences in regulation and expression of genes, their effects on social behavior, and ultimately on health, need to be investigated further. Moreover, more complex models of social affiliation may come from nonhuman primates that have been shown to demonstrate reconciliatory behavior after aggressive encounters, which are thought to be important for maintaining cooperative social hierarchies (de Waal, 2000).

Genetic Differences in Stress-Responsivity and Susceptibility to Autoimmune Disease

Evidence suggests that three contributing factors result in susceptibility to inflammatory and autoimmune disorders (Mason, 1991; Tsigos and Chrousos, 1994; Sternberg, 1995; Wick et al., 1998; Ermann and Fathman,

2001): First is the presence of host immune response genes which carry the potential for autoimmunity. Second is exposure to a proinflammatory or antigenic challenge (which may include infection) that initiates the cascade of immune reactions that ultimately result in autoimmunity. Third is a dysregulation of the immune response to which a deficiency in the HPA axis responsivity is thought to contribute.

Studies have shown that hyporeactive stress responsivity can contribute to increased susceptibility to autoimmune and proinflammatory disorders (Mason et al., 1990; Sternberg et al., 1992a; Harbuz et al., 1997; Tonelli et al., 2001; Sternberg, 2001; Webster et al., 2002; Harbuz et al., 2003). In a series of seminal studies, Sternberg et al. (1989) showed that decreased HPA axis reactivity to inflammatory stimuli results in increased susceptibility to experimental arthritis (Sternberg et al., 1989a; Sternberg et al., 1989b; Sternberg et al., 1992b). These investigators studied the development of streptococcal cell wall (SCW)-induced arthritis in female rats belonging to the genetically related Lewis/N (LEW/N) and Fischer 344/N (F344/N) strains (Sternberg and Wilder, 1989; Sternberg et al., 1989a; Sternberg et al., 1989b; Webster et al., 2002). The F344/N strain is resistant to the development of SCW-induced arthritis, while the LEW/N strain is susceptible. Interestingly, the F344/N strain mounts a significantly greater corticosterone and adrenocorticotropin response than does the LEW/N strain when challenged with a variety of stressors or with inflammatory mediators such as SCW peptidoglycan polysaccharide or interleukin-1α (IL-1α) (Sternberg et al., 1989a; Sternberg et al., 1989b; Dhabhar et al., 1995a).

Compared to the F344 strain, the Lewis strain shows a significantly greater habituation or adaptation to an acute or chronic stressor (Dhabhar et al., 1997). F344/N rats treated with the glucocorticoid receptor antagonist RU486 are rendered susceptible to SCW-induced arthritis, indicating that they do carry the immune response genes with potential for triggering autoimmunity (Sternberg et al., 1989a; Sternberg et al., 1989b). Conversely, LEW rats treated with pharmacologic doses of dexamethasone become completely resistant to the development of SCW-induced arthritis (Sternberg et al., 1989a; Sternberg et al., 1989b). Furthermore, compared to Fischer 344 (F344) rats, adrenal steroid receptors in neural and immune tissues of LEW rats show a significantly lower magnitude of activation in response to stress-induced increases in plasma corticosterone (Dhabhar et al., 1993; Dhabhar et al., 1995a). Thus, strain differences in plasma corticosterone levels also are manifest as significant differences in the extent of activation of corticosterone receptors in target tissues.

Experimental allergic encephalomyelitis (EAE) is another animal model of an autoimmune disease in which a similar immunosuppressive role for the HPA axis has been proposed (for reviews see Mason et al., 1990; Mason, 1991; Whitacre et al., 1998). The Lewis strain shows a greater susceptibility to EAE (Mason, 1991). Similar correlations between HPA axis hyporeactivity and susceptibility to autoimmune disease have been observed for autoimmune conditions in chickens (Wick et al., 1998) and mice (Lechner et al., 1996).

Complementing these animal studies, a series of elegantly conducted clinical studies (Torpy and Chrousos, 1996; Buske-Kirschbaum and Hellhammer, 2003) have shown that patients with atopic dermatitis (Buske-Kirschbaum et al., 1997; Buske-Kirschbaum et al., 1998; Buske-Kirschbaum et al., 2001) and asthma (Buske-Kirschbaum et al., 2003) show decreased HPA axis reactivity. Studies of pediatric rheumatic diseases suggest a similar HPA axis deficiency coupled with other proinflammatory hormonal biases (Chikanza et al., 2000). Differences in NK cell stress reactivity and beta(2)-adrenoreceptor upregulation on peripheral blood mononuclear cells have been observed in patients with systemic lupus erythematosus (Pawlak et al., 1999). A more complex role for sympathetic nervous system involvement in autoimmune disease also has been proposed (Kuis et al., 1996; Kavelaars et al., 1998).

Social Status

Epidemiologic studies have repeatedly linked low social status with increased disease risk and mortality rates (Adler et al., 1994; Adler and Ostrove, 1999). Sapolsky has proposed that the magnitude of chronic stress experienced by individuals of different ranks within a social hierarchy depends on the individual's personality as well as on the characteristics of social organization, such as dominance style, stability of ranks, availability of coping mechanisms, and ease of avoidance (for review see Sapolsky, 2005). Thus, despotic, top-down hierarchies maintained through aggression are more stressful for dominant animals, while those maintained through psychological intimidation are more stressful for subordinate animals (Sapolsky, 2005). Egalitarian, bottom-up hierarchies in which dominance is obtained through support from subordinate individuals are less stressful for all members. Unstable hierarchies are more stressful for dominant animals, while stable hierarchies can be more stressful for subordinate animals that have less access to food and mates. Societies that have a high availability of coping outlets (grooming, physical contact, coalition formation) are less stressful for all individuals, while those that have a low availability of coping outlets are more stressful for low-ranking individuals. Habitats that allow subordinates to easily avoid dominants are less stressful, while those that are not conducive to avoidance are generally more stressful for subordinate animals.

This highlights the fact that captive habitats that are not designed to allow subordinates to "escape" may be particularly stressful. Personality also is important in determining the influence of the social environment, regardless of rank. Individuals who perceive and react to innocuous or neutral situations as threatening and/or who are not able to muster social support are likely to experience a greater stress burden (Sapolsky, 2005). Regardless of rank, individuals who are chronically stressed as a result of their social environment show higher basal glucocorticoid hormones levels, enlarged adrenal glands, and reduced sensitivity of the HPA axis to negative glucocorticoid feedback (Sapolsky et al., 1997; Sapolsky, 2005). Such stress profiles have been associated with decreased levels of high-density lipoprotein cholesterol (Sapolsky and Mott, 1987) and decreased hippocampal volume (Uno et al., 1989).

Studies also have shown that low social status in the context of an experimental social stress paradigm is associated with a greater susceptibility to experimental viral infection (Cohen et al., 1997). Studies by Capitanio et al. have shown that social stressors that include separations and housing relocation of macaques increase susceptibility to SIV (Capitanio et al., 1998; Capitanio and Lerche, 1998) and that personality characteristics of individual animals are related to stable HPA axis response characteristics (Capitanio et al., 2004). However, other studies have shown that exposure to mild early life stressors strengthens emotional and neuroendocrine stress responses in adulthood (Parker et al., 2005). Winslow and Insel showed that oxytocin, a neurohypophyseal peptide that is thought to modulate many aspects of social behavior, has different effects depending on the social status of pairs of male squirrel monkeys. Central oxytocin administration results in increased sexual behavior and aggression in dominant males and in increased associative and marking behaviors in subordinates (Winslow and Insel, 1991). It also has been shown that individual differences in the promoter for the serotonin transporter gene interact with early rearing conditions to affect behavioral development (Champoux et al., 2002), HPA axis reactivity, and vulnerability to alcoholism (Barr et al., 2004) in nonhuman primates. Tree shrews, which are thought to provide a model for early primate behavior, have been used to study chronic social stress, which is thought to model depression in subordinate animals (Fuchs et al., 2001; Fuchs, 2005). It has been suggested that the rhesus monkey provides a particularly attractive model for aging because of the similarities between human and rhesus aging phenotypes and the close genetic relationship of this species to humans (Roth et al., 2004). Therefore, nonhuman primate models provide a significant resource for examining interactions among social environment, behavior, and health outcomes. These models, although more difficult to study than rodent models, may offer the closest approximation for humans. More research is needed to examine the health consequences of chronic social stressors and the contribution of genetic factors within these models.

Social Stressors

Rodent models of social stressors include visible burrow systems that examine group housed animals under conditions that attempt to mimic their natural habitats (Blanchard et al., 1995) and models that use cage-housed animals to induce experimentally social confrontation (Stefanski et al., 1996; Stefanski et al., 2005), social disruption (Avitsur et al., 2002a), and social isolation (Sanchez et al., 1998). Using the visible burrow system, which consists of groups of male and female rats housed in a semi-natural, visible environment, the Blanchards and their colleagues have examined the deleterious consequences of chronic social stress on numerous brain, behavioral, and physiological parameters (for review see Blanchard et al., 2001). These include differences between subordinate and dominant animals in behavior (Blanchard et al., 1993), metabolism (Tamashiro et al., 2004), HPA axis reactivity (Blanchard et al., 1995), hippocampal 5HT1A receptor levels (McKittrick et al., 1995), and corticotropin releasing factor and arginine vasopressin mRNA expression in the paraventricular hypothalamic nucleus and amygdala (Albeck et al., 1997).

Stefanski and colleagues have used a rat model of social confrontation that involves introducing intruder rats to cages of established pairs of animals (Stefanski et al., 1996; Stefanski, 2001). In this model the intruder is attacked and generally defeated. The social confrontation is allowed to proceed for hours to days, and the endocrine and immune consequences of the defeated and undefeated animals are studied (Stefanski, 2001). Studies have shown social confrontation induces a significant increase in susceptibility to metastases of injected tumor cell lines (Stefanski and Ben-Eliyahu, 1996; Stefanski, 2001), changes in blood immune cell distribution (Stefanski and Engler, 1998), T-cell maturation (Engler and Stefanski, 2003), and circulating stress hormones (Stefanski, 2000).

Sheridan and colleagues have used a model of social disruption that involves placing an aggressive retired male breeder in a cage of male mice for several consecutive stress sessions. In this model, home cage animals are attacked by the aggressive intruder and are frequently wounded. Animals that show subordinate behavior are wounded to a larger extent (Avitsur et al., 2001). Animals that show a subordinate behavioral profile also develop glucocorticoid resistance, which is measured by their lipopolysaccharide stimulated splenocyte proliferation index in vitro being resistant to suppression by corticosterone (Stark et al., 2001). These authors have shown that this stressor also alters splenocyte distribution and function (Avitsur et al.,

2002b) and increases tumor necrosis factor-alpha secretion from in vitro lipopolysaccharide stimulated splenic macrophages (Avitsur et al., 2005). The critical role of aggressive physical contact that often results in wounding in this model of social stress is highlighted by studies that show that only mice that are physically in contact with the intruder animals show the development of splenocyte glucocorticoid resistance, which is the hallmark immunological effect of this stressor (Bailey et al., 2004).

Another example of the impact of social environment involves the Watanabe heritable hyperlipidemic rabbit, which has a genetic defect in lipoprotein clearance. This defect results in severe atherosclerosis in rabbits that are raised in isolation or in an unstable social situation in which unfamiliar rabbits are paired daily. In contrast, when the rabbits are paired stably with littermates across the period of the experiment, atherosclerosis is greatly reduced (McCabe et al., 2002)—a result that demonstrates how a positive social environment that provides increased support can ameliorate the health-damaging effects of a particular genetic variant. A most interesting parallel with this animal study finding is found in one of the studies (Kaufman et al., 2004) that found increased depression among maltreated children carrying the 5HTTLPR short allele: in those with higher quality and availability of social supports the effect of the short allele to increase depression levels was ameliorated.

Stress Pathways as Mediators of Social and Behavioral Effects on Health

Physiological stress responses involving neurotransmitters and hormones are likely mediators of effects examined in almost all of the animal models described above. Therefore, it is useful to understand the concept of stress and the role that it plays in the particular model systems under examination. Numerous definitions have been proposed for the word *stress*. Each definition focuses on an aspect of an internal or external challenge, disturbance, or stimulus; on perception of a stimulus by an organism; or on a physiological response of the organism to the stimulus (McEwen, 2002; Goldstein and McEwen, 2002; Sapolsky, 2004). Physical stressors have been defined as external challenges to homeostasis and psychological stressors as the "anticipation, justified or not, that a challenge to homeostasis looms" (Sapolsky, 2005). An integrated definition states that stress is a constellation of events, consisting of a stimulus (stressor) that precipitates a reaction in the brain (stress perception) that activates physiologic fight or flight systems in the body (stress response) (Dhabhar and McEwen, 1997). The ultimate effector molecules of stress are the neurotransmitters and hormones that are released during stress, the principle mediators being norepinephrine, epinephrine, and cortisol.

It often is overlooked that a stress response has salubrious adaptive effects in the short run (Dhabhar et al., 1995b; Dhabhar and McEwen, 1996) although stress can be harmful when it is long lasting (Dhabhar and McEwen, 1997; McEwen, 1998; Glaser and Kiecolt-Glaser, 2005). Therefore, important distinguishing characteristics of stress include its duration and intensity. Acute stress has been defined as stress that lasts for a period of minutes to hours, and chronic stress as stress that persists for several hours a day for an extended period (generally months to years) (Dhabhar and McEwen, 1997). The magnitude of stress may be gauged by the peak levels of stress hormones, neurotransmitters, and other physiological changes such as increases in heart rate and blood pressure, and by the amount of time that these changes persist during and following stressor exposure. An important marker for deleterious amounts of chronic stress may be a dysregulation of the circadian corticosterone rhythm in rodents (Dhabhar and McEwen, 1997) and cortisol rhythm in humans (Sephton et al., 2000). It has been shown that moderate chronic stress experienced during UV exposure results in a significant increase in susceptibility to skin cancer (squamous cell carcinoma). This increase is mediated by a stress-induced suppression of Type 1 cytokine gene expression, a decrease in numbers of protective T cells, and an increase in numbers of suppressor T cells. Interestingly, the effects of stress on gene expression and immune cell numbers are accompanied by a disruption of the diurnal corticosterone rhythm and observed nine months after the cessation of stress (Saul et al., 2005). This indicates that stressors experienced during critical moments of immune challenge or activation may have long-term consequences.

Stress has long been suspected to play a role in the etiology of many diseases, and numerous studies have shown that stress can dysregulate or suppress immune function and hence may be detrimental to health (Herbert and Cohen, 1993; Straub and Schedlowski, 2002; Sapolsky, 2005; Glaser and Kiecolt-Glaser, 2005). Although decades of research have examined the pathological effects of stress on immune function and on health, the study of the salubrious or health-promoting effects of stress is relatively new (Dhabhar and McEwen, 1996). Much work remains to be done to elucidate the biological mechanisms mediating these bidirectional effects of stress on health and to translate basic findings regarding the adaptive effects of stress from the bench to the bedside. For a given stimulus or stressor, individual differences in genetic factors may interact with social factors to affect the degree and nature of psychological stress perception and/or the kinetics and magnitude of the physiological stress response. Therefore, when examining interactions between genes, social environment, and health, it may be critical to keep in mind the overall stress reactivity and stress status of the individual or populations being studied.

Epidemiology to Gene Knockout

Extensive epidemiological work and clinical genetics on families indicated that there must be a gene on chromosome 17(q) whose heritable mutation increased susceptibility to breast and ovarian cancer. With positional cloning, BRCA1 (BReast CAncer1) was identified (Futreal et al., 1994) and subsequently many different mutations have been described in diverse populations, including Ashkenazi Jewish, Japanese, Korean, African, and Chinese families (Katagiri et al., 1996; Okobia and Bunker, 2003; Ademuyiwa and Olopade, 2003; Lynch et al., 2004; Judkins et al., 2005; Kim et al., 2006; Song et al., 2006).

Rodent models in which the homolog of BRCA1 is knocked out (Brodie and Deng, 2001; Zan et al., 2003) have elucidated a variety of environmental, hormonal, and genetic factors that increase the penetrance of BRCA1 mutations causing breast and ovarian cancer. Such work is impossible in humans. For example, conditional inactivation of BRCA1 in the epithelial tissue of mice led to cancer development in organs other than the breast and ovary (Berton et al., 2003). Other genes have been identified and manipulated, such as ataxia telangiectasia mutated (Atm) heterozygosity and p53, demonstrating their interaction with BRCA1 to increase mammary cancer (Bachelier et al., 2005; Bowen et al., 2005). In this mouse model, even though mammary tumors were estrogen receptor negative, removing the ovaries nonetheless reduced the development of mammary tumors late in the life span. However, in a wild-type rat model, phytoestrogen-rich diets increased the BRCA1 mRNA, but not protein produced by the tumor suppressor gene. Thus, the rodent models have the potential to manipulate the environmental, hormonal, and genetic mechanisms affecting the expression of BRCA1 mutations in order to determine which are mediators and only modulate their effect on mammary tumorigenesis.

FUTURE ISSUES

Are New Animal Models Needed?

New models may not be necessary, but they could be useful. The quest for new and improved models should continue. However, promising existing models also should be nurtured, fine-tuned, and developed further.

Criteria for Animal Models Suitable for Examination of Gene-Social Environment Interactions

Animal models have a great deal to offer in furthering our understanding of the impact of interactions among social, behavioral, and genetic factors on health and can provide an important complement to clinical and community-based research. Therefore the committee makes the following recommendation:

> **Recommendation 6: Use Animal Models to Study Gene-Social Environment Interaction.** *The NIH should develop RFAs that use carefully selected animal models for research on the impact on health of interactions among social, behavioral, and genetic factors and their interactive pathways (i.e., physiological).*

The selection of the animal model should be based upon the type and complexity of the interaction to be explored. Furthermore, studies should be conducted using outbred, inbred, and wild caught animals. Appropriate animal models should be sensitive enough to register clinically relevant change in vivo; ensure that laboratory conditions are consistent with the ecological and ethological context in which the animals naturally live; recognize, account for, and preferably measure unintended physiological consequences of experimental manipulations when generating data and interpreting results; enable the examination and identification of psychological and/or physiological mediators of interactions among genes, behavior, and the social environment; enable the experimental testing of causality; and parallel human models when relevant and possible.

It probably would be advisable to establish animal housing facilities that more closely approximate each animal's natural habitat, but this would be difficult to implement. Care would need to be taken to ensure accuracy (i.e., thoroughly understand and replicate most if not all relevant ecological and ethological factors in the vivarium) and standardization across different research groups (i.e., once ecological and ethological factors are established, housing conditions designed to take them into account should be standardized across different laboratories). The standardization aspect may be a significant obstacle, because different research groups may have different opinions on what ethologically and ecologically relevant conditions are and how they should be replicated in the vivarium. However, not standardizing housing could result in significant interlaboratory variations that may make studies difficult if not impossible to replicate and compare between laboratories. In contrast, it also may be beneficial to have multiple types of environments, as an approach that would more closely mimic human living conditions (e.g., country versus city dwelling).

REFERENCES

Ademuyiwa FO, Olopade OI. 2003. Racial differences in genetic factors associated with breast cancer. *Cancer and Metastasis Reviews* 22(1):47-53.

Adler N, Ostrove J. 1999. Socioeconomic status and health: What we know and what we don't. *Annals of the New York Academy of Sciences* 896:3-15.

Adler N, Boyce T, Chesney M, Cohen S, Folkman S, Kahn R, Syme S. 1994. Socioeconomic status and health: The challenge of the gradient. *American Psychologist* 49(1):15-24.

Albeck DS, McKittrick CR, Blanchard DC, Blanchard RJ, Nikulina J, McEwen BS, Sakai RR. 1997. Chronic social stress alters levels of corticotropin-releasing factor and arginine vasopressin mRNA in rat brain. *Journal of Neuroscience* 17(12):4895-4903.

Angulo JA, Ledoux M, McEwen BS. 1991. Genomic effects of cold and isolation stress on magnocellular vasopressin mRNA-containing cells in the hypothalamus of the rat. *Journal of Neurochemistry* 56(6):2033-2038.

Avitsur R, Stark JL, Sheridan JF. 2001. Social stress induces glucocorticoid resistance in subordinate animals. *Hormones and Behavior* 39(4):247-257.

Avitsur R, Stark JL, Dhabhar FS, Padgett DA, Sheridan JF. 2002a. Social disruption-induced glucocorticoid resistance: Kinetics and site specificity. *Journal of Neuroimmunology* 124(1-2):54-61.

Avitsur R, Stark JL, Dhabhar FS, Sheridan JF. 2002b. Social stress alters splenocyte phenotype and function. *Journal of Neuroimmunology* 132(1-2):66-71.

Avitsur R, Kavelaars A, Heijnen C, Sheridan JF. 2005. Social stress and the regulation of tumor necrosis factor-alpha secretion. *Brain, Behavior, and Immunity* 19(4):311-317.

Bachelier R, Xu X, Li C, Qiao W, Furth PA, Lubet RA, Deng CX. 2005. Effect of bilateral oophorectomy on mammary tumor formation in BRCA1 mutant mice. *Oncology Reports* 14(5):1117-1120.

Bailey MT, Avitsur R, Engler H, Padgett DA, Sheridan JF. 2004. Physical defeat reduces the sensitivity of murine splenocytes to the suppressive effects of corticosterone. *Brain, Behavior, and Immunity* 18(5):416-424.

Baron RM, Kenny DA. 1986. The moderator-mediator variable distinction in social psychological research: Conceptual, strategic, and statistical considerations. *Journal of Personality and Social Psychology* 51(6):1173-1182.

Barr CS, Newman TK, Lindell S, Shannon C, Champoux M, Lesch KP, Suomi SJ, Goldman D, Higley JD. 2004. Interaction between serotonin transporter gene variation and rearing condition in alcohol preference and consumption in female primates. *Archives of General Psychiatry* 61(11):1146-1152.

Berton TR, Matsumoto T, Page A, Conti CJ, Deng CX, Jorcano JL, Johnson DG. 2003. Tumor formation in mice with conditional inactivation of BRCA1 in epithelial tissues. *Oncogene* 22(35):5415-5426.

Bielsky IF, Hu SB, Szegda KL, Westphal H, Young LJ. 2004. Profound impairment in social recognition and reduction in anxiety-like behavior in vasopressin V1a receptor knockout mice. *Neuropsychopharmacology* 29(3):483-493.

Blanchard DC, Sakai RR, McEwen B, Weiss SM, Blanchard RJ. 1993. Subordination stress: Behavioral, brain, and neuroendocrine correlates. *Behavioural Brain Research* 58(1-2):113-121.

Blanchard DC, Spencer RL, Weiss SM, Blanchard RJ, McEwen B, Sakai RR. 1995. Visible burrow system as a model of chronic social stress: Behavioral and neuroendocrine correlates. *Psychoneuroendocrinology* 20(2):117-134.

Blanchard RJ, McKittrick CR, Blanchard DC. 2001. Animal models of social stress: Effects on behavior and brain neurochemical systems. *Physiology and Behavior* 73(3):261-271.

Bowen TJ, Yakushiji H, Montagna C, Jain S, Ried T, Wynshaw-Boris A. 2005. Atm heterozygosity cooperates with loss of BRCA1 to increase the severity of mammary gland cancer and reduce ductal branching. *Cancer Research* 65(19):8736-8746.

Brodie SG, Deng CX. 2001. BRCA1-associated tumorigenesis: What have we learned from knockout mice? *Trends in Genetics* 17(10):S18-S22.

Buske-Kirschbaum A, Hellhammer DH. 2003. Endocrine and immune responses to stress in chronic inflammatory skin disorders. *Annals of the New York Academy of Sciences* 992:231-240.

Buske-Kirschbaum A, Jobst S, Psych D, Wustmans A, Kirschbaum C, Rauh W, Hellhammer D. 1997. Attenuated free cortisol response to psychosocial stress in children with atopic dermatitis. *Psychosomatic Medicine* 59(4):419-426.

Buske-Kirschbaum A, Jobst S, Hellhammer DH. 1998. Altered reactivity of the hypothalamus-pituitary-adrenal axis in patients with atopic dermatitis: Pathologic factor or symptom? *Annals of the New York Academy of Sciences* 840:747-754.

Buske-Kirschbaum A, Geiben A, Hellhammer D. 2001. Psychobiological aspects of atopic dermatitis: An overview. *Psychotherapy and Psychosomatics* 70(1):6-16.

Buske-Kirschbaum A, von Auer K, Krieger S, Weis S, Rauh W, Hellhammer D. 2003. Blunted cortisol responses to psychosocial stress in asthmatic children: A general feature of atopic disease? *Psychosomatic Medicine* 65(5):806-810.

Byrne G, Suomi SJ. 2002. Cortisol reactivity and its relation to homecage behavior and personality ratings in tufted capuchin (*Cebus apella*) juveniles from birth to six years of age. *Psychoneuroendocrinology* 27(1-2):139-154.

Cacioppo JT, Hawkley LC. 2003. Social isolation and health, with an emphasis on underlying mechanisms. *Perspectives in Biology and Medicine* 46(3 Suppl):S39-S52.

Capitanio JP, Lerche NW. 1998. Social separation, housing relocation, and survival in simian AIDS: A retrospective analysis. *Psychosomatic Medicine* 60(3):235-244.

Capitanio JP, Mendoza SP, Lerche NW, Mason WA. 1998. Social stress results in altered glucocorticoid regulation and shorter survival in simian acquired immune deficiency syndrome. *Proceedings of the National Academy of Sciences of the United States of America* 95(8):4714-4719.

Capitanio JP, Mendoza SP, Baroncelli S. 1999. The relationship of personality dimensions in adult male rhesus macaques to progression of simian immunodeficiency virus disease. *Brain, Behavior, and Immunity* 13(2):138-154.

Capitanio JP, Mendoza SP, Bentson KL. 2004. Personality characteristics and basal cortisol concentrations in adult male rhesus macaques (*Macaca mulatta*). *Psychoneuroendocrinology* 29(10):1300-1308.

Cavigelli SA, McClintock MK. 2003. Fear of novelty in infant rats predicts adult corticosterone dynamics and an early death. *Proceedings of the National Academy of Sciences of the United States of America* 100(26):16131-16136.

Champoux M, Bennett A, Shannon C, Higley JD, Lesch KP, Suomi SJ. 2002. Serotonin transporter gene polymorphism, differential early rearing, and behavior in rhesus monkey neonates. *Molecular Psychiatry* 7(10):1058-1063.

Changizi MA, McGehee RM, Hall WG. 2002. Evidence that appetitive responses for dehydration and food-deprivation are learned. *Physiology and Behavior* 75(3):295-304.

Chida Y, Sudo N, Kubo C. 2005. Social isolation stress exacerbates autoimmune disease in MRL/lpr mice. *Journal of Neuroimmunology* 158(1-2):138-144.

Chikanza IC, Kuis W, Heijnen CJ. 2000. The influence of the hormonal system on pediatric rheumatic diseases. *Rheumatic Diseases Clinics of North America* 26(4):911-925.

Coe CL, Lubach GR, Ershler WB, Klopp RG. 1989. Influence of early rearing on lymphocyte proliferation responses in juvenile rhesus monkeys. *Brain, Behavior, and Immunity* 3(1):47-60.

Coe CL, Lubach GR, Schneider ML, Dierschke DJ, Ershler WB. 1992. Early rearing conditions alter immune responses in the developing infant primate. *Pediatrics* 90(3 Pt 2):505-309.

Cohen S. 2004. Social relationships and health. *American Psychologist* 59(8):676-684.

Cohen S, Line S, Manuck SB, Rabin BS, Heise ER, Kaplan JR. 1997. Chronic social stress, social status, and susceptibility to upper respiratory infections in nonhuman primates. *Psychosomatic Medicine* 59(3):213-221.

Cohen S, Doyle WJ, Turner R, Alper CM, Skoner DP. 2003. Sociability and susceptibility to the common cold. *Psychological Science* 14(5):389-395.

Cole SW, Kemeny ME, Taylor SE. 1997. Social identity and physical health: Accelerated HIV progression in rejection-sensitive gay men. *Journal of Personality and Social Psychology* 72(2):320-335.

Cole SW, Kemeny ME, Weitzman OB, Schoen M, Anton PA. 1999. Socially inhibited individuals show heightened DTH response during intense social engagement. *Brain, Behavior, and Immunity* 13(2):187-200.

Cole SW, Kemeny ME, Fahey JL, Zack JA, Naliboff BD. 2003. Psychological risk factors for HIV pathogenesis: Mediation by the autonomic nervous system. *Biological Psychiatry* 54(12):1444-1456.

Day LB, Schallert T. 1996. Anticholinergic effects on acquisition of place learning in the Morris water task: Spatial mapping deficit or inability to inhibit nonplace strategies? *Behavioral Neuroscience* 110(5):998-1005.

de Waal FB. 2000. Primates—a natural heritage of conflict resolution. *Science* 289(5479): 586-590.

DeVries AC, DeVries MB, Taymans SE, Carter CS. 1996. The effects of stress on social preferences are sexually dimorphic in prairie voles. *Proceedings of the National Academy of Sciences of the United States of America* 93(21):11980-11984.

Dhabhar FS, McEwen BS. 1996. Stress-induced enhancement of antigen-specific cell-mediated immunity. *Journal of Immunology* 156(7):2608-2615.

Dhabhar FS, McEwen BS. 1997. Acute stress enhances while chronic stress suppresses cell-mediated immunity in vivo: A potential role for leukocyte trafficking. *Brain, Behavior, and Immunity* 11(4):286-306.

Dhabhar FS, McEwen BS, Spencer RL. 1993. Stress response, adrenal steroid receptor levels and corticosteroid-binding globulin levels-a comparison between Sprague-Dawley, Fischer 344 and Lewis rats. *Brain Research* 616(1-2):89-98.

Dhabhar FS, Miller AH, McEwen BS, Spencer RL. 1995a. Differential activation of adrenal steroid receptors in neural and immune tissues of Sprague Dawley, Fischer 344, and Lewis rats. *Journal of Neuroimmunology* 56(1):77-90.

Dhabhar FS, Miller AH, McEwen BS, Spencer RL. 1995b. Effects of stress on immune cell distribution. Dynamics and hormonal mechanisms. *Journal of Immunology* 154(10):5511-5527.

Dhabhar FS, McEwen BS, Spencer RL. 1997. Adaptation to prolonged or repeated stress-comparison between rat strains showing intrinsic differences in reactivity to acute stress. *Neuroendocrinology* 65(5):360-368.

Dobson MG, Redfern CP, Unwin N, Weaver JU. 2001. The N363S polymorphism of the glucocorticoid receptor: Potential contribution to central obesity in men and lack of association with other risk factors for coronary heart disease and diabetes mellitus. *Journal of Clinical Endocrinology and Metabolism* 86(5):2270-2274.

Engler H, Stefanski V. 2003. Social stress and T cell maturation in male rats: Transient and persistent alterations in thymic function. *Psychoneuroendocrinology* 28(8):951-969.

Ermann J, Fathman CG. 2001. Autoimmune diseases: Genes, bugs and failed regulation. *Nature Immunology* 2(9):759-761.

Faggioni R, Feingold KR, Grunfeld C. 2001. Leptin regulation of the immune response and the immunodeficiency of malnutrition. *FASEB Journal* 15(14):2565-2571.

Fahlke C, Lorenz JG, Long J, Champoux M, Suomi SJ, Higley JD. 2000. Rearing experiences and stress-induced plasma cortisol as early risk factors for excessive alcohol consumption in nonhuman primates. *Alcoholism, Clinical and Experimental Research* 24(5):644-650.

Fantuzzi G. 2005. Adipose tissue, adipokines, and inflammation. *Journal of Allergy and Clinical Immunology* 115(5):911-919.

Fish EW, Shahrokh D, Bagot R, Caldji C, Bredy T, Szyf M, Meaney MJ. 2004. Epigenetic programming of stress responses through variations in maternal care. *Annals of the New York Academy of Sciences* 1036:167-180.

Francis D, Diorio J, Liu D, Meaney MJ. 1999. Nongenomic transmission across generations of maternal behavior and stress responses in the rat. *Science* 286(5442):1155-1158.

Fuchs E. 2005. Social stress in tree shrews as an animal model of depression: An example of a behavioral model of a CNS disorder. *CNS Spectrums* 10(3):182-190.

Fuchs E, Flugge G, Ohl F, Lucassen P, Vollmann-Honsdorf GK, Michaelis T. 2001. Psychosocial stress, glucocorticoids, and structural alterations in the tree shrew hippocampus. *Physiology and Behavior* 73(3):285-291.

Futreal PA, Liu Q, Shattuck-Eidens D, Cochran C, Harshman K, Tavtigian S, Bennett LM, Haugen-Strano A, Swensen J, Miki Y, et al. 1994. BRCA1 mutations in primary breast and ovarian carcinomas. *Science* 266(5182):120-122.

Glaser R, Kiecolt-Glaser JK. 2005. Stress-induced immune dysfunction: Implications for health. *Nature Reviews. Immunology* 5(3):243-251.

Goldstein DS, McEwen B. 2002. Allostasis, homeostats, and the nature of stress. *Stress* 5(1): 55-58.

Hall WG, Swithers-Mulvey SE. 1992. Developmental strategies in the analysis of ingestive behavior. *Annals of the New York Academy of Sciences* 662:1-15.

Hammock EAD, Young LJ. 2005. Microsatellite instability generates diversity in brain and sociobehavioral traits. *Science* 308:1630-1634.

Harbuz MS, Conde GL, Marti O, Lightman SL, Jessop DS. 1997. The hypothalamic-pituitary-adrenal axis in autoimmunity. *Annals of the New York Academy of Sciences* 823: 214-224.

Harbuz MS, Chover-Gonzalez AJ, Jessop DS. 2003. Hypothalamo-pituitary-adrenal axis and chronic immune activation. *Annals of the New York Academy of Sciences* 992:99-106.

Hawkley LC, Cacioppo JT. 2003. Loneliness and pathways to disease. *Brain, Behavior, and Immunity* 17(Suppl 1):S98-S105.

Herbert TB, Cohen S. 1993. Stress and immunity in humans: A meta-analytic review. *Psychosomatic Medicine* 55(4):364-379.

House JS. 2001. Social isolation kills, but how and why? *Psychosomatic Medicine* 63(2): 273-274.

Huh GS, Boulanger LM, Du H, Riquelme PA, Brotz TM, Shatz CJ. 2000. Functional requirement for class I MHC in CNS development and plasticity. *Science* 290(5499):2155-2159.

Jacob S, McClintock MK, Zelano B, Ober C. 2002. Paternally inherited HLA alleles are associated with women's choice of male odor. *Nature Genetics* 30(2):175-179.

Jiang T, Liu S, Tan M, Huang F, Sun Y, Dong X, Guan W, Huang L, Zhou F. 2001. The phase-shift mutation in the glucocorticoid receptor gene: Potential etiologic significance of neuroendocrine mechanisms in lupus nephritis. *Clinica Chimica Acta* 313(1-2):113-117.

Jordan WC, Bruford MW. 1998. New perspectives on mate choice and the MHC. *Heredity* 81(Pt 3):239-245.

Judkins T, Hendrickson BC, Deffenbaugh AM, Scholl T. 2005. Single nucleotide polymorphisms in clinical genetic testing: The characterization of the clinical significance of genetic variants and their application in clinical research for BRCA1. *Mutation Research* 573(1-2):168-179.

Kagan J, Reznick JS, Snidman N. 1988. Biological bases of childhood shyness. *Science* 240(4849):167-171.

Kagan J, Snidman N, Julia-Sellers M, Johnson MO. 1991. Temperament and allergic symptoms. *Psychosomatic Medicine* 53(3):332-340.

Kaiser S, Sachser N. 2005. The effects of prenatal social stress on behaviour: Mechanisms and function. *Neuroscience and Biobehavioral Reviews* 29(2):283-294.

Kaiser S, Kruijver FP, Straub RH, Sachser N, Swaab DF. 2003. Early social stress in male guinea-pigs changes social behaviour, and autonomic and neuroendocrine functions. *Journal of Neuroendocrinology* 15(8):761-769.

Kalin NH, Shelton SE. 2003. Nonhuman primate models to study anxiety, emotion regulation, and psychopathology. *Annals of the New York Academy of Sciences* 1008: 189-200.

Kalin NH, Shelton SE, Rickman M, Davidson RJ. 1998. Individual differences in freezing and cortisol in infant and mother rhesus monkeys. *Behavioral Neuroscience* 112(1):251-254.

Katagiri T, Emi M, Ito I, Kobayashi K, Yoshimoto M, Iwase T, Kasumi F, Miki Y, Skolnick MH, Nakamura Y. 1996. Mutations in the BRCA1 gene in Japanese breast cancer patients. *Human Mutation* 7(4):334-339.

Kaufman J, Yang BZ, Douglas-Palumberi H, Houshyar S, Lipschitz D, Krystal JH, Gelernter J. 2004. Social supports and serotonin transporter gene moderate depression in maltreated children. *Proceedings of the National Academy of Sciences of the United States of America* 101(49):17316-17321.

Kavelaars A. 2002. Regulated expression of alpha-1 adrenergic receptors in the immune system. *Brain, Behavior, and Immunity* 16(6):799-807.

Kavelaars A, de Jong-de Vos van Steenwijk T, Kuis W, Heijnen CJ. 1998. The reactivity of the cardiovascular system and immunomodulation by catecholamines in juvenile chronic arthritis. *Annals of the New York Academy of Sciences* 840:698-704.

Kim TJ, Lee KM, Choi CH, Lee JW, Lee JH, Bae DS, Kim BG. 2006. Germline mutations of BRCA1 in two Korean hereditary breast/ovarian cancer families. *Oncology Reports* 15(3):565-569.

Kuis W, de Jong-de Vos van Steenwijk C, Sinnema G, Kavelaars A, Prakken B, Helders PJ, Heijnen CJ. 1996. The autonomic nervous system and the immune system in juvenile rheumatoid arthritis. *Brain, Behavior, and Immunity* 10(4):387-398.

Landgraf R, Frank E, Aldag JM, Neumann ID, Sharer CA, Ren X, Terwilliger EF, Niwa M, Wigger A, Young LJ. 2003. Viral vector-mediated gene transfer of the vole V1a vasopressin receptor in the rat septum: Improved social discrimination and active social behaviour. *European Journal of Neuroscience* 18(2):403-411.

Lanfear DE, Jones PG, Marsh S, Cresci S, McLeod HL, Spertus JA. 2005. Beta2-adrenergic receptor genotype and survival among patients receiving beta-blocker therapy after an acute coronary syndrome. *Journal of the American Medical Association* 294(12):1526-1533.

Lechner O, Hu Y, Jafarian-Tehrani M, Dietrich H, Schwarz S, Herold M, Haour F, Wick G. 1996. Disturbed immunoendocrine communication via the hypothalamo-pituitary-adrenal axis in murine lupus. *Brain, Behavior, and Immunity* 10(4):337-350.

LeFevre J, McClintock MK. 1991. Isolation accelerates reproductive senescence and alters its predictors in female rats. *Hormones and Behavior* 25(2):258-272.

Li J, Precht DH, Mortensen PB, Olsen J. 2003. Mortality in parents after death of a child in Denmark: A nationwide follow-up study. *Lancet* 361(9355):363-367.

Lim MM, Wang Z, Olazabal DE, Ren X, Terwilliger EF, Young LJ. 2004. Enhanced partner preference in a promiscuous species by manipulating the expression of a single gene. *Nature* 429(6993):754-757.

Lin RC, Wang WY, Morris BJ. 1999. Association and linkage analyses of glucocorticoid receptor gene markers in essential hypertension. *Hypertension* 34(6):1186-1192.

Lin RC, Wang XL, Morris BJ. 2003. Association of coronary artery disease with glucocorticoid receptor N363S variant. *Hypertension* 41(3):404-407.

Lynch HT, Rubinstein WS, Locker GY. 2004. Cancer in Jews: Introduction and overview. *Familial Cancer* 3(3-4):177-192.

Manning CJ, Wakeland EK, Potts WK. 1992. Communal nesting patterns in mice implicate MHC genes in kin recognition. *Nature* 360(6404):581-583.

Martikainen P, Valkonen T. 1996. Mortality after the death of a spouse: Rates and causes of death in a large Finnish cohort. *American Journal of Public Health* 86(8 Pt 1):1087-1093.

Martin L, Scheuerlein A, Wikelski M. 2003. Immune activity elevates energy expenditure of house sparrows: A link between direct and indirect costs? *Proceedings of the Royal Society B: Biological Sciences* 270(1511):153-158.

Mason D. 1991. Genetic variation in the stress response: Susceptibility to experimental allergic encephalomyelitis and implications for human inflammatory disease. *Immunology Today* 12(2):57-60.

Mason D, MacPhee I, Antoni F. 1990. The role of the neuroendocrine system in determining genetic susceptibility to experimental allergic encephalomyelitis in the rat. *Immunology* 70(1):1-5.

McCabe PM, Gonzales JA, Zaias J, Szeto A, Kumar M, Herron AJ, Schneiderman N. 2002. Social environment influences the progression of atherosclerosis in the watanabe heritable hyperlipidemic rabbit. *Circulation* 105(3):354-359.

McClintock MK, Conzen SD, Gehlert S, Masi C, Olopade F. 2005. Mammary cancer and social interactions: Identifying multiple environments that regulate gene expression throughout the life span. *Journals of Gerontology: Series B* 60B(Special Issue 1):32-41.

McDade TW. 2005. Life history, maintenance, and the early origins of immune function. *American Journal of Human Biology* 17(1):81-94.

McEwen BS. 1998. Protective and damaging effects of stress mediators. *New England Journal of Medicine* 338(3):171-179.

McEwen BS. 2002. *The End of Stress as We Know It*. Washington, DC: The Dana Press/Joseph Henry Press.

McKittrick CR, Blanchard DC, Blanchard RJ, McEwen BS, Sakai RR. 1995. Serotonin receptor binding in a colony model of chronic social stress. *Biological Psychiatry* 37(6): 383-393.

Meaney MJ. 2001. Maternal care, gene expression, and the transmission of individual differences in stress reactivity across generations. *Annual Review of Neuroscience* 24:1161-1192.

Meaney MJ, Szyf M. 2005. Maternal care as a model for experience-dependent chromatin plasticity? *Trends in Neurosciences* 28(9):456-463.

Miller GE, Cohen S, Rabin BS, Skoner DP, Doyle WJ. 1999. Personality and tonic cardiovascular, neuroendocrine, and immune parameters. *Brain, Behavior, and Immunity* 13(2): 109-123.

Morley JE. 1990. Appetite regulation by gut peptides. *Annual Review of Nutrition* 10: 383-395.

Morse G. 2006. Decisions and desire. *Harvard Business Review* 84(1):42, 44-51, 132.

Nelson RJ. 2004. Seasonal immune function and sickness responses. *Trends in Immunology* 25(4):187-192.

Oka K, Sakuarae A, Fujise T, Yoshimatsu H, Sakata T, Nakata M. 2003. Food texture differences affect energy metabolism in rats. *Journal of Dental Research* 82(6):491-494.

Okobia MN, Bunker CH. 2003. Molecular epidemiology of breast cancer: A review. *African Journal of Reproductive Health* 7(3):17-28.

Parker KJ, Buckmaster CL, Justus KR, Schatzberg AF, Lyons DM. 2005. Mild early life stress enhances prefrontal-dependent response inhibition in monkeys. *Biological Psychiatry* 57(8):848-855.

Pawlak CR, Jacobs R, Mikeska E, Ochsmann S, Lombardi MS, Kavelaars A, Heijnen CJ, Schmidt RE, Schedlowski M. 1999. Patients with systemic lupus erythematosus differ from healthy controls in their immunological response to acute psychological stress. *Brain, Behavior, and Immunity* 13(4):287-302.

Petitto JM, Lysle DT, Gariepy JL, Clubb PH, Cairns RB, Lewis MH. 1993. Genetic differences in social behavior: Relation to natural killer cell function and susceptibility to tumor development. *Neuropsychopharmacology* 8(1):35-43.

Petitto JM, Lysle DT, Gariepy JL, Lewis MH. 1994. Association of genetic differences in social behavior and cellular immune responsiveness: Effects of social experience. *Brain, Behavior, and Immunity* 8(2):111-122.

Petitto JM, Gariepy JL, Gendreau PL, Rodriguiz R, Lewis MH, Lysle DT. 1999. Differences in NK cell function in mice bred for high and low aggression: Genetic linkage between complex behavioral and immunological traits? *Brain, Behavior, and Immunity* 13(2):175-186.

Pitkow LJ, Sharer CA, Ren X, Insel TR, Terwilliger EF, Young LJ. 2001. Facilitation of affiliation and pair-bond formation by vasopressin receptor gene transfer into the ventral forebrain of a monogamous vole. *Journal of Neuroscience* 21(18):7392-7396.

Prendergast BJ, Nelson RJ. 2001. Spontaneous "regression" of enhanced immune function in a photoperiodic rodent *Peromyscus maniculatus*. *Proceedings of the Royal Society B: Biological Sciences* 268(1482):2221-2228.

Pressman SD, Cohen S, Miller GE, Barkin A, Rabin BS, Treanor JJ. 2005. Loneliness, social network size, and immune response to influenza vaccination in college freshmen. *Health Psychology* 24(3):297-306.

Pruessner JC, Champagne F, Meaney MJ, Dagher A. 2004. Dopamine release in response to a psychological stress in humans and its relationship to early life maternal care: A positron emission tomography study using. *Journal of Neuroscience* 24(11):2825-2831.

Ramsay CE, Hayden CM, Tiller KJ, Burton PR, Hagel I, Palenque M, Lynch NR, Goldblatt J, LeSouef PN. 1999. Association of polymorphisms in the beta2-adrenoreceptor gene with higher levels of parasitic infection. *Human Genetics* 104(3):269-274.

Rock KL, Shen L. 2005. Cross-presentation: Underlying mechanisms and role in immune surveillance. *Immunological Reviews* 207:166-183.

Ronnback A, Dahlqvist P, Svensson PA, Jernas M, Carlsson B, Carlsson LM, Olsson T. 2005. Gene expression profiling of the rat hippocampus one month after focal cerebral ischemia followed by enriched environment. *Neuroscience Letters* 385(2):173-178.

Roth GS, Mattison JA, Ottinger MA, Chachich ME, Lane MA, Ingram DK. 2004. Aging in rhesus monkeys: Relevance to human health interventions. *Science* 305(5689):1423-1426.

Sachser N, Kaiser S. 1996. Prenatal social stress masculinizes the females' behaviour in guinea pigs. *Physiology and Behavior* 60(2):589-594.

Sachser N, Durschlag M, Hirzel D. 1998. Social relationships and the management of stress. *Psychoneuroendocrinology* 23(8):891-904.

Sanchez MM, Aguado F, Sanchez-Toscano F, Saphier D. 1998. Neuroendocrine and immunocytochemical demonstrations of decreased hypothalamo-pituitary-adrenal axis responsiveness to restraint stress after long-term social isolation. *Endocrinology* 139(2):579-587.

Sanders VM, Straub RH. 2002. Norepinephrine, the beta-adrenergic receptor, and immunity. *Brain, Behavior, and Immunity* 16(4):290-332.

Sapolsky RM. 2004. *Why Zebras Don't Get Ulcers.* 3rd edition. New York: Owl Books.

Sapolsky RM. 2005. The influence of social hierarchy on primate health. *Science* 308(5722): 648-652.

Sapolsky RM, Mott GE. 1987. Social subordinance in wild baboons is associated with suppressed high density lipoprotein-cholesterol concentrations: The possible role of chronic social stress. *Endocrinology* 121(5):1605-1610.

Sapolsky RM, Alberts SC, Altmann J. 1997. Hypercortisolism associated with social subordinance or social isolation among wild baboons. *Archives of General Psychiatry* 54(12): 1137-1143.

Saul AN, Oberyszyn TM, Daugherty C, Kusewitt D, Jones S, Jewell S, Malarkey WB, Lehman A, Lemeshow S, Dhabhar FS. 2005. Chronic stress and susceptibility to skin cancer. *Journal of the National Cancer Institute* 97(23):1760-1767.

Schaefer C, Quesenberry CP Jr, Wi S. 1995. Mortality following conjugal bereavement and the effects of a shared environment. *American Journal of Epidemiology* 141(12):1142-1152.

Schwartz CE, Wright CI, Shin LM, Kagan J, Rauch SL. 2003. Inhibited and uninhibited infants "grown up": Adult amygdalar response to novelty. *Science* 300(5627):1952-1953.

Sephton SE, Sapolsky RM, Kraemer HC, Spiegel D. 2000. Diurnal cortisol rhythm as a predictor of breast cancer survival. *Journal of the National Cancer Institute* 92(12):994-1000.

Song CG, Hu Z, Yuan WT, Di GH, Shen ZZ, Huang W, Shao ZM. 2006. BRCA1 and BRCA2 gene mutations of familial breast cancer from Shanghai in China. *Zhonghua Yi Xue Yi Chuan Xue Za Zhi / Chinese Journal of Medical Genetics* 23(1):27-31.

Stark JL, Avitsur R, Padgett DA, Campbell KA, Beck FM, Sheridan JF. 2001. Social stress induces glucocorticoid resistance in macrophages. *American Journal of Physiology. Regulatory, Integrative and Comparative Physiology* 280(6):R1799-1805.

Stefanski V. 2000. Social stress in laboratory rats: Hormonal responses and immune cell distribution. *Psychoneuroendocrinology* 25(4):389-406.

Stefanski V. 2001. Social stress in laboratory rats: Behavior, immune function, and tumor metastasis. *Physiology and Behavior* 73(3):385-391.

Stefanski V, Ben-Eliyahu S. 1996. Social confrontation and tumor metastasis in rats: Defeat and beta-adrenergic mechanisms. *Physiology and Behavior* 60(1):277-282.

Stefanski V, Engler H. 1998. Effects of acute and chronic social stress on blood cellular immunity in rats. *Physiology and Behavior* 64(5):733-741.

Stefanski V, Solomon GF, Kling AS, Thomas J, Plaeger S. 1996. Impact of social confrontation on rat CD4 T cells bearing different CD45R isoforms. *Brain, Behavior, and Immunity* 10(4):364-379.

Stefanski V, Raabe C, Schulte M. 2005. Pregnancy and social stress in female rats: Influences on blood leukocytes and corticosterone concentrations. *Journal of Neuroimmunology* 162(1-2):81-88.

Sternberg E, Wilder R. 1989. The role of the hypothalamic-pituitary-adrenal axis in an experimental model of arthritis. *Progress in Neuroendocrinimmunology* 2:102-108.

Sternberg EM. 1995. Neuroendocrine factors in susceptibility to inflammatory disease: Focus on the hypothalamic-pituitary-adrenal axis. *Hormone Research* 43(4):159-161.

Sternberg EM. 2001. Neuroendocrine regulation of autoimmune/inflammatory disease. *Journal of Endocrinology* 169(3):429-435.

Sternberg EM, Hill JM, Chrousos GP, Kamilaris T, Listwak SJ, Gold PW, Wilder RL. 1989a. Inflammatory mediator-induced hypothalamic-pituitary-adrenal axis activation is defective in streptococcal cell wall arthritis-susceptible Lewis rats. *Proceedings of the National Academy of Sciences of the United States of America* 86(7):2374-2378.

Sternberg EM, Young WS, Bernardini R, Calogero AE, Chrousos GP, Gold PW, Wilder RL. 1989b. A central nervous system defect in biosynthesis of corticotropin-releasing hormone is associated with susceptibility to streptococcal cell wall-induced arthritis in Lewis rats. *Proceedings of the National Academy of Sciences of the United States of America* 86(12):4771-4775.

Sternberg EM, Chrousos GP, Wilder RL, Gold PW. 1992a. The stress response and the regulation of inflammatory disease. *Annals of Internal Medicine* 117(10):854-866.

Sternberg EM, Glowa JR, Smith MA, Calogero AE, Listwak SJ, Aksentijevich S, Chrousos GP, Wilder RL, Gold PW. 1992b. Corticotropin releasing hormone related behavioral and neuroendocrine responses to stress in Lewis and Fischer rats. *Brain Research* 570(1-2):54-60.

Straub RH, Schedlowski M. 2002. Immunology and multimodal system interactions in health and disease. *Trends in Immunology* 23(3):118-120.

Svensson E, Sinervo B, Comendant T. 2001. Condition, genotype-by-environment interaction, and correlational selection in lizard life-history morphs. *Evolution; International Journal of Organic Evolution* 55(10):2053-2069.

Tamashiro KL, Nguyen MM, Fujikawa T, Xu T, Yun Ma L, Woods SC, Sakai RR. 2004. Metabolic and endocrine consequences of social stress in a visible burrow system. *Physiology and Behavior* 80(5):683-693.

Thakkinstian A, McEvoy M, Minelli C, Gibson P, Hancox B, Duffy D, Thompson J, Hall I, Kaufman J, Leung TF, Helms PJ, Hakonarson H, Halpi E, Navon R, Attia J. 2005. Systematic review and meta-analysis of the association between {beta}2-adrenoceptor polymorphisms and asthma: A HuGE review. *American Journal of Epidemiology* 162(3):201-211.

Tonelli L, Webster JI, Rapp KL, Sternberg E. 2001. Neuroendocrine responses regulating susceptibility and resistance to autoimmune/inflammatory disease in inbred rat strains. *Immunological Reviews* 184:203-211.

Torpy DJ, Chrousos GP. 1996. The three-way interactions between the hypothalamic-pituitary-adrenal and gonadal axes and the immune system. *Baillieres Clinical Rheumatology* 10(2):181-198.

Tsigos C, Chrousos GP. 1994. Physiology of the hypothalamic-pituitary-adrenal axis in health and dysregulation in psychiatric and autoimmune disorders. *Endocrinology and Metabolism Clinics of North America* 23(3):451-466.

Uchino BN, Cacioppo JT, Kiecolt-Glaser JK. 1996. The relationship between social support and physiological processes: A review with emphasis on underlying mechanisms and implications for health. *Psychological Bulletin* 119(3):488-531.

Ukkola O, Perusse L, Weisnagel SJ, Bergeron J, Despres JP, Rao DC, Bouchard C. 2001. Interactions among the glucocorticoid receptor, lipoprotein lipase, and adrenergic receptor genes and plasma insulin and lipid levels in the Quebec Family Study. *Metabolism: Clinical and Experimental* 50(2):246-252.

Ulbrecht M, Hergeth MT, Wjst M, Heinrich J, Bickeboller H, Wichmann HE, Weiss EH. 2000. Association of beta(2)-adrenoreceptor variants with bronchial hyperresponsiveness. *American Journal of Respiratory and Critical Care Medicine* 161(2 Pt 1):469-474.

Uno H, Tarara R, Else JG, Suleman MA, Sapolsky RM. 1989. Hippocampal damage associated with prolonged and fatal stress in primates. *Journal of Neuroscience* 9(5):1705-1711.

van Rossum EF, Koper JW, Huizenga NA, Uitterlinden AG, Janssen JA, Brinkmann AO, Grobbee DE, de Jong FH, van Duyn CM, Pols HA, Lamberts SW. 2002. A polymorphism in the glucocorticoid receptor gene, which decreases sensitivity to glucocorticoids in vivo, is associated with low insulin and cholesterol levels. *Diabetes* 51(10):3128-3134.

Vidal J, Rama R. 1994. Association of the antibody response to hemocyanin with behavior in mice bred for high or low antibody responsiveness. *Behavioral Neuroscience* 108(6): 1172-1178.

Webster JI, Tonelli L, Sternberg EM. 2002. Neuroendocrine regulation of immunity. *Annual Review of Immunology* 20:125-163.

Weiss ST. 2005. Obesity: Insight into the origins of asthma. *Nature Immunology* 6(6): 537-539.

Whitacre CC, Dowdell K, Griffin AC. 1998. Neuroendocrine influences on experimental autoimmune encephalomyelitis. *Annals of the New York Academy of Sciences* 840: 705-716.

White NM. 1986. Control of sensorimotor function by dopaminergic nigrostriatal neurons: Influence on eating and drinking. *Neuroscience and Biobehavioral Reviews* 10(1):15-36.

Wick G, Sgonc R, Lechner O. 1998. Neuroendocrine-immune disturbances in animal models with spontaneous autoimmune diseases. *Annals of the New York Academy of Sciences* 840:591-598.

Williams G, Bing C, Cai XJ, Harrold JA, King PJ, Liu XH. 2001. The hypothalamus and the control of energy homeostasis: Different circuits, different purposes. *Physiology and Behavior* 74(4-5):683-701.

Winslow JT, Insel TR. 1991. Social status in pairs of male squirrel monkeys determines the behavioral response to central oxytocin administration. *Journal of Neuroscience* 11(7): 2032-2038.

Young LJ, Wang Z, Insel TR. 1998. Neuroendocrine bases of monogamy. *Trends in Neurosciences* 21(2):71-75.

Young LJ, Lim MM, Gingrich B, Insel TR. 2001. Cellular mechanisms of social attachment. *Hormones and Behavior* 40(2):133-138.

Zan Y, Haag JD, Chen KS, Shepel LA, Wigington D, Wang YR, Hu R, Lopez-Guajardo CC, Brose HL, Porter KI, Leonard RA, Hitt AA, Schommer SL, Elegbede AF, Gould MN. 2003. Production of knockout rats using ENU mutagenesis and a yeast-based screening assay. *Nature Biotechnology* 21(6):645-651.

Zehr JL, Gans SE, McClintock MK. 2001. Variation in reproductive traits is associated with short anogenital distance in female rats. *Developmental Psychobiology* 38(4):229-238.

8

Study Design and Analysis for Assessment of Interactions

A clear formulation of the concept of "interaction" and an understanding of the research designs that can be used to test for it are central to progress in assessing the impact of interactions among social, behavioral, and genetic factors on health. This chapter discusses definitions of interaction, from both statistical and biological points of view. It also describes the models of interaction that are likely to be relevant for evaluating the joint influence of multiple factors on health. Finally, the chapter considers research methods for the detection and elucidation of interactive effects and statistical issues related to the application of these methods.

DEFINITIONS OF INTERACTIONS

Statistical tests for interaction are entirely dependent on the *measurement scale*—additive or multiplicative—that is used to evaluate the effects of different factors on health. This problem is absolutely critical to the design of future studies and to the interpretation of their results. Researchers using different scales would reach different conclusions about the same underlying biological processes and make different public health recommendations depending on the scale being used. Moreover, the same data can be made to fit more than one statistical model (e.g., by analysis of the original measurement versus log-transformed data), making it very difficult to interpret the results. Thus, if the goal is to understand biology, a different way to conceptualize interaction must be found that is not dependent on statistical models. Epidemiologists have struggled with this problem and have developed an alternative conceptual framework for interaction that is not based on statistical models (Rothman and Greenland, 1998a; Rothman and Greenland, 2005). The logic of this new conceptualization is described in detail in the paper provided in Appendix E by Sharon Schwartz and summarized below; it leads to the conclusion that the *additive scale* is the only meaningful reference point for the measurement of interaction. Few investigators are aware of this new conceptualization of interaction, and hence few have used it as a basis for interpreting their findings.

Statistical Interaction

From a statistical point of view, interaction can be defined as a deviation from *conditional independence*, a state in which the effect of one factor (social, behavioral, or genetic) on health is the same within strata defined by another factor. This definition implies that an interaction is present if the effect of a social or behavioral factor on disease risk differs among individuals with different genotypes, or if the effect of a genotype on disease risk differs among individuals with different levels of a social or behavioral factor. The problem with this definition, as indicated above, is that it is entirely dependent on the measurement scale (multiplicative or additive). Ratio measures such as relative risks (RRs) or odds ratios (ORs) assess the effects of risk factors on a multiplicative scale, because they reflect the degree to which disease risk (for RR) or odds (for OR) are *multiplied* in individuals with the risk factor compared to those without. In contrast, risk differences (RD) assess the effects of risk factors on an additive scale, because they reflect how much disease risk is *added* in individuals who have the risk factor, compared with those who do not. The statistical definition of interaction differs depending on which of these measurement scales is used. For example, in the consideration of factors A and B, interaction on a multiplicative scale is defined as a different RR for factor A across strata defined by factor B, while on an additive scale, interaction is defined as a different RD for factor A across strata defined by factor B. Use of these two different measurement scales can lead to substantively different conclusions in studies of interaction.

Table 8-1 illustrates the relationship, in general, between interaction defined on the multiplicative and additive scales. The risks to four categories of individuals are considered: those who have both a risk-influencing genotype and an environmental exposure (r_{11}), the genotype but not the exposure (r_{01}), the exposure but not the genotype (r_{10}), and neither (r_{00}). On the multiplicative scale, interaction is defined by (r_{11}/r_{01}) ≠ (r_{10}/r_{00}), while on the additive scale it is defined by ($r_{11}- r_{01}$) ≠ ($r_{10}- r_{00}$). To facilitate comparison of these two measures, it is convenient to express both as RRs, with the risk in individuals with neither genotype nor exposure as the

TABLE 8-1 Epidemiologic Measures of the Effects of a High-Risk Genotype and a Social or Behavioral Risk Factor

Disease Status	High-Risk Genotype		Low-Risk Genotype	
	Social or behavioral risk factor present	Social or behavioral risk factor absent	Social or behavioral risk factor present	Social or behavioral risk factor absent
Cohort Study				
Affected	a	B	e	f
Unaffected	c	D	g	h
Risk	$r_{11} = a/(a+c)$	$r_{01} = b/(b+d)$	$r_{10} = e/(e+g)$	$r_{00} = f/(f+h)$
Relative Risk	$RR_{11} = r_{11}/r_{00}$	$RR_{01} = r_{01}/r_{00}$	$RR_{10} = r_{10}/r_{00}$	$RR_{00} = 1.0$ (referent)
Case-Control Study				
Cases	a	B	e	f
Controls	c	D	g	h
Odds Ratio	$OR_{11} = a\,h/cf$	$OR_{01} = bh/df$	$OR_{10} = eh/gf$	$OR_{00} = 1.0$ (referent)

reference category, so that $RR_{11} = (r_{11}/r_{00})$, $RR_{10} = (r_{10}/r_{00})$, $RR_{01} = (r_{01}/r_{00})$, and $RR_{00} = (r_{00}/r_{00}) = 1.0$. Table 8-1 illustrates how these RRs would be calculated in a cohort study and how ORs would be calculated in a case-control study, as estimates of these RRs. With these definitions it is easy to show, by simple algebra, that interaction measured on a multiplicative scale is defined by $RR_{11} \neq RR_{10} \times RR_{01}$, while interaction measured on an additive scale is defined by $RR_{11} \neq RR_{10} + RR_{01} - 1$. Thus, under a multiplicative scale, interaction would be tested for by determining whether or not RR_{11} is equal to the product of RR_{10} and RR_{01}, while under an additive scale, interaction would be tested for by determining whether or not RR_{11} is equal to the sum of these two RRs – 1. The only circumstances under which the same conclusion would be drawn about whether or not interaction is present from both measurement scales is when either RR_{10} or RR_{01} is equal to 1.0—that is, one (or both) of the risk factors has no effect when acting in the absence of the other. If each risk factor increases risk when acting in the absence of the other (i.e., $RR_{10} > 1.0$ and $RR_{01} > 1.0$), the RR_{11} that must be observed to declare that interaction is present will be higher under a multiplicative scale than under an additive scale, thus, if a multiplicative scale is used, interaction on an additive scale could be missed. Also, if both factors have effects when acting by themselves and no interaction is present on a multiplicative scale, interaction will always be present on an additive scale.

The importance of considering the underlying model against which gene-environment interactions are tested (i.e., additive versus multiplicative) is illustrated by a study of the Factor V Leiden variant, use of oral contraceptives (OCs), and risk of deep vein thrombosis (DVT) (Vandenbroucke et al., 1994; Austin and Schwartz, 2006). The use of OCs is associated with an approximately four-fold increased risk of DVT among women who do not carry Factor V Leiden (RR_{10} = 3.7); and, similarly, carrying Factor V Leiden is associated with a four-fold increased risk among women who do not take OCs (RR_{01} = 4.0). The risk of DVT in women who carry Factor V Leiden *and* use OCs is close to what would be expected under a model of no interaction on a multiplicative scale—that is, RR_{11} = 19.8, which is very similar to 3.7 × 4.0 = 14.8. However, the absolute incidence of DVT is higher among OC users compared to OC nonusers; hence, the same relative increase in risk of DVT associated with OC use in Factor V Leiden carriers and noncarriers translates into a larger absolute risk difference due to OC use among Factor V Leiden carriers—that is, RR_{11} (19.8) is almost three times greater than RR_{01} + RR_{10} - 1 (6.7). In short, the use of a multiplicative scale to test for interaction fails to identify an important interaction between these characteristics: excess DVT cases occur among women who are both OC users and Factor V Leiden carriers compared to what would be expected based on the sum of the individual effects of these factors (Austin and Schwartz, 2006). Such information could potentially be important, for example, in counseling Factor V Leiden carriers regarding the risks and benefits of choosing OCs for birth control as opposed to other available methods.

Newer Conceptualization in Epidemiology

To build a conceptual framework more closely tied to biology, epidemiologists begin by considering what happens at the *individual* level rather than at the level of the population. Much of this thinking is based on the counterfactual model, which defines a "cause" of disease, in an individual, as any factor without which the disease would not have occurred (Greenland and Robins, 1986; Rothman and Greenland, 1998b; Maldonado and Greenland, 2002; Rothman and Greenland, 2005). This framework assumes that multiple etiologic pathways ("sufficient causes") can lead to the same disease, and within each etiologic pathway, multiple factors can work in tandem (multiple "component causes" within a sufficient cause) to cause the disease. An interaction, then, is defined as the co-participation of two component causes within one sufficient cause, so that both factors are necessary for the sufficient cause to occur. The commissioned paper by Sharon Schwartz (see Appendix E) describes this "sufficient-component cause" model in detail, specifically as it applies to the assessment of gene-environment interactions. An abbreviated description appears here.

For an individual who is exposed to a set of risk factors, only two outcomes are possible with regard to disease occurrence: the individual either develops disease or remains unaffected. Individuals can be classified into "response types" depending on whether or not they develop disease under different exposure combinations. In the simplest case of two dichotomous risk factors, there are four possible risk factor combinations (i.e., in the case considered here: both genotype and exposure, genotype alone, exposure alone, and neither). For each risk factor combination, an individual can be either a responder (i.e., he/she develops disease) or a nonresponder (i.e., he/she remains unaffected), which means that there are $2 \times 2 \times 2 \times 2 = 16$ possible response types. For example, in one response type, individuals develop disease under all four exposure combinations (so-called Doomed). At the opposite extreme, some individuals never develop disease regardless of their genotype or exposure status (so-called Immune) (Rothman and Greenland, 1998a).

Among the 16 possible response types, 10 can be viewed as involving interaction, in the sense that an individual's response to one risk factor depends on his/her status with respect to the other. (This definition may appear similar to the statistical concept discussed above—a deviation from conditional independence—but it is totally different because it considers the response at an individual level, rather than at the level of population measures of risk.) For example, in one interaction type, an individual develops disease only if exposed to both factors (causal synergism), and in another, an individual develops disease if exposed to either factor alone, but not if exposed to both or neither (causal antagonism). All other possible interaction types are also considered, involving either risk-raising or protective effects and either synergism or antagonism.

This model, based on conceptualizing risks at an individual level, can be used to derive an expected pattern of risk in a population under different exposure conditions. To do this, the population is assumed to contain a distribution of the different response types. The average risk, or incidence proportion, among individuals with each exposure combination is computed by adding the proportions of response types who will be affected if they have that exposure combination. When the average risks are computed in this way, under the assumption that none of the 10 interaction types is present in the population, the results show that the risks are consistent with risk additivity (i.e., no interaction on an additive scale, or $(r_{11} - r_{01}) = (r_{10} - r_{00})$) (Rothman and Greenland, 1998a). This derivation demonstrates, based on a conceptual rather than a statistical argument, that tests for the presence of interaction should be based on risks measured on an additive scale. One important caveat should be noted in this regard, however: although departures from risk additivity reflect the presence of interaction types in the population, a *lack* of departure from additivity does not necessarily imply *absence* of interaction types. This is because different types of inter-

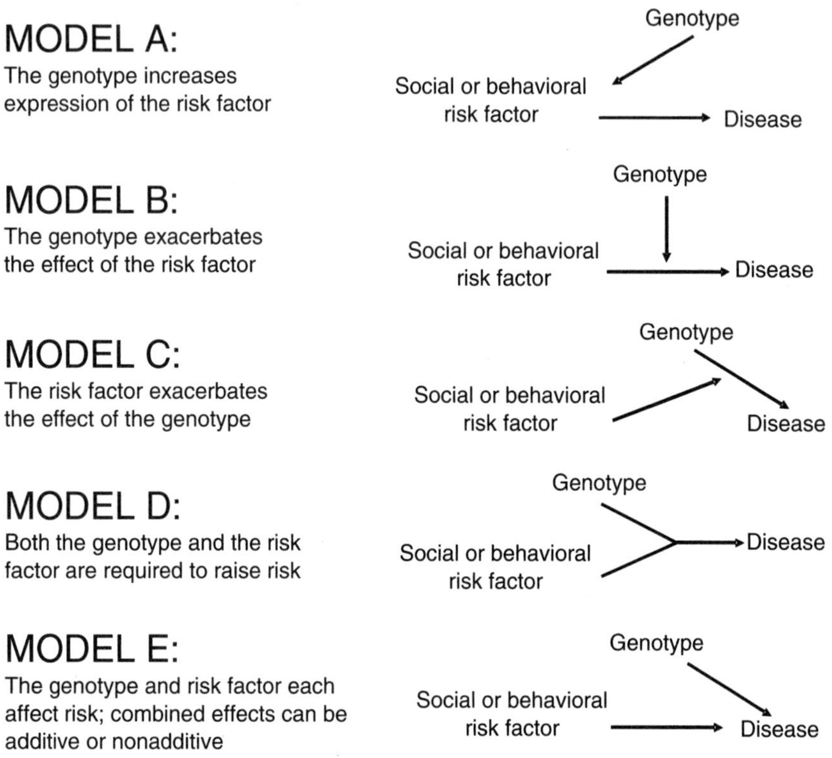

MODEL A:
The genotype increases expression of the risk factor

MODEL B:
The genotype exacerbates the effect of the risk factor

MODEL C:
The risk factor exacerbates the effect of the genotype

MODEL D:
Both the genotype and the risk factor are required to raise risk

MODEL E:
The genotype and risk factor each affect risk; combined effects can be additive or nonadditive

FIGURE 8-1 Five plausible models of interaction between a genotype and a social or behavioral risk factor.
SOURCE: Adapted from Ottman, 1996.

action could counterbalance one another. Moreover, departures from additivity may be difficult to detect because of limitations in statistical power. Also, most commonly used statistical software tests for interaction only on a multiplicative scale—for example, through the inclusion of an interaction term in a logistic regression model.

Plausible Models of Gene-Environment Interaction

Khoury et al. (1988) and Ottman (1990, 1996) have outlined several plausible models of interaction that are relevant for the consideration of the joint effects of the interaction of social, behavioral, and genetic factors on health (Figure 8-1). The following describes these models in both conceptual terms and in terms of the average risks expected in a population.

In model A, a genotype has a causal effect on a social or behavioral risk factor and through this pathway influences disease risk indirectly (Ottman, 1990; Ottman, 1996). This model differs markedly from the others because it does not involve interaction in either the statistical or biological sense. One hypothetical example would be a genetic influence on the behavior of alcohol consumption, which would be expected to affect risk for cirrhosis of the liver and other alcohol-related diseases. In this model a behavioral risk factor, alcohol consumption, is an intervening or mediating variable in the relation of a genotype to health. Although this model does not involve interaction, it is important for the consideration of the joint effects of genetic, social, and behavioral factors on health. An understanding of the intervening factors relevant to the effects of genes on health may facilitate the development of methods to prevent adverse health outcomes associated with genetic effects. For example, dietary treatment is used to prevent mental retardation in individuals with phenylketonuria, an autosomal recessive condition characterized by a deficiency of the enzyme needed to convert phenylalanine to tyrosine. Removal of phenylalanine from the diet prevents the buildup of blood levels of phenylalanine (the intervening variable), which would otherwise have a toxic effect on brain development. In the alcohol consumption example above, programs targeted to drinking behavior in individuals with a genetic susceptibility could reduce the risk of liver disease.

Model B postulates that a social or behavioral factor has a direct causal relationship to disease, and a genotype exacerbates this relationship without having any effect on disease when acting by itself. Returning to the example of alcohol drinking behavior, this model might involve a genetic influence on alcohol metabolism that leads to exacerbation of the health effects of drinking behavior. No effect of the genotype would be expected in the absence of exposure to drinking. In model C, a genotype is assumed to have a direct effect on disease risk, and the social or behavioral factor exacerbates this effect, without influencing disease risk when acting by itself. An example might involve a genetic influence on liver disease that is exacerbated by even small amounts of drinking (i.e., levels that do not influence risk in persons who do not carry the high-risk genotype). In model D, both a genotype and a social or behavioral risk factor are required to influence risk—neither affects risk in the absence of the other. Such a model might involve an alcohol-sensitivity genotype, heavy drinking, and liver disease—the assumption being that only individuals who drink heavily *and* who have a genetically mediated sensitivity to the effects of drinking develop disease. Finally, in model E both a genotype and a social or behavioral factor influence disease risk in the absence of the other factor. Although models B, C, and D involve interaction regardless of the scale of measurement, in model E the joint effect of the two factors can be consistent with interaction on an additive scale, a multiplicative scale, or neither.

The models discussed above do not explicitly incorporate protective effects of genes and social or behavioral factors, nor antagonistic rather than synergistic interactions. These are important and are likely to become increasingly so as new chemotherapeutic interventions and pharmacogenomic applications are developed. For example, one model not included in models B, C, D, and E involves a protective effect observed only in persons with a particular genotype. This type of effect was observed in a recent study, in which postmenopausal estrogen usage was protective for cognitive impairment only in women who did not carry ApoE-ε4 alleles (Yaffe et al., 2000). A risk-raising effect of carrying ≥ 1 ApoE-ε4 allele ($RR_{01} > 1$) and a protective effect of menopausal estrogen usage ($RR_{10} < 1$) were observed, and the RR in women with both the genotype and the exposure (RR_{11}) was greater than expected under an additive model, since the protective effect of the exposure was restricted to those without the high-risk genotype. Taken together, all of these models illustrate the potential complexity of evaluating interactions, even for a single genotype and a single social or behavioral factor.

RESEARCH DESIGNS FOR EVALUATING INTERACTIONS

A number of research designs for testing models of gene-environment interaction have been described (Ottman, 1994; Andrieu and Goldstein, 1998; Andrieu et al., 2001; Liu et al., 2004; Lake and Laird, 2004; Tweel and Schipper, 2004; Andrieu and Goldstein, 2004; Kraft and Hunter, 2005; Hunter, 2005; Moffit et al., 2005). These include traditional cohort and case-control designs with measured genotypes, sibling pair and case-parent triad designs, twin studies, and other approaches. The different designs offer advantages and disadvantages with respect to validity and efficiency. Population stratification, a special type of confounding in allelic association studies (Wacholder et al., 2002; Thomas and Witte, 2002), also may affect studies of gene-environment interaction, possibly leading to spurious findings. In allelic association studies, this type of confounding consists of a spurious association between an allele and disease, resulting from the existence of subgroups that vary both in allele frequency and disease occurrence. In a population "stratified" in this way, the distribution of subgroups will differ for affected and unaffected individuals, leading to a difference between the cases and controls in allele frequencies when the allele does not have a true association with disease. The importance of population stratification in studies of allelic association is controversial, but one recent study found evidence for substantial confounding that would not have been detected by standard analytical methods (Campbell et al., 2005). Campbell and colleagues (2005) found that a single nucleotide polymorphism (SNP) in the LCT gene was strongly associated with height in a sample of European Americans. This association was due to stratification: both height and the frequency of the SNP varied widely across Europe. When subjects were rematched on the basis of location of European ancestry, the apparent association was greatly diminished. Since population subgroups also may be expected to differ with respect to environmental exposures, the same type of confounding could occur in studies of gene-environment interaction. That is, different subgroups may have both different genetic backgrounds and different cultures or socioeconomically influenced patterns of behavior, creating a correlation between genotype and environmental exposure that must be controlled for. Thus the impact of stratification on different designs should be considered.

Cohort Studies

Unlike retrospective case-control studies, prospective cohort studies are immune to recall bias (i.e., different recall of exposures by cases and controls) and are expected to have minimal selection bias if the follow-up rate is high. If participation in the study is unrelated to ethnicity, the susceptibility of these types of studies to confounding due to population stratification also should be minimal. However, if this is a concern, *genomic control* methods can be used to control for stratification effects (Pritchard and Donnelly, 2001; Devlin et al., 2001; Tang et al., 2005). Genomic control methods make use of information on unlinked genetic markers throughout the genome to assess population stratification and adjust for it, if necessary. The rationale is that with population stratification, affected and unaffected individuals are likely to differ with respect to allele frequencies at many loci throughout the genome, while if an allele is truly related to disease risk, the association is likely to be restricted to a single genomic region.

In nested case-control studies, cases and controls are ascertained from within a cohort study. That is, cases are identified at follow-up and generally are matched on appropriate covariates to a group of controls, but risk factor data collected at baseline are used as predictors. Thus, again, information on exposures is collected prior to disease onset, and the problem of recall bias is avoided. However, some selection bias (i.e., differential inclusion of some subgroups of cases, because of differential survival or loss to follow-up) may occur (although usually less than in retrospective case-control studies). The main disadvantage of these studies is the large number of subjects that must be enrolled to ensure an adequate number of cases during the follow-up period for analysis. Also, for late-onset disorders, a very long follow-up period may be needed. Prospective studies of gene-environment interaction are feasible only for common disorders, and even then, collaborative studies are essential to obtain sample sizes for sufficient statistical power.

Case-Control Studies

In retrospective case-control studies, the potential for recall bias and selection bias is considerable (Kraft and Hunter, 2005). Case-control studies also are susceptible to bias due to population stratification; this should be minimized by matching on ethnicity (Wacholder et al., 2000; Wacholder et al., 2002; Cardon and Palmer, 2003; Reiner et al., 2005), and/or using genomic control methods in the analysis (Pritchard and Donnelly, 2001; Devlin et al., 2001; Tang et al., 2005). Temporality may be difficult to establish with respect to the environmental exposures; that is, it may be difficult to ensure that exposure occurred before the onset of disease. For example, an association with a behavioral factor could be due to a change in behavior in cases following the onset of disease symptoms. Liu et al. (2004) compared the traditional case-control design with several alternatives recently suggested for testing gene-environment interaction, including the case-only (described below), partial case-control, and case-parent trio designs. They found that the validity of these alternative designs was reduced by common problems such as population stratification, genotyping error, and correlation between genotype and exposure in the population. Thus, despite the potential for bias in the traditional case-control design, they concluded it was preferable to the alternatives they assessed.

Case-Only Design

In the case-only design, interaction is assessed by testing for an association between the genotype and the exposure within the cases only (Khoury and Flanders, 1996). Such an association within the cases provides evidence for interaction because it is not expected if there is no interaction on a multiplicative scale, assuming that the genotype and environmental exposure occur independently in the population (i.e., individuals are no more likely to have both genotype and exposure than would be expected based on their individual frequencies). This design has been criticized on several grounds (Gatto et al., 2004; Hunter, 2005). First, it cannot be used to assess the main effects of a genetic or environmental factor, but only the presence or absence of interaction. Second, the assumption of independence between genetic and environmental factors in the population can be difficult to assess (Gatto et al., 2004). Third, it is a valid test for interaction only on a multiplicative scale, and not on an additive scale (Ottman, 1996).

Family-Based Designs

Several family-based designs for evaluating gene-environment interaction have been described (Witte et al., 1999; Goldstein and Andrieu, 1999), including case-control studies using sibling or cousin controls and designs involving cases and their unaffected parents. Recently, Andrieu and Goldstein (2004) described an alternative design using both related and unrelated controls. An important advantage of family-based designs is the elimination of the potential for population stratification (although power for the detection of genetic effects is lower than in designs using unrelated controls) (Risch and Teng, 1998; Teng and Risch, 1999). Gauderman (2002) showed that designs using cases and their unaffected siblings were more efficient for the detection of interaction than were those using unrelated controls. In some studies, family members may be more willing to participate than unrelated subjects, leading to improved efficiency in data collection; on the other hand, enrollment of family members can be complicated and expensive and can involve difficult confidentiality issues. Also, in family-based studies the collection of information on exposure status generally is retrospective, leading to the same potential for recall bias as in case-control studies (Kraft and Hunter, 2005).

Mendelian Randomization

Mendelian randomization is another design used to examine the combined effects of genetic and nongenetic factors on disease risk (Davey Smith and Ebrahim, 2003). This design is not intended as a test of interaction; instead, it is designed to validate an association between an environmental exposure and disease risk. When a disease is found to be associated with an environmental exposure, it can be difficult to prove that the association reflects a true biological effect rather than confounding with some other factor. Mendelian randomization is a way to test such a biological relationship. If the exposure being studied is influenced by a genotype, the effect of the exposure can be studied indirectly by testing for an association of the genotype with disease. "Mendelian randomization" refers to the random assignment of the genotype according to Mendel's laws, which is assumed to be less susceptible to confounding than is measurement of the exposure. The model tested with this approach is one in which the exposure serves as an intervening variable in the relation of a genotype to disease risk; this is the same as in model A (Figure 8-1).

For example, one study used Mendelian randomization to test the validity of its findings regarding blood pressure and C-reactive protein (CRP). The authors initially found that blood pressure was associated with CRP levels, but they suspected that this finding was due to confounding (Davey Smith et al., 2005). To evaluate the validity of the association, they examined the relationship between blood pressure and a genetic polymorphism in the human CRP gene that was strongly associated with CRP levels. Blood pressure did not vary by genotype of the CRP polymorphism, suggesting the original association was due to confounding. Although a disease-genotype association may sometimes validate the idea that an environmental exposure raises disease risk, other design and analysis issues, such as population stratification, complicate studies aimed at detecting the genotype association. Also, the genotype might influence phenotypic traits other than the one being studied, thus its association with disease might not validate the relationship that it is intended to validate.

Human Laboratory Research

Human laboratory designs, in which interventions are tested in human subjects in a highly controlled laboratory setting, also can be utilized effectively to investigate the interacting effects of social, behavioral, and genetic factors on health. Such designs afford an opportunity for greater experimental control over environmental exposures and the use of interventions (within-subject designs) for exposure testing that increase statistical power. A hypothetical example would be an investigation of the effects of genetic variation in alcohol metabolizing enzymes (between subject factor) and exposure to a stress paradigm (e.g., public speaking or interpersonal stress compared to a low-stress task) (within subject factor) on alcohol sensitivity and ad lib alcohol intake. One recent human laboratory study conducted by Lerman et al. (2004) involved 71 smokers enrolled in a randomized controlled trial of buproprion treatment versus placebo for smoking cessation. The goal was to examine the degree to which abstinent smokers experience increased reward from food (possibly related to weight gain following smoking cessation), and the moderating effects of buproprion treatment and the Taq 1 polymorphism of the dopamine D2 receptor gene (DRD2). At two time points (before and after smoking cessation), subjects participated in a taste test to measure palatability of various foods followed by a behavioral economics evaluation of food reward. Carriers of the minor allele (A1) of the DRD2 polymorphism had significant increases in the rewarding value of food following smoking cessation that were not observed in noncarriers. Moreover, these effects were attenuated by buproprion treatment and predicted subsequent weight gain. Similar paradigms can be tested in animal and human laboratory models providing cross-species validation (Blendy et al., 2005). However, this approach generally is not feasible on a population basis.

Evaluation of Gene-Environment Interactions for Nonbinary Outcome Variables

In most classical epidemiologic designs, the outcome variable is binary. Other types of health outcomes include a continuous quantity (e.g., hypertension and quantitative scale of depression), count (e.g., number of brain tumors), and occurrence of an event over time (e.g., timing of cancer or heart attack). Many of the intermediate phenotypes (endophenotypes) likely to be productive in the search for gene-environment interactions are quantitative rather than binary. For these nonbinary outcomes, the basic concept of gene-environment interaction remains the same: co-participation of a genotype and environmental exposure in the causal mechanism of the outcome of interest (Siemiatycki and Thomas, 1981; Rothman and Greenland, 1998a). In an empirical analysis, the assessment of statistical interaction depends on the type of outcome variable and the corresponding statistical model; linear regression, Poisson regression, and survival analysis are the standard statistical models for continuous, count, and survival outcomes, respectively. A study of gene-environment interaction would be similar to a study of interaction in general in the sense that a gene-environment interaction study still needs to follow all the established rules of the statistical model used. However, because an interaction study tends to involve more, and oftentimes many more, parameters, issues such as sample size/power and multiple testing would become more severe. With nonbinary outcomes as with binary outcomes, analysis of statistical interaction is scale dependent. To understand the underlying biology, a conceptual framework that is not rooted solely in statistical modeling is needed, and the modern epidemiologic framework also applies here (see Appendix E).

STATISTICAL ISSUES COMMON TO ALL RESEARCH DESIGNS

Sample Size and Power

A critical design issue is the determination of the minimum sample size required to generate sufficient statistical power for a study to be able to detect an interaction. That is, a sufficient sample size is needed to ensure that an effect that is truly present can be detected in the study. A study on interactions generally requires a substantially larger sample than a study only on a main effect. Roughly speaking, for the multiplicative model, detecting an interaction requires a sample at least four times as large as a sample required for detecting a main effect (Smith and Day, 1984). Power requirements are likely to be even greater for the detection of interaction on an additive scale (Garcia-Closas and Lubin, 1999). For studies of gene-environment interactions, the methods for calculating sample size have been developed for cases in which the outcome variable is categorical and the environmental exposure is binary, ordered categorical, or continuous (Hwang et al., 1994; Foppa and Spiegelman, 1997; Garcia-Closas et al., 1999; Garcia-Closas and Lubin, 1999) and for cases in which both the outcome variable and the exposure are continuous (Luan et al., 2001).

Yang et al. (2003) discussed the use of population attributable fraction to determine sample size for case-control studies of gene-environment interaction.

The sample size required for the detection of interaction between a genetic variant and a continuous environmental exposure on a continuous outcome is determined by the magnitude of the interaction, the allele frequency, and the strength of the association between exposure and outcome. In addition, statistical power and required sample size are highly influenced by the amount of measurement error in environmental exposures and phenotypes (Wong et al., 2003). Wong et al. (2003) evaluated the effect of measurement error on the sample size needed to detect gene-environment interactions. The model examined was a simple linear regression relating a continuous exposure to a continuous outcome, where the ratio of the slopes of two genotypes served as the interaction parameter. These authors found that the sample size required for the detection of interaction under this model is determined by the magnitude of the interaction, the allele frequency, and the strength of the association between exposure and outcome in those with the common allele. They also found that statistical power and required sample size are highly influenced by the amount of measurement error in environmental exposures and phenotypes, so that studies using imprecise exposure and outcome variables need much larger samples than those that utilize repeated and more precise measures. This result suggests that investment in more precise measures may be a more cost-effective approach to studies of gene-environment interaction than investment in larger samples. Vineis proposed a greatly increased investment in validated exposure assessment procedures, including incorporating repeated measures, assessment of regression dilution bias (i.e., reduction of the differences between comparison groups because of measurement errors), and validation of novel research methods (Vineis, 2004).

Pooling samples from a number of studies may be an effective means of increasing power for studies on gene-environment interaction studies in prospective analyses. The National Cancer Institute Breast and Prostate Cancer and Hormone-Related Cohort Consortium, for example, is combining data across 10 prospective studies to examine gene-environment interactions. The combined samples consist of 6,000 breast cancer patients and 8,000 prostate cancer patients based on more than 800,000 individuals and more than 7 million years of life (Hunter, 2005). An additional advantage is that these kinds of joint efforts will facilitate prior coordination and provide relatively uniform data and analyses.

Two computer programs that estimate power and sample size are available on the Internet: POWER (dceg.cancer.gov/POWER/) and QUANTO (hydra.usc.edu/gxe/).

Multiple Comparisons

The problem of multiple comparisons arises when researchers conduct several tests simultaneously and affirm a statistically significant result from any test having a p value smaller than a critical value of, say, 0.05. This approach fails to take into consideration the increased probability of a false positive test result as the number of tests performed increases. The problem is well known and commonly addressed by procedures such as Bonferroni correction, controlling the false discovery rate (FDR) (Hochberg, 1988; Benjamini et al., 2001), and the permutation test (Edgington, 1995; Nichols and Holmes, 2002).

The multiple comparisons problem poses a formidable challenge in gene-environment interaction studies, because multiple variables will be assessed, regardless of whether or not interaction is specifically evaluated. In genome-wide association scans, thousands or hundreds of thousands of SNPs are tested to search the genome for risk-associated variants. Such tests recently have been made possible by technological advances in genotyping (Thomas et al., 2005). In studies of gene-environment interactions, the multiple-comparison problem is further exacerbated. Researchers must deal not only with a large number of SNPs, but also with environmental exposures and social and behavioral factors, as well as a large number of interactions among the SNPs and the exposures. The presence of a large number of potential gene-environment interactions dramatically increases the chance of finding false-positive results if the problem is ignored. Moreover, the statistical correction for the problem, which normally requires extremely small p values, will result in a lower probability of reporting true positive interactions. That is, the possibility of not detecting important, real associations is increased in this setting.

Multiple testing must be addressed through a number of statistical procedures. The classic Bonferroni correction tends to overcorrect the problem because it assumes tests are independent when often they are not. Several advances in procedures used to correct for multiple testing have been made in recent years. One of the seminal advances was the work of Benjamini et al. (2001) who, following the ideas of Holm (1979) and Hochberg (1988), focused on controlling the FDR (defined as the percent of statistical tests that are false positives, among those deemed significant) rather than the traditional family-wise type I error rates (FEW, the probability of making at least one false positive inference). By controlling the FDR, the researchers assure themselves that on average only perhaps about 5 percent of the total positive discoveries are false. This preserves greater power to detect true positives than the more traditional Bonferroni-type FEW procedures. Several recent variations of this method have been published (Efron and Tibshirani, 2002; Keselman et al., 2002; Sabatti et al., 2003; Storey and Tibshirani, 2003; Becker and Knapp, 2004).

Another approach uses permutation testing (Edgington, 1995; Nichols and Holmes, 2002). Here researchers take advantage of the correlation structure between the tests (testing SNPs in linkage disequilibrium will produce correlated tests of significance) in the multiple adjustment procedure. The permutation test computes significance by counting the number of ways the data can be permuted that produce results more extreme than observed (as in the Fisher's exact test). Much less power is lost in correcting for multiple testing, since researchers are automatically accounting for the exact correlation between tests and not overcorrecting by assuming that all tests are independent.

In addition to the statistical approaches to control the false positive rate, which are largely similar to those proposed for the analysis of main effects, Hunter (2005) suggested restricting the number of interactions by *biological plausibility* and *reproducibility*. The exposures will be restricted to those that plausibly interact with genetic variants in the same biological pathways. Similarly, genetic variants will be restricted to those that plausibly alter gene function. Biological plausibility, however, is relative, and whether a particular gene-environment interaction is biologically plausible is at least partially subject to interpretation. The reproducibility of gene-environment interactions from more than one study also is crucial. The assessment of reproducibility requires the prior coordination of large studies so that the results can be compared. It is important to make all results, both positive and negative, available to avoid the publication bias of suppressing "negative" results.

CONCLUSION

This chapter has examined several aspects of the concept of interaction in order to establish a foundation for discussions about the effects of interactions among specific social, behavioral, and genetic factors in their influence on health. It has been noted that statistical definitions of interaction, and their interpretation, depend on whether the effects being studied are characterized on a multiplicative scale using relative risk measures or on an additive scale using risk difference measures. This chapter has described recent developments in epidemiology that conceptualize interaction in terms of patterns of response in individuals and that advocate the use of the additive model and has characterized a series of models that can be used to delineate the potential causal effects of combinations of multiple factors.

This chapter also considered a series of research designs that can be used to investigate these interactions. These study designs include traditional epidemiological approaches such as prospective cohort studies and case-control studies, but also include less commonly used designs such as family studies and case-only studies. Each of these research designs has advantages and disadvantages, but all of them share common statistical challenges in the context of interactions. These challenges include obtaining sufficient study sample size to ensure that the statistical power is sufficient to detect interactions, the growing problem of multiple comparisons as genomic technologies dramatically increase the number of polymorphisms being studied, the biological plausibility of gene-disease associations, and the reproducibility of genetic association studies.

Testing for interactions will require the development of new, accessible statistical software for implementing tests for interaction on an additive scale. Furthermore, multisite collaborations may be required in order to assemble databases of sufficient size needed to ensure adequate statistical power for the testing of interactions. Additionally, several steps are needed to advance the science of testing interactions. Therefore the committee recommends the following:

Recommendation 7: Advance the Science of the Study of Interactions. *Researchers should base testing for interaction on a conceptual framework rather than simply the testing of a statistical model, and they must specify the scale (e.g., additive or multiplicative) used to evaluate whether or not interactions are present. If a multiplicative scale is used, consistency with an additive relation between the effects of different factors also should be evaluated. The NIH should develop RFAs for research on developing study designs that are efficient at testing interactions, including variations in interactions over time and development.*

The issues discussed in this chapter clearly illustrate the complexity of studying interactions between genetic factors and social and environmental factors and the need to consider carefully the feasibility of available research approaches. However, as we build on existing study designs and use emerging methodological and analysis techniques, it is becoming more and more possible to understand these interactions and their potential for improving public health and preventing disease on a population basis. In order to make substantive advances in this transdisciplinary work, it will be essential for genomic scientists and social scientists to learn to communicate and collaborate effectively.

REFERENCES

Andrieu N, Goldstein AM. 1998. Epidemiologic and genetic approaches in the study of gene-environment interaction: An overview of available methods. *Epidemiologic Reviews* 20(2):137-147.

Andrieu N, Goldstein AM. 2004. The case-combined-control design was efficient in detecting gene-environment interactions. *Journal of Clinical Epidemiology* 57(7):662-671.

Andrieu N, Goldstein AM, Thomas DC, Langholz B. 2001. Counter-matching in studies of gene-environment interaction: Efficiency and feasibility. *American Journal of Epidemiology* 153(3):265-274.

Austin M, Schwartz S. 2006. Cardiovascular disease. In: Costa L, Eaton D, editors. *Gene-Environment Interactions: Fundamental of Ecogenetics*. New York: John Wiley & Sons.

Becker T, Knapp M. 2004. A powerful strategy to account for multiple testing in the context of haplotype analysis. *American Journal of Human Genetics* 75(4):561-570.

Benjamini Y, Drai D, Elmer G, Kafkafi N, Golani I. 2001. Controlling the false discovery rate in behavior genetics research. *Behavioural Brain Research* 125(1-2):279-284.

Blendy JA, Strasser A, Walters CL, Perkins KA, Patterson F, Berkowitz R, Lerman C. 2005. Reduced nicotine reward in obesity: Cross-comparison in human and mouse. *Psychopharmacology* 180(2):306-315.

Campbell CD, Ogburn EL, Lunetta KL, Lyon HN, Freedman ML, Groop LC, Altshuler D, Ardlie KG, Hirschhorn JN. 2005. Demonstrating stratification in a European American population. *Nature Genetics* 37(8):868-872.

Cardon LR, Palmer LJ. 2003. Population stratification and spurious allelic association. *Lancet* 361(9357):598-604.

Davey Smith G, Ebrahim S. 2003. "Mendelian randomization": Can genetic epidemiology contribute to understanding environmental determinants of disease? *International Journal of Epidemiology* 32(1):1-22.

Davey Smith G, Lawlor DA, Harbord R, Timpson N, Rumley A, Lowe GD, Day IN, Ebrahim S. 2005. Association of C-reactive protein with blood pressure and hypertension: Life course confounding and Mendelian randomization tests of causality. *Arteriosclerosis, Thrombosis, and Vascular Biology* 25(5):1051-1056.

Devlin B, Roeder K, Wasserman L. 2001. Genomic control, a new approach to genetic-based association studies. *Theoretical Population Biology* 60(3):155-166.

Edgington ES. 1995. *Randomization Tests*. 3rd edition. New York: Marcel Dekker.

Efron B, Tibshirani R. 2002. Empirical Bayes methods and false discovery rates for microarrays. *Genetic Epidemiology* 23(1):70-86.

Foppa I, Spiegelman D. 1997. Power and sample size calculations for case-control studies of gene-environment interactions with a polytomous exposure variable. *American Journal of Epidemiology* 146(7):596-604.

Garcia-Closas M, Lubin JH. 1999. Power and sample size calculations in case-control studies of gene-environment interactions: Comments on different approaches. *American Journal of Epidemiology* 149(8):689-692.

Garcia-Closas M, Rothman N, Lubin J. 1999. Misclassification in case-control studies of gene-environment interactions: Assessment of bias and sample size. *Cancer Epidemiology, Biomarkers and Prevention* 8(12):1043-1050.

Gatto NM, Campbell UB, Rundle AG, Ahsan H. 2004. Further development of the case-only design for assessing gene-environment interaction: Evaluation of and adjustment for bias. *International Journal of Epidemiology* 33(5):1014-1024.

Gauderman WJ. 2002. Sample size requirements for matched case-control studies of gene-environment interaction. *Statistics in Medicine* 21(1):35-50.

Goldstein AM, Andrieu N. 1999. Detection of interaction involving identified genes: Available study designs. *Journal of the National Cancer Institute Monograph* (26):49-54.

Greenland S, Robins JM. 1986. Identifiability, exchangeability, and epidemiological confounding. *International Journal of Epidemiology* 15(3):413-419.

Hochberg Y. 1988. A sharper Bonferroni procedure for multiple tests of significance. *Biometrika* 75(4):800-802.

Holm S. 1979. A simple sequentially rejective multiple test procedure. *Scandinavian Journal of Statistics* 6:65-70.

Hunter DJ. 2005. Gene-environment interactions in human diseases. *Nature Reviews Genetics* 6(4):287-298.

Hwang SJ, Beaty TH, Liang KY, Coresh J, Khoury MJ. 1994. Minimum sample size estimation to detect gene-environment interaction in case-control designs. *American Journal of Epidemiology* 140(11):1029-1037.

Keselman HJ, Cribbie R, Holland B. 2002. Controlling the rate of Type I error over a large set of statistical tests. *British Journal of Mathematical and Statistical Psychology* 55(Pt 1):27-39.

Khoury MJ, Flanders WD. 1996. Nontraditional epidemiologic approaches in the analysis of gene-environment interaction: Case-control studies with no controls! *American Journal of Epidemiology* 144(3):207-213.

Khoury MJ, Adams MJ Jr, Flanders WD. 1988. An epidemiologic approach to ecogenetics. *American Journal of Human Genetics* 42(1):89-95.

Kraft P, Hunter D. 2005. Integrating epidemiology and genetic association: The challenge of gene-environment interaction. *Philosophical Transactions of the Royal Society of London. Series B: Biological Sciences* 360(1460):1609-1616.

Lake SL, Laird NM. 2004. Tests of gene-environment interaction for case-parent triads with general environmental exposures. *Annals of Human Genetics* 68(Pt 1):55-64.

Lerman C, Berrettini W, Pinto A, Patterson F, Crystal-Mansour S, Wileyto EP, Restine SL, Leonard DG, Shields PG, Epstein LH. 2004. Changes in food reward following smoking cessation: A pharmacogenetic investigation. *Psychopharmacology* 174(4):571-577.

Liu X, Fallin MD, Kao WH. 2004. Genetic dissection methods: Designs used for tests of gene-environment interaction. *Current Opinion in Genetics and Development* 14(3):241-245.

Luan JA, Wong MY, Day NE, Wareham NJ. 2001. Sample size determination for studies of gene-environment interaction. *International Journal of Epidemiology* 30(5):1035-1040.

Maldonado G, Greenland S. 2002. Estimating causal effects. *International Journal of Epidemiology* 31(2):422-429.

Moffit TE, Caspi A, Rutter M. 2005. Strategy for investigating interactions between measured genes and measured environments. *Archives of General Psychiatry* 62:473-481.

Nichols TE, Holmes AP. 2002. Nonparametric permutation tests for functional neuroimaging: A primer with examples. *Human Brain Mapping* 15(1):1-25.

Ottman R. 1990. An epidemiologic approach to gene-environment interaction. *Genetic Epidemiology* 7:177-185.

Ottman R. 1994. Epidemiologic analysis of gene-environment interaction in twins. *Genetic Epidemiology* 11:75-86.

Ottman R. 1996. Gene-environment interaction: Definitions and study designs. *Preventive Medicine* 25:764-770.

Pritchard JK, Donnelly P. 2001. Case-control studies of association in structured or admixed populations. *Theoretical Population Biology* 60(3):227-237.

Reiner AP, Ziv E, Lind DL, Nievergelt CM, Schork NJ, Cummings SR, Phong A, Burchard EG, Harris TB, Psaty BM, Kwok PY. 2005. Population structure, admixture, and aging-related phenotypes in African American adults: The Cardiovascular Health Study. *American Journal of Human Genetics* 76(3):463-477.

Risch N, Teng J. 1998. The relative power of family-based and case-control designs for linkage disequilibrium studies of complex human diseases I. DNA pooling. *Genome Research* 8(12):1273-1288.

Rothman KJ, Greenland S. 1998a. Concepts of interaction. In: Rothman KJ, Greenland S. *Modern Epidemiology*. 2nd edition. Philadelphia, PA: Lippincott-Raven. Pp. 329-342.

Rothman KJ, Greenland S. 1998b. *Modern Epidemiology*. 2nd edition. Philadelphia, PA: Lippincott-Raven.

Rothman KJ, Greenland S. 2005. Causation and causal inference in epidemiology. *American Journal of Public Health* 95(Suppl 1):S144-S150.

Sabatti C, Service S, Freimer N. 2003. False discovery rate in linkage and association genome screens for complex disorders. *Genetics* 164(2):829-833.

Siemiatycki J, Thomas DC. 1981. Biological models and statistical interactions: An example from multistage carcinogenesis. *International Journal of Epidemiology* 10(4):383-387.

Smith PG, Day NE. 1984. The design of case-control studies: The influence of confounding and interaction effects. *International Journal of Epidemiology* 13(3):356-365.

Storey JD, Tibshirani R. 2003. Statistical significance for genomewide studies. *Proceedings of the National Academy of Sciences of the United States of America* 100(16):9440-9445.

Tang H, Quertermous T, Rodriguez B, Kardia SLR, Zhu X, Brown A, Pankow JS, Province MA, Hunt SC, Boerwinkle E, Schork NJ, Risch NJ. 2005. Genetic structure, self-identified race/ethnicity, and confounding in case-control association studies. *American Journal of Human Genetics* 76(2):268-275.

Teng J, Risch N. 1999. The relative power of family-based and case-control designs for linkage disequilibrium studies of complex human diseases. II. Individual genotyping. *Genome Research* 9(3):234-241.

Thomas DC, Witte JS. 2002. Point: Population stratification: A problem for case-control studies of candidate-gene associations? *Cancer Epidemiology, Biomarkers and Prevention* 11(6):505-512.

Thomas DC, Haile RW, Duggan D. 2005. Recent developments in genomewide association scans: A workshop summary and review. *American Journal of Human Genetics* 77(3):337-345.

Tweel I, Schipper M. 2004. Sequential tests for gene-environment interactions in matched case-control studies. *Statistics in Medicine* 23(24):3755-3771.

Vandenbroucke JP, Koster T, Briet E, Reitsma PH, Bertina RM, Rosendaal FR. 1994. Increased risk of venous thrombosis in oral-contraceptive users who are carriers of factor V Leiden mutation. *Lancet* 344(8935):1453-1457.

Vincis P. 2004. A self fulfilling prophecy: Are we underestimating the role of the environment in gene-environment interaction research? *International Journal of Epidemiology* 33(5):945-946.

Wacholder S, Rothman N, Caporaso N. 2000. Population stratification in epidemiologic studies of common genetic variants and cancer: Quantification of bias. *Journal of the National Cancer Institute* 92(14):1151-1158.

Wacholder S, Rothman N, Caporaso N. 2002. Counterpoint: Bias from population stratification is not a major threat to the validity of conclusions from epidemiological studies of common polymorphisms and cancer. *Cancer Epidemiology, Biomarkers and Prevention* 11(6):513-520.

Witte JS, Gauderman WJ, Thomas DC. 1999. Asymptotic bias and efficiency in case-control studies of candidate genes and gene-environment interactions: Basic family designs. *American Journal of Epidemiology* 149(8):693-705.

Wong MY, Day NE, Luan JA, Chan KP, Wareham NJ. 2003. The detection of gene-environment interaction for continuous traits: Should we deal with measurement error by bigger studies or better measurement? *International Journal of Epidemiology* 32(1):51-57.

Yaffe K, Haan M, Byers A, Tangen C, Kuller L. 2000. Estrogen use, APOE, and cognitive decline: Evidence of gene-environment interaction. *Neurology* 54(10):1949-1954.

Yang Q, Khoury MJ, Friedman JM, Flanders WD. 2003. On the use of population attributable fraction to determine sample size for case-control studies of gene-environment interaction. *Epidemiology* 14(2):161-167.

9

Infrastructure

Research that is conducted to elaborate the impact of interactions among social, behavioral, and genetic factors on human health places several demands on the research infrastructure. This infrastructure includes, in addition to laboratory space and equipment, the human infrastructure (e.g., education and training), data, and incentives and rewards. Some aspects of infrastructure are largely affected by the actions of the National Institutes of Health (NIH), while others are largely driven by university actions, although the two domains are inextricably related. For example, NIH supports training, but the universities actually provide the training; research tools are needed by university researchers, while NIH policies and practices may dictate what tools are funded. This chapter examines three aspects of infrastructure: education, data, and incentives and rewards. The discussion explores ways in which existing mechanisms can be focused to strengthen the infrastructure and examines potential new mechanisms that could be developed.

EDUCATION

The foundation of the research enterprise is the education of its researchers. Ideally, appropriate training would occur before launching a research career. The committee believes that the responsibility for education and training is shared among our universities (and high schools) and NIH and other funders of research training. For example, the National Science Foundation (NSF) calls for a more explicit involvement in precollege education. A partnership between NSF and NIH could help ensure the seamless development of a talent pool that could address biomedical research topics or other topics that require a fundamental grounding in math and science.

However, since the advances in genomics have been recent—and the challenge of incorporating genetic research with behavior and social factors is even more recent—it is likely that there are many current researchers who have gaps in their scientific training. Therefore, by and large, the recommendations offered here are directed to NIH and aimed at the college level and beyond.

NIH is the major source of funding for researchers in the biomedical and behavioral arenas and is poised to contribute to the training of a cadre of researchers who could address the issues described in this report. As the pace setter for the biomedical research enterprise, NIH is central to the infrastructure issues for this research, especially in the realm of education and training.

NIH provided about $704 million in 2004 in support of research training through the National Research Service Act (NRSA) (NIH, 2004b). It is generally agreed that postdoctoral training received in conjunction with research grants serves at least as many—and perhaps twice as many—as postdoctoral training through the NRSA (NRC, 2000). Since NIH is the dominant source of funding for the training of researchers in these fields, the NIH policies are fundamental to the ability of the United States to advance research on transdisciplinary issues such as those addressed in this report.

The need for transdisciplinary research to address the study of gene-environment interactions was discussed earlier. As a beginning approach to fostering the development of transdisciplinary research on the impact of interactions among social, behavioral, and genetic factors on health, the committee believes that NIH should consider holding a conference for interested individuals. Such a conference would assist universities in sharing their best practices in interdisciplinary and transdisciplinary research and would foster the exchange of knowledge and practices. It will be challenging, but important, to ensure that participants in such a conference share specific strategies that others could adopt or modify; the conference should not simply provide another forum devoted to encouraging the goal of collaboration.

Also, since the challenge of educating across boundaries is not exclusive to health, it might be timely for the NSF or a private foundation (e.g., the Pew Charitable Trust) to bring together educators from many fields that have developed interdisciplinary and transdisciplinary programs specifically in order to educate across boundaries and help students learn how to work in transdisciplinary teams. The Science Education Partnership Awards

(NIH, 2005c) are an example of an excellent outreach effort to support science at the K-12 level. Although this program does not focus on transdisciplinary research, other programs, such as the Genome Science Education Program (SEPA, 2005), could emphasize the transdisciplinary aspects.

Early Career Education

At the initiation of careers, fellowship support is very important. Therefore, NIH could advertise individual pre- and postdoctoral awards specifically for transdisciplinary research on the impact of interactions of social, behavioral, and genetic factors on health and provide easy links to the institutions and investigators who already are working in a transdisciplinary manner. This would not involve creating any new mechanisms, or even necessarily identifying additional funds. However, it would require NIH to make support in this area a priority and to take active steps to ensure that potential applicants are aware of NIH's interests. As NIH identifies universities that are conducting transdisciplinary research effectively, such universities could be urged to advertise specific opportunities at their sites for postdoctoral work. Additionally, these universities could be funded to support innovative outreach efforts in the topical areas of interest. In general, the committee believes that NIH could apply its existing training mechanisms specifically to the transdisciplinary topic addressed here. In other cases, modifications of existing mechanisms would make them more valuable in this area.

Although postdoctoral training is common in biology, it is less so in the social sciences. Therefore, it is important that opportunities at the postdoctoral level are available in order to expose social scientists to broad, transdisciplinary training. Also, since postdoctoral fellows may devote two or three years to their disciplinary training, NIH could consider extending training beyond three years for those who are reaching beyond their traditional boundaries and would be likely to contribute as researchers in the areas of social, behavioral, and genetic factors and health. In general, there is concern that individuals must be well grounded in a discipline, but also able to work and communicate with other disciplines. A slight extension of the training period might serve this focus well, and there may be value in continuing to support mechanisms that support disciplinary training, while also providing the means to extend skills to those in complementary scientific areas (see Box 9-1).

NIH initiated a T90 grant in 2004 to support transdisciplinary training (NIH, 2004a). Although these projects have been under way for only about one year, it would be useful to assess what has been learned from these early experiences and craft a T90 specifically for training in the impact of

---

BOX 9-1
Institute for Public Health Genetics,
University of Washington, Seattle

The Institute for Public Health Genetics (IPHG) at the University of Washington provides graduate education, opportunities for interdisciplinary research, and policy workshops that integrate genomics with the public health sciences disciplines (epidemiology, biostatistics, environmental health sciences, and health services), and with pharmacogenetics, bioethics, social sciences, law, public policy and health economics. The mission of the Institute is to "provide broad, interdisciplinary training for future public health professionals, to facilitate research in public health genetics, and to serve as a resource for continuing professional education" (Brochure). Specifically, the IPHG offers an accredited masters of public health (M.P.H.), a doctorate (Ph.D.) in Public Health Genetics, and a transcripted graduate certificate, all of which include this interdisciplinary training (IPHG, 2005).

---

interactions among social, behavioral, and genetic factors on health. This program is fairly new, and while some may believe that it is premature to extend it without understanding the elements that lead to its success, the concept is clearly in alignment with the issues addressed by this committee. Therefore, the committee urges NIH to take every opportunity to learn from this cohort of projects and to extend them, while incorporating into new T90 projects the elements that have contributed to the program's success.

New programs require time to become organized and to enroll and educate trainees. Also, a lengthy period of time is required to observe the impact on the trainees' careers and, subsequently, assess their impact on the field. Thus, the committee urges NIH to develop and use intermediate indicators for such programs in order to facilitate transdisciplinary training efforts, rather than wait the years that it might take to conduct a definitive assessment of impact. Intermediate indicators might include the level of interest in the program, success in recruiting top students, successful completion of the training program, and continued interest in transdisciplinary research.

Established Faculty

Transdisciplinary research requires the development of "professionals that can interact synergistically" (IOM, 2003). Therefore, to develop a cadre of researchers who can participate in transdisciplinary research, faculty members who are at a particular level of accomplishment in a particular field must be presented with a realistic opportunity to extend their skills into new or changing fields. To assist established faculty in broadening their skills, NIH could revisit the senior fellowship (F33) concept to determine whether it might be used as a mechanism to provide salary support for a defined release period (e.g., 30 percent to 50 percent). This support could be specifically used in structured education for disciplines involved in researching the impact of interactions among social, behavioral, and genetic factors on health. The award also would need to provide a modest institutional stipend. An advantage of this approach would be that researchers would be encouraged to seek out existing expertise in fields that they themselves were lacking. NIH would not have to identify the fields or the individuals, but, instead, would provide support if researchers in one field (e.g., social sciences) were to propose a structured study in another field (e.g., genetics). Although it is appealing to envision a new cohort of researchers who are trained from the earliest stages in transdisciplinary research, such a process requires time to develop. This kind of program could be part of a "toolbox" of approaches that would help to support the need for the continuous extension of abilities in these complex and changing fields.

A more limited approach to extending skills could occur through the short course approach. In this way, NIH could assist researchers at all stages in broadening their skills by supporting a short course that focuses on studying the impact of interactions among social, behavioral, and genetic factors on health. Mechanisms exist (e.g., the T35) for this, but care is needed to ensure that such a course takes advantage of lessons learned from similar activities. It would be important for the short course to address the theoretical, statistical, and ethical aspects of this work and to ensure that participants are already strong in one (or more) of the arenas. In other words, such a short course should not be narrowly constructed in ways that would allow all of the students to be geneticists seeking to learn about social factors or, as another example, to be sociologists seeking to understand genetics. NIH has had experience in providing support in the past to areas of focus such as population behavior and Alzheimer's disease (Bachrach and Abeles, 2004). The importance of bringing together investigators to collaborate on the study of the impact of social, behavioral, and genetic influences on health is no less compelling.

It certainly can be argued that education that prepares researchers to work across fields needs to start early—perhaps at the undergraduate level. Universities have considerable latitude in how they construct courses and degree programs in order to allow, for example, social scientists to be exposed genetics and biologists to be exposed to cultural studies. Some universities offer degrees that explicitly encourage students to draw from more than one field. NIH has provided limited programs for undergraduates—typically those that encourage undergraduate programs to support the growth of a diverse cohort of biomedical researchers. It has not played a significant role in guiding undergraduate education, and no compelling reason appears to indicate that the situation should be any different in this area. However, universities are capable of adjusting their course work to meet changing scientific needs—and this is clearly within their purview.

The report *Facilitating Interdisciplinary Research* by the National Academy of Sciences/National Academy of Engineering/Institute of Medicine (NAS/NAE/IOM) involved government, university, and industry members participating in a broad discussion of ways to foster interdisciplinary research, and offers many useful recommendations (NAS/NAE/IOM, 2004). (See Appendix B for the complete set of recommendations.) The challenges in supporting interdisciplinary research are not unique to the social, behavioral, and genetic aspects of health, and the lessons that can be learned from these other fields can be shared so that not all tools need to be developed de novo. There is, however, no clear single way for the transmission of this knowledge to take place.

MECHANISMS OF SUPPORT

NIH uses a variety of mechanisms to support research, each with its own advantages. The core mechanism for supporting research is the R01, the Research Project Grant or what is known as the "individual investigator award," which reflects the value that is placed on the work of sole investigators. The challenge, however, is to reconcile the historic focus on the work of the individual with the needs of team science. This is the heart of the cultural challenge to research today. Should the R01 be changed to reflect teams? Or, should the R01 continue to play a pivotal role as the means to support an individual scientist's work?

There always will be value in the work of the individual scientist, and not all scientific questions require the mustering of a team. However, when the R01 is portrayed as the highest form of achievement—the gold standard—it then undermines the valuation of the team approach. How can the scientific community best value both the work of the individual and the work of the team? How can NIH best value the different types of approaches needed for answering different scientific questions?

There is a benefit to clearly identifying the expectations for a given support mechanism and, therefore, it is most likely that success would come from creating a new mechanism that specifically supports team research. Given a new identifying number, such a mechanism might be used only when a team is needed and when the Principal Investigator (PI) is a team leader with some number of co-PIs who clearly are equal collaborators. Such a mechanism would differ from program projects in which the subprojects, although oriented around a common theme, are fairly independent.

Furthermore, this new mechanism would be different from a Core Grant (P30) for a center that provides elements that are shared by other, individually funded projects. The new mechanism would not be a center mechanism, but rather would fund a specific research project that requires a team of people working together to conduct the research. It is important that this mechanism be given the same status as the R01 if the goal is to support and reward team science. It also is important that all the members of the team receive appropriate recognition, which would not eliminate the role of the PI or minimize the leadership that is required to bring a team together to ensure successful project functioning. Since, as described in the 2004 NAS report, it may take time to develop a team for transdisciplinary research, it may be helpful to construct a mechanism that allows one to two years for a developmental phase, followed by three to five years for the support of the research following administrative review. This would help to ensure that the team has established a well-functioning structure and has access to the data or populations that are needed. Such an approach is consistent with the NAS report (2004) suggestion that an allowance should be made "for the longer startup time required by some IDR programs."

Private foundations have used different mechanisms for supporting complex teams of investigators. The MacArthur Foundation has supported networks devoted to specific topics. These awards typically are highly selective of the individuals involved, flexible in structure, and well funded (see Box 9-2). Another transdisciplinary effort was that conducted under the auspices of the Family Research Consortium III, described in Box 9-3 below.

Whatever approaches NIH decides to take, the value of flexibility and sufficient funding should be incorporated into them.

Just as the topic of social, behavioral, and genetic influences on health is broad, so too must be the approaches taken to support the diverse workforce that can address these topics. Therefore, the committee makes the following recommendation:

Recommendation 8: Expand and Enhance Training for Transdisciplinary Researchers. *The NIH should use existing and modified training tools both to reach the next generation of researchers and to enhance the training of current researchers. Approaches include individual fellowships (F31, F32) and senior fellowships (F33), transdisciplinary institutional grants (T32, T90), and short courses.*

DATA

Infrastructure also involves the tools that researchers use. In the area of social, behavioral, and genetic effects on health, there is a significant need for datasets that provide information across these disciplines and that would allow for the testing of interactions. Datasets used to study such interac-

---

BOX 9-2
The MacArthur Network Model

The research networks that have been established by the John D. and Catherine T. MacAthur Foundation over the last two decades are successful examples of interdisciplinary collaboration. As described by the foundation, these research networks function as "research institutions without walls" devoted to topics related primarily to human and community development. According to the description provided by the foundation:

"They are Foundation-initiated projects that bring together highly talented individuals from a spectrum of disciplines, perspectives, and research methods. The networks explore basic theoretical issues and empirical questions that will increase the understanding of fundamental social issues and are likely to yield significant improvements in policy and practice" (www.macfound.org/site/c.lkLXJ8MQKrH/ b.948165/k.E3C/Domestic_Grantmaking__Research_Networks.htm).

An example of a currently ongoing network is the Network on Socioeconomic Status and Health, established in 1997 and chaired by Nancy Adler of the University of California, San Francisco (www.macses.ucsf.edu). The mission of the Network on Socioeconomic Status and Health is to enhance the understanding of the mechanisms by which socioeconomic factors affect the health of individuals and their communities. The network's research agenda is designed to inform both policy and practice, to stimulate additional research in diverse fields, to contribute data to discussions of economic and social policy, and to provide a basis for social and medical interventions that will foster better health among individuals and communities.

To achieve their mission, the network's investigators are drawn from a diverse range of fields including psychology, sociology, psychoimmunology, medicine, epidemiology, neuroscience, biostatistics, and economics. Their research is organized around an integrated, transdisciplinary conceptual model of the environmental and psychosocial pathways by which socioeconomic status (SES) alters the

---

tions typically are large, difficult to collect, and costly. Therefore, it is important to support such datasets as a research tool to be shared among a wide audience of researchers. Three ways that NIH could foster the development of such datasets are described below. All three would have value, but each has different costs and benefits.

Existing Datasets

First, it is important that NIH undertake a systematic review of existing and ongoing datasets to determine their current usefulness for transdisciplinary research that is aimed at assessing the interactions among social, behavioral, and genetic factors on health. Furthermore, this review should examine how the datasets could be made more useful with supplemental

performance of biological systems, thereby affecting disease risk, disease progression, and ultimately, mortality.

In its first phase, the network undertook a variety of studies focusing on the social, psychological, and biological processes involved in "social gradients" in health and disease. For example, the Network added new measures to waves of data collection in the Whitehall Study of British Civil Servants, a longitudinal study that has shown a persistent influence of SES on health well into old age. The group also has added new psychosocial measures to the 15-year follow-up wave of the CARDIA study, a multisite, longitudinal project funded by the National Heart, Lung, and Blood Institute, and has added the collection of biomarkers to characterize "allostatic load" in ancillary studies at the project's Oakland and Chicago sites.

The network also initiated a large study of work environment and health across 15 plants of a large industrial company. It is using data on administrative and physical status, supplemented with new surveys, to assess psychosocial and environmental factors affecting allostatic indicators and health. Data collected from these and other studies will enable the group to test its integrative, transdisciplinary model of the pathways by which SES alters biological systems and health (description taken from the Network's website: www.macses.ucsf.edu).

The success of the MacArthur Research Network model rests on several factors: first, it has facilitated the integration (or "*consilience*") of knowledge, concepts, and methods across social and biological disciplines by carefully selecting a group of scholars who have demonstrated the willingness and capacity to overstep disciplinary boundaries in their previous research. Second, the network has been willing to invest in innovative, high-risk/high-reward projects initiated by the group's members. These projects have ranged in size from small pilot projects to more ambitious undertakings (such as the collection of new data piggy-backed onto large-scale ongoing studies). Third, the network has been involved in mentoring a cadre of junior investigators who have attended the meetings of the Network over the years, and who have benefited from collaborating with the network investigators who have been funded by the network in carrying out exploratory research projects.

data collection. In making this assessment, care should be given to identify datasets:

- that are especially valuable for specific health outcomes,
- in which there is sufficient social variation, and
- in which the linkage to genetic factors can be plausibly explored with genetic measures that could be added to an existing project.

It is possible that some datasets that already include biological and genetic measures could be augmented to include social and behavioral variables. Not every dataset will ultimately be determined to be valuable for transdisciplinary research. However, to the extent that existing datasets can be augmented, there are efficiencies that should be exploited.

BOX 9-3
The Family Research Consortium III

The Family Research Consortium (FRC) III was a multisite, 3-year postdoctoral training program that promoted interdisciplinary collaborative research and training for the study of ethnic/racial diversity, family process, and child and adolescent mental health. Research partners and postdoctoral students came from a variety of disciplines. Trainees attended yearly Summer Institutes designed to bring together members of the consortium as well as approximately 100 scholars from various universities worldwide. Students also attended a 6-week intensive summer training program during their first year, winter meetings focused on particular topics, and worked with one of the consortium faculty at his/her home institution. During their 3-year term, students were required to collaborate with at least two faculty members at different sites. The success of this effort led to funding for FRC IV which includes scholars from sociology, demography, developmental psychology, anthropology, economics, statistics, public health, and pediatrics.

One example of a dataset that could be useful for the kind of transdisciplinary research described in this report is the National Longitudinal Survey of Youth and Child Supplement. This dataset includes extensive information on social factors, developmental measures of children, data on family members (for a subset of participants), and a host of other measures. It does not, however, include biomarkers (Bureau of Labor Statistics et al., 2002). Therefore, NIH could explore the possibility of enhancing health measures and adding biomarkers. The large size, the representativeness of the sample, its longitudinal nature, and the wealth of existing data in this survey argue for a careful review of its potential regarding the impact of interactions on health.

It could be valuable to collect biologic samples at any point in time during the course of a longitudinal study when the markers are stable over time, such as is the case with the HapMap. If the biologic measure is quite variable over time—and especially if the time sensitivity is associated with other behaviors of interest—then it would be necessary to collect the specimens at specific times. Similarly, social measures that are stable (e.g., parental education) can be collected at virtually any point, but some measures are subject to considerable recall error, which makes the timing of their collection important.

The development of complex datasets must involve careful consideration of the stability of measures, the importance of different levels of stability, and the patience of research subjects to continue participation, among other factors.

The AddHealth survey currently includes measures of social, behavioral, and genetic characteristics (Udry, 2003). In the recently funded Wave IV data collection, biomarkers (e.g., glycosylated hemoglobin, C-reactive protein levels, blood pressue, lipids, etc.) will be collected, and DNA will also be collected from all 17,000+ participants. These new data, added to the rich longitudinal social environment data already available from adolescence on this sample that is now aged 25-31 will make the AddHealth dataset a valuable resource for transdisciplinary research in the coming years. Care should be taken to ensure that this uniquely valuable dataset is sustained and made available to researchers. The inclusion of parents and the oversampling of twins and siblings make this an especially valuable dataset. Of course, the focus on adolescents restricts the types of health questions that can be addressed, but each dataset has its own strengths and weaknesses. In this case, the strengths are exceptional, and a high priority should be placed on continuing the study and the excellent access that is available to it.

With the increased use of existing datasets comes the challenges that are associated with data sharing, privacy, confidentiality, and the scope of informed consent. Chapter 10 examines these issues. However, it is worth a brief discussion here. The sharing of data is a powerful tool for ensuring that the benefits of large investments in complex datasets are realized and that such datasets are not unduly restricted to a small number of researchers. Careful consideration must be given to the understanding that participants have about who will have access to their information, under what circumstances, and perhaps for what purposes. However, it is difficult—if not impossible—to envision all of the specific ways in which the data could be used. Researchers and funding agencies should give careful attention to how participants are informed of the potential sharing of their information (i.e., data and/or biological samples), the protections in place to guard their privacy, and the uses to which these data might be put. Although there is movement toward greater sharing of data, the need still exists to be attentive to and involved in this fast-moving field.

Another valuable NIH role could be the development of a guide that includes measures of key concepts in data collection about the impact of interactions among social, behavioral, and genetic factors on health. This is not a new idea, but it is one that has proved useful in other fields (NIH, 2005b) and is one that also could help introduce researchers in disparate fields to the methods used by their colleagues. It is not uncommon for researchers to realize the need for measures from another field and, therefore, to add data elements that are either not state-of-the-art or not appropriate for the specific circumstances. To the extent that such a guide includes a discussion of the underlying concepts being measured or the circumstances in which specific measures were or were not appropriate, it could be useful to researchers.

In some cases, the measures are either static (e.g., the genome) or can be recalled (e.g., education), such that supplementation is feasible. However, these additions must be scientifically compelling and not simply feasible. A particularly challenging issue in data collection is the collection of biological specimens, as well as their storage, sharing, and characterization. Also, collecting DNA at one point in time does not provide the breadth of gene expression data over time that might be necessary in order to understand the interplay of genetic and environmental factors over time.

Development of such a guide also could also include a review that would seek to identify broadly agreed-upon measures in different scientific sectors. These measures would aid researchers who are interested in adding biomarkers to a behavioral study, providing, for example, some specific guidance about preferable measures and the logistics of collecting biological samples. Similarly, guidance for geneticists about better or worse ways to collect social and behavioral data could have widespread value and could be a valuable contribution regardless of which approach is taken to data collection. Such a guide would facilitate discussion of the concepts underlying frequently used measures and specifically address them in the context of contributing to the understanding of the impact of interactions among social, behavioral, and genetic factors on health. Although such a review would be a significant undertaking, it could provide a useful guide to a wide array of researchers.

New Datasets

A second approach to strengthening the data infrastructure is to design specific studies of social, behavioral, and genetics factors that influence specific health outcomes. Health conditions or diseases could be identified for which there is a suspected or known genetic contribution, for which behavioral factors are likely to be involved, and for which hypotheses have been formed regarding the role of social factors. Given the relationship of some social factors, such as race, ethnicity, and social support, to a variety of health conditions (see Chapters 2 and 5), the number of most likely candidate studies could be narrowed. Such studies also could focus on topics for which the best methodological tools already exist or can be developed fairly easily. An advantage of this approach would be that the datasets would be specifically focused on a given condition. However, it is likely that a large sample size would be required, and this could make such studies costly and perhaps difficult to construct.

Finally, it also is possible that true advancement in this field requires a major new cohort study. Some refer to this as a "last cohort" concept. The benefits of such a study lie in the ability to craft specific measures of relevant concepts and to ensure that the periods of data collection are appropriate for the scientific questions being asked. A study to assess the impact of interactions of social, behavioral, and genetic factors on health could require hundreds of thousands—perhaps even a million—subjects and would involve a large part of the scientific community. This number of subjects would be far larger than any other existing U.S. cohort, but the sheer size of such a cohort would make it a valuable tool for many different exploratory projects, even those that were not conceptualized at the outset. Although there would be benefit in having such statistical power, the cost and the time required for data collection might seem daunting.

Because there is still the need to identify topics that are likely to benefit from understanding the interplay of social, behavioral, and genetic factors on health, it was not clear to the committee that a last cohort approach would be necessary, at this time, to advance transdisciplinary research linking these domains. Understanding the interplay of these factors and health likely will progress through the building up of each of these key areas and most likely will require substantial investment in understanding linkages, developing measures, and carefully selecting subjects. This is not to say that a last cohort approach would not bring considerable insight and statistical power to many issues related to our understanding of health and disease, but it is not clear that this is an ideal strategy for understanding the interactions among the levels discussed here. Such a major investment in a research effort must build upon the basis of a skilled research workforce. It is not clear that strategies exist for optimal training in transdisciplinary skills, and it is not apparent that the necessary rewards and incentives are in place to support successful transdisciplinary research on a massive scale.

Replication

Another area in which NIH could be involved concerns the replication of research results. Scientific fields advance when findings are reproducible and when replication is a routine, expected stage of research. One role that NIH could play would be to ensure that the costs of replication are viewed as legitimate—either as a part of the original study, as an add-on to a study, or as a separate project. For example, supplements can be made available if there are findings that warrant replication and the costs cannot be absorbed within the basic costs of a research project. Projects that are essentially replications of other findings may need to be funded, yet they often lack the appeal (or innovativeness) of new projects. Without replication, however, fields can zig-zag from "finding" to "finding" without developing a critical mass of reproducible results. In a field such as the transdisciplinary study of health, the demands of replication may be greater than for other, more discrete, fields because studies are more difficult to mount and more expensive to replicate.

A number of data issues and concerns need to be addressed in order to facilitate research that is aimed at explicating the impact on health of interactions among social, behavioral, and genetic factors. Therefore, the committee makes the following recommendation:

> **Recommendation 9: Enhance Existing and Develop New Datasets.** *The NIH should support datasets that can be used by investigators to address complex levels of social, behavioral, and genetic variables and their interactive pathways (i.e., physiological). This should include the enhancement of existing datasets that already provide many, but not all, of the needed measures (e.g., the National Longitudinal Survey of Youth, ADDHealth) and the encouragement of their use. Furthermore, NIH should develop new datasets that address specific topics that have high potential for showing genetic contribution, social variability, and behavioral contributions—topics such as obesity, diabetes, and smoking.*

INCENTIVES AND REWARDS—NIH AND ACADEME

According to the 2004 NAS/NAE/IOM report on interdisciplinary research, several key conditions are required for the conduct of effective interdisciplinary research. The committee believes that these same conditions are necessary for success in transdisciplinary research—a primary recommendation of this current report. The conditions identified by the 2004 report include "sustained and intense communication, talented leadership, appropriate reward and incentive mechanisms (including career and financial rewards), adequate time, seed funding for initial exploration, and willingness to support risky research." Although such aspects of university functioning are not within the NIH's purview, they may affect its ability to find scientists who can conduct the kind of transdisciplinary research that is envisioned here. The 2004 report is thorough and detailed, and the committee believes that its recommendations are crucial to the successful implementation of interdisciplinary and transdisciplinary research. The purpose of discussing certain key points of the 2004 report in the following section of this report is to emphasize their relevance and importance to the topic under consideration—assessing the interactions among social, behavioral, and genetic factors on health.

Researchers generally work in a university environment (only about 10 percent of NIH research funds are expended at independent research institutes, and an additional 10 percent of the NIH budget is expended in the NIH intramural program [IOM, 1998]). Furthermore, universities are the sites for virtually all professional training. Therefore, it is useful to consider how the structures and reward systems of universities may influence the incentives and rewards that are available for working in the area of transdisciplinary research.

One example is the hiring, promotion, and tenure (P/T) process, a key aspect of success in the university research setting. It is widely acknowledged that the P/T process rewards individual initiative and products of research, such as grant proposals, research projects, and publications. Although being a PI on a research project or a senior, lead, or sole author on a paper is a clear sign of scholarly achievement, there may be other indicators as well. Participation in team projects for transdisciplinary research may be discouraged for those who have not yet achieved tenure, but there would be value in providing junior faculty ways to become engaged in transdisciplinary research from the early stages of their careers. The criteria for promotion and tenure will, of course, affect the hiring process, because presumably an institution seeks to hire people it believes will thrive and grow in that instituition's environment.

Because the review process for P/T shares some attributes with the NIH peer review process, there may be some common observations about how these processes can support transdisciplinary research. One would be the importance of having someone participate in the P/T review who is experienced in transdisciplinary research, who understands the challenges and metrics for success, and who is able to evaluate contributions made by individuals to team projects. Because the P/T process is so critical to the career pathway of academic researchers it might be valuable for leading university associations (e.g., the American Association of Universities, the National Association of State Universities and Land Grant Colleges, and the American Association of Medical Colleges) to jointly develop models that would ensure that work in such important and critical fields is adequately rewarded before faculty reach the senior—or tenured—level.

One of the suggestions made in the 2004 NAS/NAE/IOM report is that academic institutions should "increase recognition of co-principal investigators' research activities during promotion and tenure decisions." As NIH explores new approaches to acknowledging multiple investigators on team projects, the next step would be for universities to use that information in ways that would ensure that the impact of the incentives and rewards are felt at the campus level. Interestingly, in the guidance for the new clinical (Clinical and Translational Science Award) awards, NIH specifically calls on universities to put forward PIs with broad institutional authority, including authority over promotion:

> . . . that the program director have authority, perhaps shared with other high-level institutional officials, over requisite space, resources, faculty appointments, protected time, and promotion (NIH, 2005a).

This is a sign that NIH acknowledges the need for institutions to align their policies with the goals of supporting transdisciplinary research, even though these are policies and procedures that are under the control of the university, not NIH.

Universities also should consider the incentives that come with sharing funding with researchers, departments, and colleges. The NAS/NAE/IOM report (2004) suggests that academic institutions "experiment with administrative structures that lower administrative and funding walls between departments and other kinds of academic units." Furthermore, the report recommends that "institutions should develop equitable and flexible budgetary and cost-sharing policies that support IDR." Among examples provided is the suggestion that institutions "credit a percentage of all projects' indirect costs to support the infrastructure of research activities that cross departmental and school boundaries."

There is wide variability in how recovered indirect costs are shared within universities, in part because universities face an array of demands and constraints on their use of such funds. In many institutions, some portion of the indirect cost recovery may be returned to investigators, departments, colleges, or other components to be used for research purposes. Even small amounts of such funding can either advance or retard the development of transdisciplinary teams. If, for example, the only recovery made is to the PI, then there is a clear disadvantage to being a participant on a team project. On the other hand, if the recovery is divided according to the involvement of individuals in the project, then there is a greater incentive to engage in teamwork. Such sharing could also minimize individual departments having to work to keep faculty participating only on projects within the department and could help encourage faculty members to engage in opportunities to work across organizational boundaries.

Another example comes from the experiences of nontraditional structures within universities, such as transdisciplinary research centers or institutes. If all of the faculty members involved in a project reside within such an institute, there may not be any issue regarding how incentives flow. But, if individuals have home departments or colleges, as well as center or institute affiliations, there can be problems of attributing "credit" and providing rewards. The amount of incentive funds could be increased to ensure that some share went to such centers or institutes, or the overall amount of funding could be held steady, with the proportions adjusted to allow for incentives to accrue to such centers. Universities should examine their practices to ensure that, to the extent funds are distributed as incentives, they do not disadvantage the structures that may be key to conducting transdisciplinary research. Simply put, universities cannot embrace the concept of transdisciplinary research without reviewing their policies and procedures in order to ensure that they facilitate such work rather than penalize it.

Support for transdisciplinary research also will require incentives from NIH. Although the most powerful incentive is the financial support that is provided for such work, there are other types of incentives as well. One incentive is the credit that accrues to those who participate in such projects. The recent NIH announcement of plans to recognize multiple PIs represents a significant advancement in providing external recognition for members of research teams. This is a very important and valuable step in the enhancement of team science, and it is a step that clearly is needed for transdisciplinary research. However, it is important that in seeking to appropriately recognize those who are conducting team research, systems not minimize the value of the leadership that such teams require. The development of complex proposals and the leadership for projects has been historically recognized in the role of PI. That role still is important, even in settings in which there may be several collaborators who are critically important to the project. If NIH were also to recognize such collaborators, it would be easier for institutions to see how those roles are being played, not just on their own campus but on others as well.

In reflecting on the P/T discussion, it was noted that it is obvious that those on an investigator's campus know what their roles have been on research projects. However, it may be challenging in P/T review to understand how those roles are experienced by other researchers at a similar career stage. Team members need to be recognized for support received from NIH (e.g., through Computer Retrieval of Information on Scientific Projects). The use of such data is the responsibility of the university, but NIH (and other funders) could help by making the appropriate data easily available.

The federal laboratories and industry have organizational structures that differ from those in universities in that they tend to be more problem focused than discipline focused. This structure may foster research that brings together different skills through a team approach to address a problem, efforts that could provide insights for universities and, possibly, avoid unwarranted advocacy for particular disciplines, independent of their contribution to identified problems. In fact, the 2004 NAS report recommends that "universities may benefit by incorporating many IDR [interdisciplinary research] strategies used by industrial and national laboratories, which have long experience in supporting IDR." Interestingly, some industries report that they prefer to hire trainees early in their careers before they have become too focused on independent activity and the anticipation of individual rewards. Industry frequently needs individuals who can work well in teams, a skill that is not necessarily fostered through a lengthy commitment to an academic career.

However, the larger obligations of universities to provide a broad intellectual environment and grounding for the next generation of scholars would not make a totally problem-oriented focus plausible as a general approach to institutional structure. The many calls for transdisciplinary research approaches to contemporary problems may mean that it is timely for universities to consider how to approach a balance of needs. Some of the deliberation in this area should involve considering how best to educate the next generation to contribute to the academy, as well as to industry and government.

PEER REVIEW

Scientific peer review of applications is a key step in supporting any area of research. It is not uncommon to hear investigators lament that transdisciplinary projects have difficulty in peer review. This reality, or even this impression, undermines the willingness of researchers to take on these important and difficult scientific areas. Transdisciplinary research is a challenge for the review of applications. The involvement of multiple disciplines means that review groups need to reflect that diversity. There is great value in having more than one person representing a field on a review group, and a significant number should represent transdisciplinary research experiences and skills.

The need to have multiple people in multiple fields, not to mention the need to include those who are skilled in systems approaches to analysis and disease endpoints, could rapidly escalate the number of reviewers needed overall. It is probably not enough to simply place people from different disciplines on a review group. Rather, it is important to take specific steps to ensure that reviewers will be able to appreciate the transdisciplinary nature or goals of a proposal. For example, selecting reviewers who actually do transdisciplinary research would be one important step. If the project members are expected to function as team, then real-life experience with successful team science would be essential. This is far different from recruiting reviewers who have knowledge of the specific elements of work that the team will address, but no experience in working in teams.

Another strategy for the NIH could be to focus review criteria to specifically address transdisciplinary aspects (so that truly transdisciplinary projects were clearly valued). Establishing a specific mechanism for such team science could assist in focusing review groups on the requirements for such research. Also, review group members might spend time in advance of the specific review of proposals learning about one another's disciplines, discussing the meaning of transdisciplinary inquiry, or even presenting a summary of the contributions of a field other than their own. Expanding the skill set of reviewers (and presumably program and review staff at the same time) would lead to the development of more reviewers who are aware of the conceptual and technical issues involved in a particular type of research.

In any emerging field there is the risk of a dearth of qualified reviewers or an over-reliance on a few reviewers. NIH could take modest steps that would help to improve the quality of review and that would be helpful to its own program and review staff as well. Although the tools might vary, the goal is to ensure that transdisciplinary work is fairly reviewed and truly valued. It is not sufficient to place individual scientists with expertise in the elements of a complex, integrative project on review groups unless there are already members with experience working in a transdisciplinary setting or with a special initiative within the review group to build that perspective prior to conducting the review. The gravitational pull to individual perspectives must be actively countered.

If NIH establishes a goal of supporting projects that are addressing transdisciplinary issues or bringing transdisciplinary teams to bear on a project, then funding decisions need to reflect that goal. The value of adhering to peer review assessments is obviously quite strong, but it is not incompatible with also making programmatic decisions in support of some defined areas. If a field or approach is truly cutting edge, it may present a challenge to peer review, but if it is to be advanced, then NIH should consider making modest use of such programmatic decisions. It is not clear that this necessarily requires specific solicitations for such research (although Requests for Applications and Program Announcements have a role), but may simply involve placing high programmatic relevance on such projects when funding decisions are made.

In many respects the tools needed to advance this field are not novel, but they need to be systematically applied toward the goal of fostering a type of research that has inherent scientific challenges and that faces specific institutional hurdles. In other words, it is the determination to use the available tools more than the need to rely on the development of new administrative tools that will allow this field to grow. Therefore, the committee makes the following recommendation:

Recommendation 10: Create Incentives to Foster Transdisciplinary Research. *The NIH and universities should explore ways to create incentives for the kinds of team science needed to support transdisciplinary research. Areas to address include (1) hiring, promotion, and tenure policies that acknowledge the contributions of collaborators on transdisciplinary teams; (2) peer review that includes reviewers who have experience with inter- or transdisciplinary research and are educated about the complexity and challenges involved in such research; (3) mechanisms for peer review of*

research grants that ensure the appropriate evaluation of trans-
disciplinary research projects; and (4) credit for collaborators in
teams, such as NIH acknowledgement of co-investigators and uni-
versity sharing of incentive funds.

CONCLUSION

The infrastructure needed to support research on the impact of interactions among social, behavioral, and genetic aspects on health will require transdisciplinary teams of researchers. This infrastructure may be construed as a matrix of training, tools, and incentives that are applied by universities and by funders, most specifically NIH. As discussed in this chapter, steps could be taken in each of these arenas, some easy and some difficult. However, it is unlikely that there are simple responses to this complex challenge. If the university community and funding agencies would come together to share experiences, many strategies that already exist to address aspects of this challenge could be communicated and recorded. If a full range of approaches was applied rigorously to the current problem, considerable progress could undoubtedly be made. However, this may require new funding mechanisms from NIH, and it may challenge universities to address fundamental practices such as the P/T process.

The infrastructure needed may take the form of new ways to train and educate researchers as well as ways to fund that training. The incentives to conduct research are influenced by both university and funding agency policies and practices and need to reflect the value of team science. Finally, there are tools that are needed to conduct research, not just the concrete tools of equipment and facilities, but the data that are key to this complex area of research.

REFERENCES

Bachrach CA, Abeles, RP. 2004. Social science and health research: Growth at the National Institutes of Health. *American Journal of Public Health* 94(1):22-28.

Bureau of Labor Statistics, U.S. Department of Labor, National Institute for Child Health and Human Development. 2002. *Children of the National Longitudinal Survey of Youth (NLSY79), 1979-2002.* [Computer File]. Columbus, OH: Center for Human Resource Research, Ohio State University.

IOM (Institute of Medicine). 1998. *Scientific Opportunities and Public Needs: Improving Priority Setting and Public Input at the National Institutes of Health.* Washington, DC: National Academy Press.

IOM. 2003. *Who Will Keep the Public Healthy? Educating Health Professionals for the 21st Century.* Washington, DC: The National Academies Press.

IPHG (University of Washington Institute for Public Health Genetics). 2005. *Institute for Public Health Genetics.* [Online]. Available: depts.washington.edu/phgen/about/about_ intro.shtml [accessed March 30, 2006].

NAS/NAE/IOM (National Academy of Sciences/National Academy of Engineering/Institute of Medicine). 2004. *Facilitating Interdisciplinary Research*. Washington, DC: The National Academies Press.

NIH (National Institutes of Health). 2004a. *Training for a New Interdisciplinary Research Workforce*. (NIH Award No. RFA-RM-04-015). Bethesda, MD: NIH.

NIH. 2004b. *NIH Awards (Competing and Non-Competing) by Fiscal Year and Funding Mechanism Fiscal Years 1994-2004*. [Online]. Available: grants2.nih.gov/grants/award/trends/fund9404.htm. [accessed February 15, 2006].

NIH. 2005a. *Institutional Clinical and Translational Science*. (NIH Award No. RFA-RM-06-002). Bethesda, MD: NIH.

NIH. 2005b. *Methodology and Measurement in the Behavioral and Social Sciences*. (NIH PA No. PA-05-090). Bethesda, MD: NIH.

NIH. 2005c. *NCRR Science Education Partnership Award (SEPA)*. (NIH PA No. PAR-05-068). Bethesda, MD: NIH.

NRC (National Research Council). 2000. *Addressing the Nation's Changing Needs for Biomedical and Behavioral Scientists*. Washington, DC: National Academy Press.

SEPA (Science Education Partnership Award). 2005. *Genome Science Education Program*. [Online]. Available: www.ncrrsepa.org/program/year/2001/Genome.htm [accessed March 30, 2006].

Udry J. 2003. *The National Longitudinal Study of Adolescent Health (Add Health), Waves I & II, 1994-1996; Wave III, 2001-2002*. [Machine-Readable Data File and Documentation]. Chapel Hill, NC: Carolina Population Center, University of North Carolina at Chapel Hill.

10

Ethical, Legal, and Social Implications

Earlier in this report, the committee addressed the challenges involved in identifying how individual human genes interact with other genes and with social and behavioral factors over time to affect human health. Research that elucidates how social, behavioral, and genetic influences interact to impact health may reveal findings that demonstrate beneficial effects on individuals and their health while other findings on interactions may show harmful effects. This lack of consistency may lead to differing perceptions of the value of research on interactions, which in turn may affect the willingness of researchers to do this work; funders to support it; care providers to act on existing evidence; and the population to embrace the findings. At its best, such findings could ensure that public health practice and medical care are attuned to the complex of factors that are affecting a patient, or an individual might be able to use such information as motivation for his/her own health-promoting behavior. On the other hand, such findings could lead to stigmatization and could have negative effects on the ability of individuals or groups to receive appropriate health care and insurance coverage. Consequently, it is important that transdisciplinary research on the impact on health of interactions among social, behavioral, and genetic factors also encompasses investigations that improve our understanding of how individuals make use of this information and how policymakers and the public interpret such research.

Efforts to address the implications of this type of knowledge are not new. For example, environmental regulation is focused to a large degree on the protection of health. Some of the more difficult issues in that arena concern whose health is to be protected—that of the average person in the population of interest and/or the health of high-risk individuals—as well as how and at what cost. Over the last two decades, much attention has been paid to the social and ethical implications of genetic and genomic information (Murray et al., 1996; Walters and Palmer, 1997; Rothstein, 1997; Rothstein, 2003; Mehlman, 2003). Indeed, the Human Genome Project occasioned the first decision by an institute of the National Institutes of Health (NIH) to designate specific funds to explore the social implications of a project. In this arena, the focus has been broader, ranging from effects on health to discrimination in work and insurance to notions of personal responsibility, including health and criminal law. More recently, these areas of inquiry have begun to merge in consideration of environmental genomics[1] and pharmacogenomics[2] (Need et al., 2005), both of which are concerned explicitly with interactions. Discussion in the following section builds upon all these discourses, with an emphasis on the implications of the interactions between genetic susceptibility and social and behavioral factors.

Another very important area in the ethical, legal, and social implications realm is that of the granting and licensing of intellectual property rights on discoveries related to genetics. A recent National Research Council report (NRC, 2006) explores this issue in depth, concluding that "the patent landscape, which is already becoming complicated in areas such as gene expression and protein-protein interactions, could become considerably more complex and burdensome over time." For a thorough and detailed examination of the very complex issues in this area, the committee refers readers to the NRC report entitled *Reaping the Benefits of Genomic and Proteomic Research: Intellectual Property Rights and Innovation in Public Health.*

CONVEYING COMPLEX SCIENTIFIC FINDINGS ACCURATELY

The picture that emerges from the study of the impact of the interactions among social, behavioral, and genetic factors on health is one of complexity. Even single gene disorders such as familial hypercholesterolemia (Austin et al., 2004a; Austin et al., 2004b) are anything but simple. Such disorders may involve hundreds of different mutations, most with

---

[1]*Environmental genomics* is defined as understanding how individuals differ in their susceptibility to environmental agents and how these susceptibilities change over time. Environmental genomics includes both the ways in which environmental factors cause genetic damage as well as the ways in which genetic variation affects responses to environmental exposures.

[2]*Pharmacogenetics* is the "branch of genetics that studies the ways in which genetically determined variations affect responses to drugs in humans or laboratory organisms" (Wordnet 2.0, 2003).

reduced penetrance. Many have pleiotropic effects. Sickle cell disease, which is caused by a single mutation but has many manifestations, is an even starker example of complexity in the face of apparent simplicity. Furthermore, the "common disease, common variant" hypothesis (Zondervan and Cardon, 2004) suggests that the most common variants in the genome have only modest effects on disease susceptibility (relative risks of 1.5 to 2), so that interactions among social, behavioral, and genetic factors may have major effects on health only for specific subgroups. Moreover, to date, the overwhelming majority of reported genetic associations have not been replicated in subsequent studies (Hirschhorn et al., 2002). Social and behavioral factors are even more difficult to measure than genetic variation. Nor are phenotypic effects readily predictable simply by characterizing the relevant genetic sequences, behaviors, and social environments, either together or individually. Network theory teaches that living systems are remarkably resistant to change and that the perturbation of one part tends to lead to a countervailing response by another in order to promote stability (Barabasi, 2002).

In contrast to this complexity, claims about scientific findings are at times simplistic and even exaggerated. The reasons for this tendency are many. The language of science plays a role. Terms such as "the gene for disease X" obscure distinctions between normal gene function—the normal variation in most genes that is present in the population—and the role of specific deleterious mutations that can cause abnormal function and disease predisposition. Furthermore, the scientific method itself is reductionist, seeking to isolate the impact of a particular factor on an outcome of interest. Finally, scientists face economic and social pressures to emphasize the significance of their findings in easily understandable terms that may have the effect of distorting the subtleties and uncertainties of the results (Holtzman et al., 2005).

These difficulties are compounded by those outside the scientific community who often are ill equipped to challenge what are perhaps overstated scientific claims. The media understandably prefers straightforward messages, while concepts of relative risk are notoriously difficult to understand. The legal system continues to struggle, in both regulatory settings and the courtroom, with the enormous disjunction between its methods of truth finding and those of science (Rothstein, 1999).

Failures to convey the limitations and complexity of scientific findings are significant because beliefs about the causation of health and disease affect the allocation of responsibility and resources, and this has ethical and social implications. Given the consequences of identifying clear causal explanations, the drive for simplification is strong. People generally seek simple explanations for events in their lives. Tort law is premised to a large degree on the notion that no more than a few factors can be held legally "responsible" for injuries to people and property. The attraction of reductionism and the search for a limited number of causes also contribute to the prominence of determinism—the idea that once a particular factor is known, biological and even social consequences follow more or less inexorably. The trend toward deterministic thinking has been particularly prominent regarding genetics, dating back at least to the eugenics movement of the nineteenth and early twentieth centuries (Kevles, 1985; Duster, 1990), but it extends throughout science and society. The history of how biological information has been used to put people at a disadvantage still looms large in the public's mind. The first step to countering the resulting fear of science is conveying accurately scientific findings and the difficulties involved in predicting the responses of complex systems.

POLICY DOES NOT INEXORABLY FOLLOW FROM SCIENTIFIC DISCOVERIES

Greater understanding of the influence of interactions among social factors, behavior, and genetic variation on health and disease pulls us in two directions (Shostak, 2003). Focusing on a person's unique physiological and genetic makeup focuses attention on the individual and his/her unique susceptibilities. Acknowledging the role of behavior and social location, however, directs attention to the situation in which the individual lives, including the social factors that influence and constrain that person's situation and his/her health-related behaviors. The question is whether and how to intervene to improve health, given this complexity.

For the purposes of this discussion, it is assumed that it will not be possible to alter particular gene sequences in an individual, at least not in the near future. Thus, any efforts to improve health and well-being in the population as a whole will necessarily depend on using pharmacologic and other medical interventions as well as on changing the social environments and individual health behaviors. Opportunities to alter these nongenetic factors in useful ways may exist at many levels, from the individual, to the family and community, to larger—even global—approaches. However, the array of realistic possibilities is constrained by a host of factors, such as the individual's personal and financial assets and cultural beliefs; the availability of resources; legal rules; and concerns about issues such as discrimination. The goals of intervention may vary because notions of health change over time and differ by cultural settings. Moreover, health-promoting actions can complement or compete with other goods at both the personal and societal levels, including individual priorities and values as well as commercial interests. Indeed, the matrix of factors that affect the application of scientific knowledge about social, behavioral, and genetic interactions and the values at stake is every bit as complex as the science that we seek to understand.

An example may be useful here. It is known that individuals who have one copy of certain mutations in the gene that codes for alpha-1-antitrypsin (A1AT) (i.e., are heterozygous) are more susceptible to lung damage when exposed to certain inhalants, ranging from chemicals typically used and produced in industry to smog and tobacco smoke (Ranes and Stoller, 2005). On its face, it seems obvious that such individuals should not experience these potential harmful exposures. But questions about how to achieve this goal quickly arise.

One might think that people with mutations in A1AT would simply choose to avoid being in harmful environments. However, a great deal of evidence demonstrates that knowledge of risk does not lead inexorably to health-promoting behavior change (Marteau and Lerman, 2001), and at times it may lead to harmful responses. The possible explanations for these apparently suboptimal outcomes are many. In some cases, susceptible individuals simply choose to ignore the risk of toxic exposures. Some argue that protecting susceptible individuals by providing health care if they become ill or by cleaning the environment creates "moral hazard"—the possibility that predisposed people would engage in socially undesirable, unhealthy activities because they are insulated from the consequences. The argument in this case would be that people with mutations in A1AT do not avoid exposing themselves to risk because they know they will receive treatment if they become ill.

Some decisions not to avoid potentially harmful exposures, however, result from trade-offs that are made with other goals. Some people with these mutations may find that they can earn a living wage only if they live in a smoggy city or work in sites with harmful fumes. They can be faced with choosing between optimizing their health and meeting their immediate needs and those of their families. Also, the personal protective equipment that could ameliorate some of the risk to such susceptible individuals can be onerous and expensive. However, no matter what the reason for lack of avoidance, it does seem likely that most people do not choose ill health as a matter of preference. Moreover, relatively little research has been done to show how to increase health-promoting behavior in these type of situations.

Nor is it clear that protecting only those who have greater risk is necessarily the best policy. Exposures to smoke and toxic fumes are potentially harmful to a large part of the population, not just to those who are particularly susceptible. Reducing such exposures, then, could improve the health of the public generally, not just those members of the public with mutations in A1AT or other susceptibilities. As a result, environmental regulation has taken a variety of approaches, sometimes requiring individual protective measures, but frequently trying to reduce exposures for everyone. This has led to noticeable improvements in air and water quality over the past 50 years, with benefits going beyond good health to those as simple as the pleasure of having blue skies and clean water (Grodsky, 2005).

Policies regarding who should bear the costs of behavioral choices and environmental exposures are mixed as well. In the individual health insurance market, people who smoke or who work in hazardous jobs pay higher premiums. At the same time, both the federal and many state governments regulate the extent to which insurers can use some types of information, particularly information about genetic predispositions, in their underwriting. Employers are concerned with health care costs because they pay higher premiums if their workers have large claims. Over the last 20 years, the ability of employers to exclude workers who may have high health care costs has been limited by laws such as the Americans with Disabilities Act (ADA) (42 U.S.C.A. §§ 12101 et seq. (2006)), which forbids discrimination against workers with disabilities so long as they can fulfill the essential elements of the job with reasonable accommodation, and cases such as *Automobile Workers v. Johnson Controls* (499 U.S. 187 111 S.Ct. 1196, 113 L.Ed.2d 158 (1991)), which held that Johnson Controls could not exclude women from the potentially fetotoxic workplace. Thus, Terri Seargent, who was essentially asymptomatic, successfully claimed that she was fired because of the costs of enzyme replacement for her A1AT deficiency (Clayton, 2001).

This body of law, however, recently has been undercut by cases such as *Chevron v. Echazabal* (536 U.S. 73, 122 S.Ct. 2045, 153 L.Ed.2d 82 (2002)), in which the Supreme Court upheld regulations issued under the ADA that permitted employers to refuse to hire workers whose underlying medical conditions make them more likely to be made ill by the toxic workplace. Finally, although society often tries to encourage its members to avoid risky behavior, it has chosen not to require people to bear all of the consequences of their actions. Instead, reflecting a belief that a civil society should provide basic care for its citizens, our health care system provides a substantial, if spotty, safety net against catastrophic illness for many of its members, even when those diseases result in part from personal behaviors.

Expressed another way, risks to individual health of whatever sort—genetic, behavioral, or social—raise a set of common questions, as illustrated below. For these purposes, we assume that a threshold level of scientific validity has been met demonstrating that a particular factor influences disease risk.[3]

---

[3]What this level might be can itself be contested. Does the likelihood of the truth of a particular scientific outcome need to be more probable than not, clear and convincing, beyond a reasonable doubt, or have a probably of less than 0.05?

- Who decides whether it is known that a particular individual has a specific risk factor, whether social, behavioral, or genetic, or a combination thereof? Does the individual have the exclusive right to make decisions about whether to find out about his/her risk status, or can third parties require testing or make testing a condition for receiving employment or other goods?
- People may have more control over access to some sorts of personal risk information than to others. For example, the fact that an individual smokes cigarettes is difficult to hide, while whether that person has a genetic variant that affects the metabolism of that smoke may not be apparent without a specific test.
- If the fact that a person has a particular risk factor is known, who should be able to *obtain access* to this knowledge? Options include the individual, the government, and private entities such as employers or insurers.
- If the fact that a person has a particular risk is known, who gets to *act upon* that information? Can a third party force the individual to ameliorate the risk, perhaps by denying employment to the person or requiring him/her to use special protective equipment? Can an insurer permissibly charge higher premiums?
- What are the costs of acting on the risk information, and who will bear those costs? The answers to this inquiry can be complex. For example, excluding particular individuals from certain opportunities or social goods may benefit some entities, such as employers, while arguably harming the individual as well as impinging on social norms of equality. It also is important to recognize that most costs are shared, albeit to varying degrees, and all, in the final analysis, are borne by the citizenry.

In some ways, traits such as the A1AT deficiency present a relatively simple case in the United States, because these mutations are present primarily in Caucasians and cause disorders—emphysema and liver damage—that are not particularly stigmatizing. Questions about appropriate interventions almost certainly will become more vexing as more is learned about the impact of interactions among social, behavioral, and genetic variation on behavioral itself. For example, it was recently reported that individuals with low levels of monoamine oxidase A (MAOA) who were subjected to severe child abuse are more likely to engage in a variety of antisocial behaviors (Caspi et al., 2002). These results could raise a host of questions, ranging from whether these children need special protection during childhood to whether they should be monitored for antisocial behavior more closely as adults, all of which have serious implications for civil liberties. Even assuming that the findings of Caspi et al. will be replicated in the future, any intervention would be overly broad, because the majority of children in the high-risk group (low MAOA + abuse) in that study exhibited no behavior problems, which often is the cause for complex phenotypes.

At times, particular genetic alleles are more frequent in individuals of a certain geographic or historical origin. For example, mutations that cause cystic fibrosis are more common in populations of Northern European ancestry than in those of Asian or African origin (Nussbaum et al., 2004). Similarly, behaviors and social environments and practices vary among cultural groups. Because it often is difficult to ascertain these variables for any particular person, it can be tempting to use more readily available social groupings, such as race or ethnicity, as proxies for variations in all these domains. (See Chapter 5 for a more detailed discussion of race/ethnicity and sex/gender.) Using categories such as race as a proxy, however, can have adverse effects. For example, in the late 1980s and early 1990s, after it became clear that penicillin prophylaxis could be lifesaving for children with sickle cell disease, a number of states decided to screen only non-Caucasian newborns for hemoglobinopathies, with the reasoning that focused screening would be more cost-effective because these mutations are most prevalent in populations that arose in equatorial areas. Most states subsequently abandoned this strategy for several reasons, not the least of which is that some affected children were missed.

One reason for incomplete ascertainment is that hemoglobinopathies occur in many populations in this country. More generally, states have faced difficulties in defining which children were to be tested. Different strategies were used, including visual determination of the race of the mother and/or the child or asking the mothers to identify their race. No matter what was tried, affected children were missed, including some whose ancestry meant that they were more likely to have inherited these mutations. It also has become increasingly clear that race is not a stable category, but rather is a social construct whose definition changes over time. The problems with targeted newborn screening for hemoglobinopathies were all the more challenging because they occurred in the context of the longstanding history of race discrimination in this country and the more recent history in the 1970s of discrimination against those with sickle cell trait (Reilly, 1977). The memories of these events never have been too far from the surface.

This example, while focusing on genetic variation, illustrates some of the difficulties that can be presented by interventions targeted at groups of people. Risk factors, be they social, behavioral, or genetic, can be both overinclusive and underinclusive—some individuals will be singled out for further attention who would never have become ill, while others who are actually at risk will not receive beneficial assistance. These problems of over- and underinclusiveness are exacerbated when the criteria for targeting are not fully concordant proxies for the actual risk factors. For example, even though more men than women ride motorcycles, it would make little sense to teach only men about the importance of wearing helmets, because most men do not ride, while some women do. Moreover, the use of historically disfavored groups as proxies for genetic variation, behavior, or social environment creates the risk of reinforcing old prejudices and stereotypes. Targeted intervention may well be appropriate at times, but such programs should be undertaken only after careful consideration of the social consequences and after weighing other alternatives.

It is beyond the scope of this report to make recommendations regarding the application of knowledge of social, behavioral, and genetic interactions in forming policy. However, the array of factors that must be considered in deciding how to use this knowledge is very broad and extends far beyond the science itself. Often, a variety of social responses are ethically and socially acceptable. Thus, the idea that social policy follows inexorably from scientific discovery is every bit as misplaced as the notion of scientific determinism itself.

To address difficulties in how individuals and groups understand complex scientific findings, as well as the potential impact such findings could have on policy development, the committee makes the following recommendations:

Recommendation 11: Communicate with Policymakers and the Public. *Researchers should (1) be mindful of public and policymakers' concerns, (2) develop mechanisms to involve and inform these constituencies, (3) avoid overstating their scientific findings, and (4) give careful consideration to the appropriate level of community involvement and the level of community oversight needed for such studies.*

Recommendation 12: Expand the Research Focus. *The NIH should develop RFAs for research that elucidate how best to encourage people to engage in health-promoting behaviors that are informed by a greater understanding of these interactions, how best to effectively communicate research results to the public and other stakeholders, and how best to inform research participants about the nature of the investigation (gene-environment interactions) and the uses of data following the study.*

ETHICAL IMPLICATIONS FOR RESEARCH

Institutional Review Boards (IRBs) perform several roles in overseeing research regarding interactions, but it is important at the outset to identify one area in which they may not act. Although they are required to weigh the risks and benefits of research protocols for research participants, they are specifically precluded from considering "possible long-range effects of applying knowledge gained in the research (for example, the possible effects of the research on public policy) as among those research risks that fall within the purview of its responsibility" (45 CFR § 46.111(a)(2) (2006)). (For an in-depth analysis of issues regarding protection of research participants, please see the report *Responsible Research: A Systems Approach to Protecting Research Participants* [IOM, 2003].) Such factors, which we have seen can be implicated by research regarding the effect of interactions among social, behavioral, and genetic factors on health, must be considered elsewhere, if at all.

Privacy and Security

IRBs are responsible for ensuring, where appropriate, the protection of the research participants' privacy and the protection of the data regarding the participants (45 CFR § 46.111(a)(7) (2006)). Studying interactions among variations in social, behavioral, and genetic factors requires the collection of information about relevant DNA variants as well as clinical or other phenotypic information, which often includes sensitive personal information about behavior and social factors. The risk to research participants, were such information to be accessed by people and institutions outside the study, could be substantial. Indeed, fear that sensitive or stigmatizing information will be uncovered or revealed is a common reason people give for declining to participate in research (Schwartz et al., 2001). Some protection from disclosure is provided by laws such as the Privacy Rule promulgated under the Health Insurance Portability and Accountability Act of 1996 (HIPAA)(45 CFR Parts 160 and 164 (2006)). However, IRBs should direct investigators to take additional administrative and technological steps to prevent the unwarranted release of data. The first approach in this regard is to provide adequate *security* for the data, which can involve storing data on computers that are kept in locked facilities with limited access by personnel, allowing no connection to the Internet, and using methods of encryption. The second draws upon the model provided by HIPAA. In this approach, investigators who share their data with others must execute data use agreements that ensure that the recipients will comply with all the restrictions that apply to the individual or institution that collected the information initially. The third approach is to obtain additional legal protections against disclosure. The most important of these protections are the still relatively underutilized Certificates of Confidentiality, which can be obtained from the U.S. Department of Health and Human Services (Cooper et al., 2004; Office of Extramural Research, 2005).

Several different methods, which vary in their impact on the utility of the data, may be taken to secure data. Irretrievably removing identifiers provides the greatest security, but if this is done correctly, it may require eliminating many variables. Furthermore, this method precludes following study participants prospectively for disease incidence by adding new clinical data. A different strategy may be to adopt one-way encryption so that neither investigators nor database managers can identify individual research participants, even though new clinical information can be added.

It also may be possible to code the information and maintain a key that makes it possible to go back to particular individuals in order to obtain additional specific data pertinent to new hypotheses, to invite them to participate in new research projects, perhaps exploring preventive or therapeutic interventions, or even to provide them with clinically meaningful research findings. Retaining a key, however, presents additional challenges. Strict limits on access to the key would be necessary to avoid seriously compromising security. Criteria and a process of review need to be developed to justify recontacting individuals for more information or to invite further research participation.

Despite individuals' concerns about their privacy, pressure is mounting from many quarters to increase the availability of data. Pharmaceutical companies are being asked to report details of all their clinical trials (Herxheimer, 2004). Investigators and funding agencies around the world are proposing expanded data sharing policies (Arzberger et al., 2004). Countries and funders are creating new, very large datasets with genomic and phenotypic data to be made broadly available. The Data Quality Act enables entities that dislike particular regulatory decisions to question the science on which they are based (Rosenstock, 2006). To date, no clear consensus has emerged about exactly what data need to be shared or how individual privacy is to be protected, although at least some writers have recognized that the latter is an issue. Given this uncertainty and the power of datasets that include rich phenotypic, genomic, and environmental information, it is particularly important that IRBs attend to questions of how fully data can be encrypted or de-identified and what research participants need to be told about the research.

Disclosure of Results

One of the most contentious issues posed by maintaining a mechanism for personal contact involves the question of whether research participants should receive individual results. Although some argue that this information should be offered as a matter of right (Council for International Organizations of Medical Sciences, 2002; Shalowitz and Miller, 2005), strong arguments have been made that it is better to reveal individual research results, if at all, only under very limited circumstances. Routine disclosure fuels the therapeutic misconception (Appelbaum et al., 1982)—the mistaken belief that research is directed toward the same goal as clinical care, namely the best interest of the patient. The purpose of research, instead, is to create generalizable knowledge. The research process typically proceeds by fits and starts. Because many initial findings cannot be replicated, particularly in areas as complex as the impact of interactions among social, behavioral, and genetic factors on health, a practice of routine disclosure often would provide misplaced reassurance or create unwarranted fear. On a more practical level, most research is conducted in laboratories that are not approved under the Clinical Laboratory Improvement Amendments and their regulations (42 USCA § 263a (2006) and 42 CFR Part 493(2006)). Therefore, research laboratories may not use the rigorous sample handling and tracking procedures used in clinical laboratories. This increases the risk that results would be attributed to the wrong person or be incorrectly reported. Finally, some people may not welcome this information. Numerous studies demonstrate that while many people express interest in learning about individual risks, fewer actually pursue testing once it is available (Bowen et al., 1999).

As a result of these problems, most commentators favor limits on the disclosure of individual research results. The National Bioethics Advisory Commission, for example, proposed that individual research results could ethically be revealed only if an ethics committee or other review body concluded that "a) the findings are scientifically valid and confirmed, b) the findings have significant implications for the subject's health concerns, and c) a course of action to ameliorate or treat these concerns is readily available" (National Bioethics Advisory Commission, 1999). The availability of effective prevention also may suffice to justify disclosure. This threshold rarely would be met in research involving the impact of interactions among social, behavioral, and genetic factors on health because the relative risks are almost always relatively modest, and because behavioral and social factors can be difficult to quantify. In any event, it is critical that investigators and IRBs define the criteria for the disclosure of individual results at the outset of the project.

Even when individual research results are not revealed to participants, it often is desirable to inform them periodically about general research findings. This can be accomplished by routine mailings, by presentations at meetings of patient organizations, or by creating websites, which may or may not be password protected, that participants can visit. Informing research participants about the progress of the project will enable them to talk more effectively with their clinicians about seeking testing or other interventions once the research findings become sufficiently robust to be incorporated into clinical care.

Community Involvement

The desirability and limits of including lay oversight and some level of community involvement in research protocols were recently reviewed in the National Research Council/Institute of Medicine's report *Ethical Considerations for Research on Housing-Related Health Hazards Involving Children* (NRC/IOM, 2005). Lay involvement can take many different forms, ranging from membership on IRBs to community-based participatory research, in which laypeople and investigators jointly define every aspect of the project. Including participants can improve research by identifying issues or risk factors that would not have been considered by investigators, improving recruitment and communication, and increasing transparency. Community advisory groups, which represent an intermediate level of involvement, increasingly act as conduits of research to the larger group of research participants (Coriell Institute for Medical Research, 2006). At the same time, greater lay involvement is time intensive for both investigators and laypeople. To date, little data exist regarding its efficacy.

In considering what level of lay involvement is appropriate in studies of the impact of interactions among social, behavioral, and genetic factors on health risks of the research, the practicability of inclusion should be taken into account. More active involvement may be desirable, for example, when study results have the potential to stigmatize individuals or groups, as might be the case in studies that explore genetic and environmental influences on antisocial behaviors. It is important to recognize that all types of differences—social, behavioral, and genetic—can be potential sources of stigma, and, where implicated, they may warrant greater lay involvement. The risk to individuals or groups may be even greater when the research participants are in some way vulnerable within the larger society.

One of the most vexing problems facing investigators is deciding what to do when research involving samples and clinical information collected for one purpose suggests new hypotheses. For example, researchers may have focused initially on cardiovascular disease risk, but new findings may suggest that the exploration of factors contributing to Alzheimer's disease also may be fruitful (e.g., the work currently being conducted on apolipoprotein E isoforms). Consultation with the community may provide insight into whether this new direction is consistent with the original intent of the participants.

Lay involvement may be more obviously required when defined political structures exist within the group from which research participants are drawn. The paradigmatic example in the United States is research involving Native Americans because they are members of sovereign nations; however, in that setting, care must be taken to ensure the representativeness of the process and of those who purport to speak on behalf of the participants

(Council for International Organizations of Medical Sciences, 2002; Sharp and Foster, 2002). Efforts to solicit public involvement in research design and dissemination also may be warranted even when community groupings are less well defined. Strategies will differ in each context, but will typically involve tapping into local social networks within the larger group.

Informed Consent

The last decade has seen an enormous amount of debate regarding the ethical and legal requirements of informed consent for the use of medical information and human biological materials for research (Clayton et al., 1995; Knoppers, 1997; National Action Plan for Breast Cancer, 1997; National Bioethics Advisory Commission, 1999). Among the issues that are often addressed in current consent forms are the types of research that may be conducted; the risks and benefits, both personal and social, that may result from the research; who is going to hold and have access to these resources; what privacy and security protections are going to be used; under what conditions, if any, individuals may be recontacted either to obtain further consent or to be provided specific health-related results; and the possibility that intellectual property may be developed. Particularly in light of evidence that research participants often are not truly informed, more work needs to be done to learn how to communicate this information effectively. These issues merit particular attention in studies of the interactions among social, behavioral, and genetic factors on health in order to ensure that participants truly understand what is at stake in the research.

Given the sensitivity of research and its implications involving interactions among the factors under discussion, it is of primary importance to address the issues of data sharing and informed consent. Therefore, the committee recommends the following:

> **Recommendation 13: Establish Data-Sharing Policies That Ensure Privacy.** *IRBs and investigators should establish policies regarding the collection, sharing, and use of data that include information about (1) whether and to what extent data will be shared; (2) the level of security to be provided by all members of the research team as well as the research and administrative process; (3) the use of state-of-the-art security for collected data, including, but not limited to, NIH's Certificates of Confidentiality; (4) the use of formal criteria for identifying the circumstances under which individual research results will be revealed; and (5) how, before sharing data with others, recipients must agree to use data only in ways that are consistent with those agreed to by the research participants. Furthermore, if a mechanism to identify individual research partici-*

pants is retained in the database, IRBs and investigators should consider whether to contact participants prior to initiating research on new hypotheses or other new research.

Recommendation 14: Improve the Informed Consent Process. *Researchers should ensure that informed consent includes the following: (1) descriptions of the individual and social risks and benefits of the research; (2) the identification of which individual results participants will and will not receive; (3) the definition of the procedural protections that will be provided, including access policies and scientific and lay oversight; and (4) specific security, privacy, and confidentiality protections for protect the data and samples of research participants.*

REFERENCES

Appelbaum PS, Roth LH, Lidz C. 1982. The therapeutic misconception: Informed consent in psychiatric research. *International Journal of Law and Psychiatry* 5(3-4):319-329.

Arzberger P, Schroeder P, Beaulieu A, Bowker G, Casey K, Laaksonen L, Moorman D, Uhlir P, Wouters P. 2004. Science and government. An international framework to promote access to data. *Science* 303(5665):1777-1778.

Austin MA, Hutter CM, Zimmern RL, Humphries SE. 2004a. Familial hypercholesterolemia and coronary heart disease: A HuGE association review. *American Journal of Epidemiology* 160(5):421-429.

Austin MA, Hutter CM, Zimmern RL, Humphries SE. 2004b. Genetic causes of monogenic heterozygous familial hypercholesterolemia: A HuGE prevalence review. *American Journal of Epidemiology* 160(5):407-420.

Barabasi, AL. 2002. *Linked: How Everything Is Connected to Everything Else and What It Means*. Cambridge, MA: Perseus Publishing.

Bowen DJ, Patenaude AF, Vernon SW. 1999. Psychosocial issues in cancer genetics: From the laboratory to the public. *Cancer Epidemiology, Biomarkers and Prevention* 8(4 Pt 2):326-328.

Caspi A, McClay J, Moffitt TE, Mill J, Martin J, Craig IW, Taylor A, Poulton R. 2002. Role of genotype in the cycle of violence in maltreated children. *Science* 297(5582):851-853.

Clayton EW. 2001. Through the lens of the sequence. *Genome Research* 11(5):659-664.

Clayton EW, Steinberg KK, Khoury MJ, Thomson E, Andrews L, Kahn MJ, Kopelman LM, Weiss JO. 1995. Informed consent for genetic research on stored tissue samples. *Journal of the American Medical Association* 274(22):1786-1792.

Cooper ZN, Nelson RM, Ross LF. 2004. Certificates of confidentiality in research: Rationale and usage. *Genetic Testing* 8(2):214-220.

Coriell Institute for Medical Research. 2006. *Coriell Cell Repositories*. [Online]. Available: locus.umdnj.edu/ccr/ [accessed March 24, 2005].

Council for International Organizations of Medical Sciences. 2002. *International Ethical Guidelines for Biomedical Research Involving Human Subjects*. Geneva: Council for International Organizations of Medical Sciences.

Duster, T. 1990. *Backdoor to Eugenics*. New York: Routledge.

Grodsky JA. 2005. Genetics and environmental law: Redefining public health. *California Law Review* 93(1):171-270.

Herxheimer A. 2004. Open access to industry's clinically relevant data. *British Medical Journal* 329(7457):64-65.

Hirschhorn JN, Lohmueller K, Byrne E, Hirschhorn K. 2002. A comprehensive review of genetic association studies. *Genetics in Medicine: Official Journal of the American College of Medical Genetics* 4(2):45-61.

Holtzman NA, Bernhardt BA, Mountcastle-Shah E, Rodgers JE, Tambor E, Geller G. 2005. The quality of media reports on discoveries related to human genetic diseases. *Community Genetics* 8(3):133-144.

IOM (Institute of Medicine). 2003. *Responsible Research: A Systems Approach to Protecting Research Participants*. Washington, DC: The National Academies Press.

Kevles, DJ. 1985. *In the Name of Eugenics: Genetics and the Uses of Human Heredity*. New York: Knopf.

Knoppers, BM. 1997. *DNA Sampling: Human Genetic Research—Ethical, Legal, and Policy Aspects*. The Hague: Kluwer Law International.

Marteau TM, Lerman C. 2001. Genetic risk and behavioural change. *British Medical Journal* 322(7293):1056-1059.

Mehlman MJ. 2003. *Wondergenes: Genetic Enhancement and the Future of Society (Medical Ethics Series)*. Bloomington, IN: Indiana University Press.

Murray TH, Rothstein MA, Murray RF. 1996. *The Human Genome Project and the Future of Health Care (Medical Ethics Series)*. Bloomington, IN: Indiana University Press.

National Action Plan for Breast Cancer. 1997. *Consent Form for the Use of tissue for Research*. [Online]. Available: www.4woman.gov/napbc/catalog.wci/napbc/consent.htm [accessed December 12, 2005].

National Bioethics Advisory Commission. 1999. *Research Involving Human Biological Materials: Ethical Issues and Policy Guidance*. Rockville, MD: National Bioethics Advisory Commission.

NRC (National Reserach Council). 2006. *Reaping the Benefits of Genomic and Proteomic Research: Intellectual Property Rights and Innovation in Public Health*. Washington, DC: The National Academies Press.

NRC/IOM. 2005. *Ethical Considerations for Research on Housing-Related Health Hazards Involving Children*. Washington, DC: The National Academies Press.

Need AC, Motulsky AG, Goldstein DB. 2005. Priorities and standards in pharmacogenetic research. *Nature Genetics* 37(7):671-681.

Nussbaum RL, McInnes RR, Willard HF. 2004. *Thompson & Thompson Genetics in Medicine*. 6th edition. Philadelphia, PA: Saunders.

Office of Extramural Research. 2005. *Certificates of Confidentiality Kiosk*. [Online]. Available: grants1.nih.gov/grants/policy/coc/ [accessed December 12, 2005].

Ranes J, Stoller JK. 2005. A review of alpha-1 antitrypsin deficiency. *Seminars in Respiratory and Critical Care Medicine* 26(2):154-166.

Reilly P. 1977. *Genetics, Law, and Social Policy*. Boston, MA: Harvard University Press.

Rosenstock L. 2006. Protecting special interests in the name of "good science." *Journal of the American Medical Association* 295(20):2407-2410.

Rothstein M. 1999. The impact of behavioral genetics on the law and the courts. *Judicature* 83(3):116-123.

Rothstein MA. 1997. *Genetic Secrets: Protecting Privacy and Confidentiality in the Genetic Era*. New Haven, CT: Yale University Press.

Rothstein MA. 2003. *Pharmacogenomics Social, Ethical, and Clinical Dimensions*. Hoboken, NJ: Wiley-Liss.

Schwartz MD, Rothenberg K, Joseph L, Benkendorf J, Lerman C. 2001. Consent to the use of stored DNA for genetics research: A survey of attitudes in the Jewish population. *American Journal of Medical Genetics* 98(4):336-342.

Shalowitz DI, Miller FG. 2005. Disclosing individual results of clinical research: Implications of respect for participants. *Journal of the American Medical Association* 294(6): 737-740.

Sharp RR, Foster MW. 2002. Community involvement in the ethical review of genetic research: Lessons from American Indian and Alaska Native populations. *Environmental Health Perspectives* 110(Suppl 2):145-148.

Shostak S. 2003. Locating gene-environment interaction: At the intersections of genetics and public health. *Social Science and Medicine* 56(11):2327-2342.

Walters L, Palmer JG. 1997. *The Ethics of Human Gene Therapy*. New York: Oxford University Press.

Wordnet 2.0. 2003. *Definition of Phamacogenetics*. [Online]. Available: dictionary. reference.com/search?q=pharmacogenetics [accessed January 19, 2005].

Zondervan KT, Cardon LR. 2004. The complex interplay among factors that influence allelic association. *Nature Reviews Genetics* 5(2):89-100.

11

Conclusion

As discussed throughout this report, human health is determined by the interaction of several factors, including the social environment, genetic inheritance, and personal behaviors. Socioeconomic status, race/ethnicity, social networks/social support, and the psychosocial work environment all have been shown to affect health outcomes (Chapter 2). These social determinants influence health at multiple levels throughout the life course. In addition to the vast array of social determinants that influence health, a person inherits a complete set of genes from each parent that contributes both directly and indirectly to the pathogenesis of disease. Genes have been identified for relatively uncommon, simple Mendelian patterns of disease inheritance, such as Tay-Sachs disease and cystic fibrosis, and recently research has begun to explore genetic susceptibility to disease as the consequence of the joint effects of many genes, each with small-to-moderate effects, often interacting among themselves and with the environment (Chapter 3). Behaviors also have been shown to affect health (Chapter 4). For example, tobacco use, obesity, and physical inactivity are the greatest preventable causes of morbidity and mortality in the United States (Mokdad et al., 2004). Furthermore, complex traits, such as sex/gender and race/ethnicity, pose both a challenge and an opportunity in our search for a better understanding of environmental, genetic, and behavioral interactions as determinants of health (Chapter 5).

As this report demonstrates, research has documented associations between social factors and health, behaviors and health, and genetics and health. Yet, researchers are only now beginning to study in earnest the potential interactions between genetic and social environmental factors that are likely to be contributing to a large fraction of disease in most populations. Key to the success of research on these interactions is the conduct of such research in a collaborative and transdisciplinary manner, which "implies the conception of research questions that transcend the individual departments or specialized knowledge bases because they are intended to solve . . . research questions that are, by definition, beyond the purview of the individual disciplines" (IOM, 2003). Furthermore, more comprehensive, predictive models of etiologically heterogeneous disease are needed, and this requires the development and implementation of new modeling strategies and the use of profiling approaches. In order to ensure that findings are applicable beyond a small population, research must be conducted in diverse groups and settings (Chapter 6). Animal models, which are explored in Chapter 7, have a great deal to offer in understanding the effects of interactions of social, behavioral, and genetic factors on health.

A clear formulation of the concept of interaction, and an understanding of research designs that can be used to test for it, are central to progress in assessing the impact on health of interactions among multiple factors. This report discusses several steps that are needed to advance the science of testing interactions (Chapter 8). These include new, accessible statistical software for implementing tests for interaction on an additive scale and research on developing study designs that are efficient at testing interactions, including variations in interactions over time and development.

Transdisciplinary research on the impact on health of interactions among social, behavioral, and genetic factors places several demands on the research infrastructure, including the need for education and training of researchers, the enhancement and development of appropriate datasets, and the creation of incentives and rewards that will encourage investigators to move beyond the single discipline approach to research. Approaches that the National Institutes of Health can use to address these barriers include providing individual and senior fellowships, transdisciplinary institutional grants, short courses, and datasets that can be enhanced to provide the necessary information. The development of new datasets for topics that have high potential for showing interactions also would be valuable. Other incentives that foster the transdisciplinary research discussed in this report address hiring, promotion and tenure policies, peer review, and the allocation of credit for collaborative research (Chapter 9).

Finally, research that elucidates how social, behavioral, and genetic factors interact to influence health raises important ethical and legal issues, including those involving how individuals and groups understand and use complex scientific findings, as well as the potential impact such findings might have on policy development (Chapter 10).

Furthermore, studying interactions among variations in social, behavioral, and genetic factors requires the collection of information that could entail significant risk to research participants if it is inappropriately accessed. This report offers recommendations for communicating with policymakers and the public, for expanding the research focus to include research on how best to encourage people to engage in health-promoting behaviors, for the establishment of data-sharing policies that ensure privacy, and for improving the informed consent process.

The intent of this report is to encourage and facilitate the growth of research on the impact of interactions among social, behavioral, and genetic factors on health that will further our understanding of disease risk and aid in the development of effective interventions to improve the health of individuals and populations. This report has resulted from collaboration that has occurred between scientists from the social and the biological worlds, and it provides a template for how their theories and methods can be integrated to advance knowledge. It is timely and important because it sets out an agenda for research that is needed to advance the science of gene-environment interactions in explaining individual and population health and health disparities.

REFERENCES

IOM (Institute of Medicine). 2003. *Who Will Keep the Public Healthy? Educating Health Professionals for the 21st Century*. Washington, DC: The National Academies Press.

Mokdad AH, Marks JS, Stroup DF, Gerberding JL. 2004. Actual causes of death in the United States, 2000. *Journal of the American Medical Association* 291(10):1238-1245.

A

Methodology: Data Collection and Analysis

INTRODUCTION

The committee reviewed a broad array of information while considering the issues associated with assessing the impact on health of interactions among social, behavioral, physiological, and genetic factors. Sources of information included primary scientific literature in sociology, psychology, genetics, gene-environment interactions, and public health; books; scientific reviews; news articles; presentations from researchers, and representatives from the sponsor. Compilations of this background material commenced in December of 2004 and ended in February of 2006, shortly after the committee held its final meeting.

To answer questions that were posed to the committee in the statement of task, members of the committee relied on their own areas of expertise supplemented by various methods of information gathering that are described in more detail below.

Literature Review

The committee and Institute of Medicine (IOM) staff used an extensive online bibliographic search to compile a reference database of peer-reviewed literature relevant to the topic of the impact of interactions among social, behavioral, physiological, and genetic factors on health. The online bibliographic search was conducted using relevant databases (Box A-1) that included EMBASE, LexisNexis, Medline, PsychINFO, Science Direct, and

BOX A-1
Online Databases

EMBASE (Excerpta Medica) database is a major biomedical and pharmaceutical containing more than 9 million records from 1974 to the present from over 4,000 journals; approximately 450,000 records are added annually. More than 80 percent of recent records contain full author abstracts. This bibliographic database indexes international journals in the following fields: drug research, pharmacology, pharmaceutics, toxicology, clinical and experimental human medicine, health policy and management, public health, occupational health, environmental health, drug dependence and abuse, psychiatry, forensic medicine, and biomedical engineering/instrumentation. EMBASE is produced by Elsevier Science.

LexisNexis provides access to full-text information from more than 5,600 sources, including national and regional newspapers, wire services, broadcast transcripts, international news, and non-English-language sources; U.S. federal and state case law, codes, regulations, legal news, law reviews, and international legal information; and business news journals, company financial information, Securities and Exchange Commission filings and reports, and industry and market news. It is produced by Reed Elsevier, Inc.

MEDLINE is the U.S. National Library of Medicine's premier bibliographic database, containing citations from the mid-1960s to the present and covering the fields of medicine, nursing, dentistry, veterinary medicine, the health care system, and the preclinical sciences. PubMed provides online access to more than 12 million MEDLINE citations. MEDLINE contains bibliographic citations and author abstracts from more than 4,600 biomedical journals published in the United States and 70 other countries. PubMed includes links to many sites providing full-text articles and other related resources. This database can be accessed at www.ncbi.nlm.nih.gov/PubMed.

Sociological Abstracts: This online search was carried out throughout the entire course of the study.

To begin the process of identifying peer-reviewed literature, the IOM staff conducted a general bibliographic search on topics that were relevant to interactions among genes and the social environment, and behavioral and physiological factors. IOM staff then categorized these references according to their subject matter and developed reference lists of key citations that were provided to the committee for review. After discussing the reference lists with the committee, areas in which additional information was needed were determined.

As the study progressed, searches of peer-reviewed literature continued regularly. Additional references were identified by reviewing the reference lists of major primary literature, key reports, relevant websites, and text-

PsycINFO is a bibliographic database of psychological literature with journal coverage from the 1800s to the present and book coverage from 1987 to the present. It contains more than 1,900,000 records, including citations and summaries of journal articles, book chapters, books, and technical reports, as well as citations to dissertations, all in the field of psychology and psychological aspects of related disciplines. Journal coverage includes full-text article links to 42 American Psychological Association journals including peer-reviewed international journals. PsycINFO is produced by the American Psychological Association.

Science Direct is a full-text journal database that indexes more than 1,800 scientific, technical, and medical peer-reviewed journals and contains more than 59 million abstracts and more than two million full-text scientific journal articles. Subject coverage includes biological sciences; business management and accounting; computer science; earth and planetary sciences; engineering and technology; environmental science; materials science; mathematics; medicine; physics and astronomy; psychology; and social science.

Sociological Abstracts indexes the international literature in sociology and related disciplines in the social and behavioral sciences from 1963 to the present. This bibliographic database contains citations (from 1963) and abstracts (only after 1974) of journal articles, dissertations, conference reports, books, book chapters, and reviews of books, films, and software. Approximately 1,700 journals and 900 other serials published in the United States and other countries in more than 30 languages are screened yearly and added to the database bimonthly. The Sociological Abstracts database contained approximately 600,000 records in 2003. A limited number of full-text references are available. Sociological Abstracts is prepared by Cambridge Scientific Abstracts.

books. Throughout the process, committee members, workshop presenters, and IOM staff supplied references and suggested key terms and authors relevant to the study. The IOM staff maintained a searchable database that was categorized to allow searches by keyword, type of literature (e.g., journal article), date, or other criteria. Reference lists of articles obtained were regularly updated and provided to the committee and consultants, who requested full text of the journal articles and other resources as needed for their information and analysis.

After many months of reviewing the rapidly expanding literature available, the final count of articles was more than one thousand. Two-thirds of the articles obtained were published after the year 2000, a reflection of the fact that interest in studying the impact of interactions among social, behavioral, physiological, and genetic factors on health continues to increase.

Commissioned Papers

In the statement of task the committee was asked to "develop case studies (e.g., obesity, stress, smoking) that will: demonstrate how the interactions of the social environment and genetics affect health outcomes; illustrate the methodological issues involved in measuring the interactions; elucidate the research gaps; point to key areas necessary for integrating social, behavioral, and genetic research; and suggest mechanisms for overcoming barriers." The committee chose to address this task by obtaining commissioned papers on sickle cell disease and obesity that would focus on the points illustrated in the statement of task. Myles S. Faith, Ph.D., and Tanya V.E. Kral, Ph.D., from the University of Pennsylvania School of Medicine, were identified as the foremost experts with the specialized knowledge necessary to write the commissioned paper on obesity. Dr. Faith and Dr. Kral provided the committee with a paper entitled "Social Environmental and Genetic Influences on Obesity and Obesit-Promoting Behaviors: Fostering Research Integration," which can be found in Appendix C. The committee identified Robert J. Thompson, Jr., Ph.D., from Duke University, as having the necessary knowledge and expertise to prepare the paper on sickle cell disease. He provided the committee with a paper entitled "The Interaction of Social, Behavioral, and Genetic Factors in Sickle Cell Disease," which can be found in Appendix D.

The committee also determined the need for a detailed analysis of genetic interactions and the current state of the science in this area. Sharon Schwartz, Ph.D., at Columbia University's Mailman School of Public Health, was asked to write this paper and provided the committee with a paper titled "Modern Epidemiologic Approaches to Interaction: Applications to the Study of Genetic Interactions," which can be found in Appendix E. Steve Cole, Ph.D., at the University of California at Los Angeles David Geffen School of Medicine, also provided a commissioned paper on immunology that was designed to increase the committee's understanding of the impact of social and genetic variation on immune function and the state of the science of this area.

Information from all four commissioned papers was used to invigorate committee deliberations and enhance the quality of the report.

Public Workshops

The committee held a total of five meetings over the course of the project. The purpose of these meetings was to address the study charge, review the data collected, and develop the report and recommendations. The first three meetings held by the committee included data-gathering

BOX A-2
Open Agenda for Meeting 1: March 28-29, 2005

Institute of Medicine
Committee on Assessing Interactions Among Social,
Behavioral, and Genetic Factors in Health
500 5th St., NW
Washington, D.C. 20001

Open Session: March 28, 2005

10:00	Welcome and Introduction **Dan G. Blazer, M.D.,** *Committee Chair*
10:15	Sponsor Presentation of Charge **Ronald Abeles, Ph.D.,** *OBSSR*
10:45	Discussion and Clarification of Charge

sessions, which were open to the public. These were held on March 28-29, 2005, June 16-17, 2005, and September 29-30, 2005.

In preparation for the data-gathering sessions, the committee discussed areas in which there were gaps in the knowledge of committee members. Once the gaps were identified, the committee developed a set of questions that needed to be answered in order for the committee to adequately address the statement of task. The committee then identified potential speakers with the appropriate level of expertise to address the questions and invited them to participate in open session workshops.

The first committee meeting, held March 28-29, 2005, in Washington, D.C. (Box A-2), included a presentation of the charge to the committee by Ronald Abeles of the Office of Behavioral and Social Sciences Research (OBSSR) and an open discussion of the statement of task with representatives from each of the sponsors from the National Institutes of Health, including Ronald Abeles and Deborah Olster from OBSSR, Colleen Mcbride from the National Human Genome Research Institute, and Brian Pike from the National Institute of General Medical Sciences.

The second committee meeting, held June 16-17, 2005, in Washington, D.C. (Box A-3), was the first of the two open data-gathering sessions. During this meeting, the committee heard presentations from seven speakers who provided overviews of social variables, genetics variables, gene

BOX A-3
Open Agenda for Meeting 2: June 16-17, 2005

Institute of Medicine
Committee on Assessing Interactions Among Social,
Behavioral, and Genetic Factors in Health

Washington Terrace Hotel
1515 Rhode Island Ave., NW
Washington, D.C. 20005
Director's Room (second floor)

Open Session: June 16, 2005

9:30   Welcome and Introductions
            Dan Blazer M.D., Ph.D.
            Committee Chair
            J.P. Gibbons Professor of Psychiatry
            Duke University Medical Center

9:45   Overview of Social Variables and Their Measurement
            Ana Diez Roux, Ph.D.
            Associate Professor of Epidemiology
            Associate Director of the Center for Social Epidemiology and
                Population Health
            University of Michigan School of Public Health

10:05   Conceptualizing Social Variables to Facilitate and Promote
                Gene/Environment Research
            Eileen Crimmins, Ph.D.
            Edna M. Jones Professor of Gerontology and Sociology
            University of Southern California

10:25   Discussion

11:00   Overview of Genetic Variables and Their Measurement
            Sharon Kardia, Ph.D.
            Director, Public Health Genetic Programs
            Associate Professor, Department of Epidemiology
            University of Michigan School of Public Health

11:20	Discussion

11:45	LUNCH

1:00-2:45   PANEL ON EPIGENETICS

1:00   Gene Expression over Time
Ming D. Li, Ph.D.
Associate Professor/START Center Genetic Professorship
Head, Program in Genomics and Bioinformatics on Drug
* Addiction*
University of Texas Health Science Center at San Antonio

1:20   Epigenetic Phenomenon: How to Approach Mechanisms by
    Which Social Variables Influence Gene Expression
Arthur Beaudet, M.D.
Chair, Department of Molecular and Human Genetics
Baylor College of Medicine

1:40   Genetics of Ethnic Populations
Sharon Kardia, Ph.D.
Director, Public Health Genetic Programs
Associate Professor, Department of Epidemiology
University of Michigan School of Public Health

2:00   Implications of Genetics of Ethnic Populations for Common
    Disease
Keith Whitfield, Ph.D.
Associate Professor of Biobehavioral Health
Pennsylvania State University

2:20   Animal Models
John Sheridan, Ph.D.
Professor, College of Medicine and Public Health
Associate Director, Institute for Behavioral Medicine Research
Ohio State University

2:45   Discussion

3:30   Workshop Adjourns

BOX A-4
Open Agenda for Meeting 3: September 29-30, 2005

Institute of Medicine
Committee on Assessing Interactions Among Social,
Behavioral, and Genetic Factors in Health

Keck Building, Room 100
500 5th St., NW
Washington, D.C. 20001

Open Session: September 29, 2005

9:00   Welcome and Introductions
        Dan Blazer, M.D., Ph.D.
        Professor of Psychiatry
        Duke University
        Committee Chair

9:15   Cultural Influences on Health
        Margaret Lock, Ph.D.
        Professor in Social Studies in Medicine
        McGill University

9:45   Discussion

10:00   Effects of Psychological Stress on Health
        Sheldon Cohen, Ph.D.
        Professor of Psychology
        Carnegie Mellon University

10:30   Discussion

11:00   Gene-Environment Interactions: Definitions and Study Design
        Ruth Ottman, Ph.D.
        Professor of Epidemiology
        Columbia University

11:30   Discussion

12:00   Workshop Adjourns

expression over time, epigenetics, genetics of ethnic populations, and animal models.

The third committee meeting, held on September 29-30, 2005, in Washington, D.C. (Box A-4), was the second main data-gathering session open to the public. During this meeting, the committee heard presentations from three speakers who provided overviews of cultural influences on health, the effects of psychological stress on health, and gene-environment interactions. The remaining two committee meetings were closed to the public in order to permit committee deliberation and report writing. They were held in November of 2005 and January of 2006.

B

Recommendation from the National Academy of Sciences/ National Academy of Engineering/ Institute of Medicine Report *Facilitating Interdisciplinary Research*[1]

RECOMMENDATIONS

On the basis of its findings, the committee offers the following recommendations. They are listed by category of people and organizations involved in interdisciplinary research, education, and training. The committee does not necessarily urge interdisciplinary research activities for all institutions and individuals, but, for parties that are interested in implementing or improving such activities, the committee provides the following recommendations.

The majority of the recommendations the committee makes to facilitate interdisciplinary research are "incremental"; however, the committee provides suggestions for "transformative" changes for those institutions willing to experiment with new approaches. Most of these are described briefly here in the section entitled "academic institutional structures," but very specific ideas are provided in Chapter 9 that expand upon these recommendations.

Students

S-1:   Undergraduate students should seek out interdisciplinary experiences, such as courses at the interfaces of traditional disciplines that address basic research problems, interdisciplinary courses that address societal problems, and research experiences that span more than one traditional discipline.

---

[1]These recommendations were developed by the Committee on Facilitating Interdisciplinary Research and were published in NAS/NAE/IOM. 2004. *Facilitating Interdisciplinary Research*. Washington, DC: The National Academies Press.

S-2: Graduate students should explore ways to broaden their experience by gaining "requisite" knowledge in one or more fields in addition to their primary field.

Postdoctoral Scholars

P-1: Postdoctoral scholars can actively exploit formal and informal means of gaining interdisciplinary experiences during their postdoctoral appointments through such mechanisms as networking events and internships in industrial and nonacademic settings.

P-2: Postdoctoral scholars interested in interdisciplinary work should seek to identify institutions and mentors favorable to interdisciplinary research (IDR).

Researchers and Faculty Members

R-1: Researchers and faculty members desiring to work on IDR, education, and training projects should immerse themselves in the languages, cultures, and knowledge of their collaborators in IDR.

R-2: Researchers and faculty members who hire postdoctoral scholars from other fields should assume the responsibility for educating them in the new specialties and become acquainted with the postdoctoral scholars' knowledge and techniques.

Educators

A-1: Educators should facilitate IDR by providing educational and training opportunities for undergraduates, graduate students, and postdoctoral scholars, such as relating foundation courses, data gathering and analysis, and research activities to other fields of study and to society at large.

Academic Institutions' Policies

I-1: Academic institutions should develop new and strengthen existing policies and practices that lower or remove barriers to IDR and scholarship, including developing joint programs with industry and government and nongovernment organizations.

I-2: Beyond the measures suggested in I-1, institutions should experiment with more innovative policies and structures to facilitate IDR, making appropriate use of lessons learned from the performance of IDR in industrial and national laboratories.

I-3:   Institutions should support interdisciplinary education and training for students, postdoctoral scholars, researchers, and faculty by providing such mechanisms as undergraduate research opportunities, faculty team-teaching credit, and IDR management training.

I-4:   Institutions should develop equitable and flexible budgetary and cost-sharing policies that support IDR.

Team Leaders

T-1:   To facilitate the work of an IDR team, its leaders should bring together potential research collaborators early in the process and work toward agreement on key issues.

T-2:   IDR leaders should seek to ensure that each participant strikes an appropriate balance between leading and following and between contributing to and benefiting from the efforts of the team.

Funding Organizations

F-1:   Funding organizations should recognize and take into consideration in their programs and processes the unique challenges faced by IDR with respect to risk, organizational mode, and time.

F-2:   Funding organizations, including interagency cooperative activities, should provide mechanisms that link interdisciplinary research and education and should provide opportunities for broadening training for researchers and faculty members.

F-3:   Funding organizations should regularly evaluate, and if necessary redesign, their proposal and review criteria to make them appropriate for interdisciplinary activities.

F-4:   Congress should continue to encourage federal research agencies to be sensitive to maintaining a proper balance between the goal of stimulating IRD and the need to maintain robust disciplinary research.

Professional Societies

PS-1: Professional societies should seek opportunities to facilitate IDR at regular society meetings and through their publications and special initiatives.

Journal Editors

J-1:   Journal editors should actively encourage the publication of IDR research results through various mechanisms, such as editorial-board membership and establishment of special IDR issues or sections.

Evaluation of IDR

E-1: IDR programs and projects should be evaluated in such a way that there is an appropriate balance between criteria characteristic of IDR, such as contributions to creation of an emerging field and whether they lead to practical answers to societal questions, and traditional disciplinary criteria, such as research excellence.

E-2: Interdisciplinary education and training programs should be evaluated according to criteria specifically relevant to interdisciplinary activities, such as number and mix of general student population participation and knowledge acquisition, in addition to the usual requirements of excellence in content and presentation.

E-3: Funding organizations should enhance their proposal-review mechanisms so as to ensure appropriate breadth and depth of expertise in the review of proposals for IDR, education, and training activities.

E-4: Comparative evaluations of research institutions, such as the National Academies' assessment of doctoral programs and activities that rank university departments, should include the contributions of interdisciplinary activities that involve more than one department (even if it involves double-counting), as well as single-department contributions.

Academic Institutional Structure

U-1: Institutions should explore alternative administrative structures and business models that facilitate IDR across traditional organizational structures.

U-2: Allocations of resources from high-level administration to interdisciplinary units, to further their formation and continued operation, should be considered in addition to resource allocations of discipline-driven departments and colleges. Such allocations should be driven by the inherent intellectual values of the research and by the promise of IDR in addressing urgent societal problems.

U-3: Recruitment practices, from recruitment of graduate students to hiring of faculty members, should be revised to include recruitment across department and college lines.

U-4: The traditional practices and norms in hiring of faculty members and in making tenure decisions should be revised to take into account more fully the values inherent in IDR activities.

U-5: Continuing social science, humanities, and information-science-based studies of the complex social and intellectual processes that make for successful IDR are needed to deepen the understanding of these processes and to enhance the prospects for the creation and management of successful programs in specific fields and local institutions.

C

Social Environmental and Genetic Influences on Obesity and Obesity-Promoting Behaviors: Fostering Research Integration

Myles S. Faith, Ph.D. and Tanja V. E. Kral, Ph.D.[*]

Weight and Eating Disorders Program

SECTION 1: INTRODUCTION

Obesity is one of the most pressing public health disorders in the United States and other westernized societies. Its prevalence is increasing world-wide and it is associated with concerning medical comorbidities, most notably the metabolic syndrome and type 2 diabetes [1-4]. Hence, innovative research that elucidates the causes of obesity has become an increasingly important focus for the National Institutes of Health. A challenge to this mission, however, is that fact that obesity is a "complex disorder." For most individuals in the population, obesity results from multiple genetic and environmental factors that may interact with, or may be correlated with, each other. Genes operate additively and through gene-gene interactions to influence body weight [5].

The topic of genetic and social environmental influences on obesity, and how they interact, is a unique topic for which conceptual frameworks are scarce. Research within each domain appears to have advanced largely within independent "camps," each of which has undergone major advances in the past decade. Research into the genetics of human obesity has become increasingly sophisticated with respect to molecular technologies, biostatistics, and efficient design strategies; however, as illustrated in this report, these studies generally did not measure specific aspects of the social environment. Research into social environmental influences on obesity has expanded its scope

[*]University of Pennsylvania School of Medicine.

of coverage from interpersonal variables to potential consequences of a broader "toxic environment;" however, these studies generally did not collect DNA or use genetically informative designs. Hence, there appears to be room for greater scientific synergy between the domains.

There are two overarching aims to the present report: (a) to review evidence for genetic and social-environmental influences on obesity, respectively, and the types of methodologies used to establish these associations, and (b) to consider opportunities for greater methodological synergy between the two domains. The report strives to foster ideas for new research that bridge genetic and social-environmental research, as they relate to obesity and obesity-promoting behaviors. Conceptual frameworks that posit potential interactions or covariation among genetic and social environmental factors are proposed.

SECTION 2: ORGANIZATIONAL FRAMEWORK OF THIS REPORT

Figure C-1 presents the conceptual framework around which the present report is organized. The model posits that genetic and social-environmental factors promote obesity through their independent influences on intermediary behavioral variables. These intermediary phenotypes may induce a positive energy balance (i.e., greater energy intake than expenditure) that, when sustained, promotes obesity. Although physiological variables are not depicted in the model, they clearly are central to energy balance regulation and the putative behavior phenotypes listed in the figure. The model is intended to reflect much of the current literature, in that correlations or interactions among the social environment and genetic factors are not explicitly posited. However, as reviewed in this report, certain studies challenge this assumption and suggest that expansions of this model may help guide future research. The final section of this report suggests additional research that would test interactions and correlations among genetic and social-environmental variables.

The following section of the report, Section 3, addresses putative social-environmental influences on obesity-promoting behaviors and obesity, corresponding to pathways b and c in Figure C-1. Section 4 addresses evidence for selected refined behavioral traits that have been associated with obesity in some studies, corresponding to the "putative behavioral phenotypes" noted in the figure. Section 5 addresses putative genetic influences on obesity-promoting behaviors and obesity, corresponding to pathways *a* and *c* in the figure. Section 6 addresses evidence for potential interactions among genetic, social, environmental, and behavioral influences on obesity. The data presented in this section challenge the premise that genetic and environmental factors do not interact or cannot influence each other. Section 7 suggests additional research questions and designs that

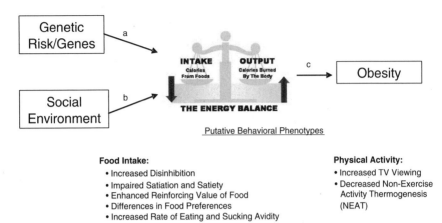

FIGURE C-1   Conceptual model relating genetic and social-environmental factors to obesity. In this figure, the effects of genetic and social-environmental factors, respectively, are posited to operate through putative behavioral phenotypes that promote positive energy balance. Although not depicted, genetic and social-environmental factors are posited to impact on physiological variables as well.

might test new questions concerning the interplay between genes, social environment, behavior, and obesity.

It should be noted that the term "obesity," used throughout this report, was not necessarily measured in the same way across all the reviewed studies. Most studies defined obesity based on the body mass index (BMI; kg/m²), which is a reasonable proxy measure of total body fat, at least in population studies. Guidelines by the National Heart, Lung, and Blood Institute stipulate a BMI between 25.0 and 29.9 as "overweight," and greater than 30.0 as "obese." More refined body composition measures were used in some studies.

Given the range of topics covered in this report, a table of contents for the major report sections and subsections is provided for the reader (Table C-1).

SECTION 3: SOCIAL-ENVIRONMENTAL INFLUENCES ON OBESITY AND OBESITY-PROMOTING BEHAVIORS

For the purposes of this report, a broad definition of "social environment" is used. Specifically, as defined by Barnett and Casper [6], "Human social environments encompass the immediate physical surroundings, social relationships, and cultural milieus within which defined groups of

TABLE C-1 Organizational Sections of Summary Report and Accompanying Pages

Section Number and Topic	*Starting Page Number*
1. Introduction	236
2. Organizational Framework of This Report	237
3. Social-Environmental Influences on Obesity and Obesity-Promoting Behaviors	238
3a. Macroenvironmental Influences	240
3b. Microenvironmental Influences	242
4. Refined Behavioral Traits Associated with Obesity	245
4a. Eating Traits	245
4b. Physical Activity and Sedentary Behavior	251
5. Genetic Influences on Obesity and Obesity-Promoting Behaviors	253
5a. Genetic Influences on BMI and Fat Mass	253
5b. Genetic Influences on Food Intake	257
6. Evidence for Interactions Among Social Environmental, Genetic, and Behavioral Factors as They Relate to Obesity	262
6a. Social Environment as a Potential Moderator Variable	263
7. Opportunities for Future Research That Would Enlighten Relationships Between Genetics and the Social Environment	266
8. Conclusion	272

people function and interact. Components of the social environment include built infrastructure; industrial and occupational structure; labor markets; social and economic processes; wealth; social, human, and health services; power relations; government; race relations; social inequality; cultural practices; the arts; religious institutions and practices; and beliefs about place and community. [. . .] Social environments can be experienced at multiple scales, often simultaneously, including households, kin networks, neighborhoods, towns and cities, and regions."

This section reviews evidence for potential social-environmental influences on obesity and obesity-promoting behaviors, corresponding to paths b and c in Figure C-1. The social-environmental variables include two "macroenvironmental" variables and two "microenvironmental" variables. Macroenvironmental factors operate across larger communities or populations, specifically, exposure to components of the "toxic environment" and socioeconomic status (SES); "microenvironmental" factors, on the other hand, refer to smaller groups of individuals or family members, specifically, the "social facilitation" of overeating that occurs in group settings and parent-child feeding dynamics. The social-environmental variables reviewed below are not necessarily independent of each other, but are presented individually for ease of presentation.

3a. Macroenvironmental Influences

The two macroenvironmental factors reviewed below are (i) exposure to the "toxic environment" and (ii) SES. These particular factors are reviewed because there is a reasonable database providing information on these variables and because of their potential relevance for obesity prevention.

i. Exposure to the "Toxic Environment"

Brownell coined the term "toxic environment" [7, 8], referring to a pervasive series of social and economic changes that have occurred in the United States during in the past several decades. Brownell argues that these changes have caused the rising obesity prevalence, even though strong causal inferences cannot be easily made from these observational trends. These changes are outlined in detail elsewhere [9-12], but include the increased portion sizes and the "super-sizing" of commercially available foods, the proliferation of fast-food restaurants, the reduced cost of fast-food products, the increasing access to energy-dense foods in schools, the increased use of labor saving devices that reduce physical activity, and reduced opportunities for physical activity in schools and at safe playgrounds.

Data have been published that are consistent with the notion that some of these changes may have contributed to the rising obesity prevalence. As reviewed elsewhere [13], for example, data on national food supply and utilization from the U.S. Marketing System indicate that the overall energy availability per capita in the United States increased by 15 percent between 1970 and 1994, a period during which there was also an increase in per capita availability of dietary fat, increased consumption of added fats (commonly found in snack or confectionary foods), reduced milk intake, and increased soft-drink intake. During this period, there was an increased number of households with two or more television sets, home video recorders, and home computers.

Despite these findings, several caveats are warranted. First, although these aforementioned findings are *consistent* with a causal influence (i.e., pathways b and c in Figure C-1), evidence for a causal relationship per se is limited [13]. Much of the evidence comes from observational studies that could not control for potential confounding factors or did not directly test associations between participant weight status and exposure to putative environmental risk factors. Second, specific aspects of the "toxic environment" that have the greatest impact on obesity are unknown [13]. Third, findings from certain studies did not support expected predictions. For example, in a cohort of over 7,000 children who were 36 to 59 months of age and from low-income families, child obesity status was not associated with access to playgrounds, proximity to fast-food restaurants, or neighborhood crime level [14].

Finally, it has not been tested whether exposure to the toxic environment is related to genotype. That is, individuals with obesity-predisposing genes may be particularly responsive to the effects of such a "toxic" environment. In addition, certain individuals may be more likely to seek out or expose themselves to aspects of the toxic environment. The topic of gene-environment correlations as a topic for additional research is discussed further in Section 7.

ii. Socioeconomic Status (SES)

Several studies (e.g., [15-17]) have documented an inverse relationship between SES and obesity in previous years. In a recent review, Ball and colleagues [18] examined 34 articles to test the hypothesis that persons from lower SES strata are at increased risk of weight gain. Their hypothesis was supported for predominantly non-African American samples, but not for African American samples. Reviewing relevant studies, they found little support for a relationship between SES and weight gain among African Americans. In contrast, depending on the particular indicator for SES that was used (i.e., occupational status, education, and income), they found that lower SES was associated with an increased risk of weight gain in non-African American individuals. Specifically, the authors found an inverse association between occupational status and weight gain for men and women. When SES was assessed using education as the indicator, the relationship became less strong (particularly among men). Using income level as the particular indicator for SES, findings for associations between weight gain and SES were inconsistent for both men and women. Finally, the authors noted a differential rate of weight gain by SES and attributed that finding to an early onset of weight gain in a person's life, when parental SES may still be influential.

Prospective analyses of the National Longitudinal Survey of Youth [19] found that children from lower SES families were more likely to have been overweight during the prior year than children from higher SES families. Negative associations between obesity status and household income and parental education were found even when controlling for ethnicity and other demographic variables.

Several mechanisms could underlie the link between low SES and obesity. Factors such as limited access to resources, poor knowledge of nutrition and health, increased exposure to fast-food outlets, and limited physical activity due to deprived or unsafe neighborhoods [20, 21] have been suggested to influence energy intake and energy expenditure and, consequently, body weight. For instance, in an ecological study of 267 postal districts in Melbourne, Australia, families living in the poorest SES strata had 2.5 times the exposure to fast-food outlets and thus increased access to relatively inexpensive, calorically dense foods compared to families from the wealthiest SES strata [22].

The relationship between SES and obesity may also be influenced by differential costs of less or more nutritious foods. For instance, in a series of elegant analyses, Drewnowski documented that the cost of healthy, nutrient-dense foods such as fruits and vegetables were reliably more expensive than more energy-dense, less nutritious foods [23-25]. Possibly for this reason, the availability of fruits and vegetables in adolescents' homes was shown to be greater among families from high compared to low SES strata [26]. These data suggest that families from lower SES strata have overall fewer monetary resources to purchase more nutrient-dense, healthy foods [23, 25, 27].

Reduced access to recreational facilities or parks in deprived neighborhoods also may contribute to diminished energy expenditure and thus increased body weight in individuals of lower SES [28].

In summary, lower SES may contribute to the onset of obesity in that it provides an environment which promotes the intake of calorically dense foods while it reduces the need or the opportunity for physical activity.

3b. Microenvironmental Influences

The two microenvironmental influences reviewed in this section are social facilitation of eating and parental feeding practices. These particular factors are reviewed because there is a reasonable database providing information on these variables and, in regards to feeding practices, because of its potential relevance for obesity prevention.

i. Social Facilitation of Eating

There is reliable evidence that total energy intake at meals is increased significantly when eating in the presence of other people, a phenomenon termed "social facilitation" [29]. This phenomenon would be represented by pathway b in Figure C-1. De Castro [30] studied 63 adults who maintained a 7-day continuous food diary and recorded the number of people present at each meal. Results indicated that energy intake during meals that were eaten alone was significantly lower compared to energy intake during meals that were consumed in the presence of others. This was observed for total energy intake (410 vs. 591 kcals), carbohydrate intake (190 vs. 241 kcals), fat intake (157 vs. 230 kcals), and protein intake (65 vs. 100 kcals). Satiety ratings were 30 percent greater following meals eaten with others compared to meals eaten alone.

Additional analyses of de Castro's data indicated that the social facilitation effect was greater for meals consumed in the presence of a spouse, family member, or friend compared to less familiar or unknown companions, suggesting that enhanced social interactions and discussions were the underlying mechanisms [31]. Indeed, de Castro and de Castro [30] argued that physiological signals that relate to appetite and meal size can be overridden by social interactions. Specifically, they found that reported total energy intake at meals was positively correlated with time since prior meal consumption, but only for meals eaten alone. When others were present at meals, there was no longer a significant association, suggesting that postprandial meal regulation may be "disrupted by the presence of other people" (p. 246).

Laboratory studies have also demonstrated this social facilitation phenomenon. Edelman et al. [32] showed that overweight and normal-weight subjects consumed more lasagna when eating in groups of 4 or 5 persons compared to when eating alone, and that there was no significant difference between the weight groups in terms of this phenomenon. Klesges et al. documented the social facilitation effect in a restaurant setting, with the effect being more pronounced for women than men. Kimm and Kissileff [33] also demonstrated the social facilitation of eating in a cafeteria setting.

The mechanism underlying social facilitation of eating has been termed "time-extension" [29, 34] and has received the most empirical support. Specifically, the presence of people at a meal serves to lengthen meal time which, in turn, promotes further energy intake. The point is important to the present paper because, as presented in Section 5, there is evidence that the tendency to eat with others may be genetically influenced. Thus, the fact that some individuals are more likely to eat in the presence of others may not be a random event; rather, eating in the presence of others may be a trait that is influenced by genes that indirectly promote social facilitation of eating at meals.

ii. Parental Feeding Practices: Breast-Feeding vs. Bottle-Feeding

An area of active research concerns parental feeding practices and parent-child feeding dynamics that might promote a positive energy balance and overweight in young children. Review of this literature reveals two specific feeding practices that are prospectively associated with increased body weight and weight gain in infants and children. These practices are, first, bottle-feeding as opposed to breast-feeding, and, second, parental use of restrictive child feeding practices. With respect to breast-feeding practices, prospective epidemiology studies have shown that childhood and adolescent obesity rates were reduced among infants who were breast-fed as opposed to never breast-fed [35] and among infants who were breast-fed for longer compared to shorter durations [36, 37]. In one seminal study, the prevalence of overweight was studied in 8,186 girls and 7,155 boys, 9 to 14 years of age, who were participating in a national growth and development study [38]. Among children who were mostly or exclusively breast-fed during the first 6 months of life, compared to children who were mostly or exclusively formula-fed, the odds ratio for being overweight was 0.78. This held true when controlling for maternal BMI and other variables reflecting SES and lifestyle activities. It should be noted that not all studies replicated this significant association [39], and that one study found the association to be true in non-Hispanic white families but not African American families [40].

The mechanisms for the apparent protective effect of breast-feeding on overweight development were unknown, although recent data implicate parental feeding patterns as a possible factor. Specifically, mothers who breast-fed their infants were less restrictive in their feeding practices (as measured by self-report questionnaire) than mothers who bottle-fed their infants [41]. As discussed in the next section, restriction of child eating may impede a child's ability to self-regulate food intake and instead teach a child to eat in response to external cues [42]. Whether or not this is the actual mechanism needs to be clarified in future research.

iii. Parental Feeding Practices: Restrictive Feeding Practices

An extensive literature has examined which parental feeding practices, if any, are associated with increased child food intake during meals and increased weight status [43]. Investigators have measured feeding practices by parent-report questionnaires, direct observation, or analysis of videotapes, with the most common assessment tool being the parent-report Child Feeding Questionnaire [44]. A recent review of this literature concluded that, across the range of parental feeding domains that have been studied, only restriction of child eating was consistently associated with increased child total energy intake and weight status [43]. Parents who restrict their children's access to foods tend to have heavier children. No other feeding domains were associated with childhood obesity, including use of food to calm infants and children, feeding on schedule, pushing child to eat more, and provision of structure during feeding, or using food as a reward [45, 46].

Several mechanisms by which parental restriction may promote increased child energy intake and body weight have been proposed. First, restrictive feeding practices may impede on a child's ability to adhere to internal hunger and satiety cues (i.e., impaired self-regulation) and thereby teach children to eat in response to external cues (e.g., portion size, time of day). Among preschool children, the ability to self-regulate food and energy intake across meals was poorer among children whose parents reported elevated efforts to control child eating [42]. Second, restricting children's access to foods may have the counterproductive effect of making those "forbidden" foods more desirable [47]. Third, restriction of foods may teach children to eat in the absence of hunger, that is, to continue eating despite being full when food is available [48].

At the same time, the body of evidence suggests that parental restriction of child eating is elicited, at least in part, by a child's increased body weight [43, 49]. Indeed, in one study, the association between restrictive feeding practices and increased child weight gain was only seen in children who were born at high risk for obesity [49]. As in other realms of child development, there appears to be a bidirectional association such that parental restriction of child eating partially is *elicited* by child's weight, which in turn may exacerbate further child weight gain. This also suggests a possible gene-environment correlation such that genes and environmental conditions that promote childhood obesity are interrelated. The topic of gene-environment correlations is discussed in Section 7.

4. REFINED BEHAVIORAL TRAITS ASSOCIATED WITH OBESITY

This section reviews refined behavioral traits that have been associated with obesity in cross-sectional or prospective investigations. As such, it addresses the putative behavioral phenotypes listed in Figure C-1. Obesity results from an imbalance between energy input and energy output. The daily energy surplus that is necessary to promote weight gain is small; specifically, Hill et al. [50] estimated that a sustained daily energy surplus above a person's daily energy requirements as small as 100 kcal/day is sufficient to promote weight gain. For this reason, it is desirable to identify refined behavioral traits that are related to positive energy balance and obesity. Identifying such intermediary traits may help elucidate the pathways through which the social environment and/or genes promote obesity.

4a. Eating Traits

In the 1970s and early 1980s there was much interest in identifying an "obese eating style" [51-57] which differentiates lean and obese individuals' eating behavior. It has been argued that intraindividual differences in various eating behaviors may underlie the disparity in energy intake and body weight among both groups. In light of the recent obesity epidemic, the search for distinctive patterns of food intake among individuals with differing body sizes continues to be of great importance. Following is a description of selected eating traits which may represent behavioral phenotypes of obesity.

i. Externality and Dietary Disinhibition

During the late 1960s results from a series of experiments conducted by Schachter and colleagues [58-60] suggested that the eating behavior of obese individuals is greatly influenced by the immediate (food) environment. In particular, the eating behavior of obese individuals was believed to be controlled by external cues related to the perception of time, taste and sight of food, and the number of highly palatable food cues present [53, 60, 61], rather than by internal physiological cues of hunger.

Subsequent studies [62, 63] failed to replicate consistent differences between lean and obese individuals in their responsiveness to external food-related and non-food-related cues. These studies found large intraindividual variability among individuals across all weight groups in their response to external cues. However, this early research on "external eating" developed into a more promising line of research on the trait of dietary "disinhibition."

Disinhibition refers to the loss of self-imposed cognitive control of eating behavior in response to external or emotional stimuli, and is the behavioral trait that most consistently differentiates between obese and nonobese individuals [64]. Obese subjects show greater disinhibition scores than do nonobese individuals [65, 66] and degree of disinhibition is strongly associated with energy intake [64, 67], weight status and weight gain [68, 69], weight fluctuations [65], binge eating [70], and body fat [71].

In summary, dietary disinhibition, a characteristic that associated with external eating, may represent a behavioral phenotype which is relevant to obesity and obesity-related traits.

ii. Impaired Satiation

In recent years there has been much debate over whether obesity is the result of impairment in the regulation of energy intake. One way to study food and energy intake in individuals is to examine satiation (or intrameal satiety). Satiation refers to the process leading to the termination of eating. It is assessed by measuring food and energy intake during a single meal which subjects consumed ad libitum.

To date only a limited number of studies is available that investigated the effects of dietary manipulation on satiation in both normal-weight and overweight/obese subjects. A study conducted by Bell and Rolls [72] was designed to examine the effects of energy density across three levels of dietary fat on intake in both lean and obese women. Results demonstrated that the energy density of the meals significantly affected subjects' energy intake across all levels of dietary fat. The response to the dietary manipulation was similar between lean and obese women. All women consumed approximately 20 percent less energy in the condition of low energy density compared to high energy density.

Likewise, studies which examined the effects of varying the portion size [73, 74] or the portion size and the energy density of food [75] on subjects' ad libitum intake found no significant difference in the eating response of lean and obese individuals. Both groups consumed significantly more energy when the portion size or the portion size and the energy density of food were increased. A longitudinal study [76] conducted in children analyzing nutritional data from nationally representative databases (i.e., CSFII 94-96; NFCS 77-78) found that portion sizes of commonly consumed foods were positively related to children's energy intake and body weight.

As outlined above, laboratory studies for the most part failed to detect significant differences between lean and obese individuals in their response to the dietary manipulation of the energy density and/or portion size. One of the great difficulties in accurately assessing food intake in obese populations has been their altered eating behavior when being monitored. As several studies on self-reported food intakes have indicated, obese individuals underreport their intakes to a greater extent than do lean individuals [77]. The measured energy intakes of obese subjects in a controlled laboratory setting may likewise be compromised by the fact that their food intake is being monitored.

Despite these null findings, there is some evidence that when self-selecting their diets obese individuals tend to consume overall greater amounts of foods that are higher in energy density than do their lean counterparts. In a study conducted by Westerterp-Plantenga et al. [78] obese women reported consuming larger portions and an overall greater percentage of their total energy intake from foods that are higher in energy density than did lean women.

There is some evidence [79] of a difference in the pattern of cumulative intakes within a meal between lean and obese individuals. While lean individuals showed a decrease in their eating rate over the course of a meal, obese and latent obese as well as restrained subjects [80] failed to do so. The authors suggested that this difference in the pattern of cumulative intakes over the course of a meal may indicate that lean and obese individuals experience satiation differently.

Recent findings from neuroimaging studies confirmed intrameal differences between lean and obese individuals. It has been shown that the hypothalamic response following glucose ingestion was significantly delayed (~4-9 min) in obese individuals compared to their normal-weight counterparts [81]. These findings suggest that obesity may be associated with an abnormal neuronal activity in certain regions of the brain [82], some of which are believed to cause a delayed response in satiation over the course of a meal.

In summary, the finding of a potentially delayed satiation in obese individuals is of interest in that it may point to differences in the experience of hunger and fullness between lean and obese individuals. Innovative research designs need to be developed to further study satiation as a possible phenotype for obesity.

iii. Impaired Satiety

Another approach to examine energy intake regulation among individuals is the study of satiety. Satiety, defined as the effects of a food or a meal after eating has ended [83], can be studied by administering a fixed amount of a given food or nutrient (preload) and, after a predetermined delay, measure its effects on subsequent intake (test meal).

Among adults, there is conflicting evidence that obese individuals experience satiety differently and compensate for energy less accurately than do lean individuals. Data generated from an experiment that was designed to compare effects of carbohydrate and fat on eating behavior in lean and obese individuals [84] suggest that obese restrained females show a relative insensitivity to the satiating power of fat in that they did not adjust their energy intake as well as did their lean counterparts after the ingestion of a high-fat preload. Outcomes from other investigations [85, 86], however, failed to detect differences in caloric compensation (i.e., satiety) among individuals with differing body sizes.

Studies have found that young children have the ability to adjust food intake at test meals in response to preloads, although compensation often is incomplete and differs between children. Johnson and Birch [42] found that children with poorer caloric compensation abilities tended to be heavier than children with better compensation abilities. On the other hand, other studies have failed to detect this same association in young children [87]. Thus, whether or not this trait reliably relates to a child's proneness for obesity remains to be further investigated.

In summary, the degree to which an individual is able to compensate for energy may represent an eating trait that distinguishes the lean from the obese. It is possible that a predisposition for obesity moderates developmental changes in compensation ability as environmental factors start to override internal feelings of hunger and satiety.

iv. Increased Reinforcing Value of Food

The reinforcing value of food can be defined as the extent to which an individual will work for a given food or food group when an alternative commodity (e.g., money) is concurrently available. Typically assessed on a computer keyboard that required "bar presses" on the keyboard, the reinforcing value of food represents the highest amount of work (i.e., bar presses) an individual will emit to earn access to food. Thus, the measure represents "drive" or hedonic motivation for foods (i.e., food reward). The paradigm is based on behavioral economics theory, which builds upon an extensive animal literature and research in the additions [88]. In a series of controlled studies, Epstein and colleagues have found that obese individuals score higher on measures of food reward than nonobese individuals [89-91]. This trait has proven to be one of the more consistent behavioral phenotypes that relates to weight status and, as described in Section 6, has even been linked to specific genes related to dopamine pathways.

v. Differences in the Eating Style

In 1962, Ferster and colleagues put forward the idea that obese individuals take larger bites and eat faster than do normal-weight individuals and that the obese would eat less if they ate more slowly [92]. Subsequent experiments tested potential differences between lean and obese individuals in their eating style, including rate of eating, bite size, and the amount and rate of chewing.

Early work by Dodd et al. [52] found that obese individuals ate more, ate at a faster rate, and took in larger bites than nonobese individuals. Some investigators [93] confirmed that obese individuals ate faster than lean individuals, however, others [57, 94] did not replicate this finding. The conflicting outcomes may have been due to methodological issues related to how the rate of eating during a meal was manipulated, as well as the failure to control for meal size in early studies.

An interesting finding has been the difference in the rate of sucking in infants who were born at high or low risk for obesity based on maternal pre-pregnancy BMI. That is, at 3 months of age, infants born at high risk for obesity displayed greater nutritive sucking rates on an artificial nipple than did infants born at low risk for obesity [95]. Moreover, among all infants, increased sucking rate was predictive of increased weight gain during the first two years of life [95, 96].

In adults, the rate of eating appears to be related to food consumption in both obese and nonobese individuals. Spiegel et al. [97] tested the effects of bite size on ingestion rate, satiation and meal size, and found that decreasing the bite size of test foods was associated with a lower ingestion rate for the whole meal. Interestingly, this decrease in the rate of eating was offset by an increase in meal duration such that overall meal sizes did not differ across conditions. This result was found true for both lean and obese individuals.

Spiegel [98] gave lean and obese men access to a buffet-style meal during which they could choose between different flavors, kinds of foods, and make their own sandwiches. Results showed that lean and obese subjects did not differ significantly in their average bite size of different foods, local ingestion rate (g ingested/min), chew efficiency (g ingested/chew), and chew frequency (chews/s). However, obese men consumed more energy per minute than lean men, a difference that was due to the higher energy density of the foods consumed by obese subjects, in particular the greater energy density of the sandwiches. Thus, there may be an interrelation between a greater rate of eating and the tendency to eat more energy-dense foods among obese individuals. Lean and obese individuals may respond similarly to the physical properties of foods, while they may be differing in the food preferences and food choices they make, which can promote a positive energy balance.

vi. Potential Differences in Food Preferences

While there has been much interest in identifying distinct taste qualities that are more or less preferred by lean versus obese populations, these studies have yielded mixed results. For instance, studies conducted to examine differences in taste preferences between lean and obese individuals [99, 100] suggested that obesity is associated with an overall heightened preference for high-fat stimuli. Others, however, could not confirm these findings [101, 102] in that they failed to find differences between lean and obese individuals for overall pleasantness scores or liking for foods with different predominant taste qualities.

Another difficulty is to find evidence for the conception that sensory preferences influence food choices in both lean and obese individuals. Epidemiologic data have documented a positive relationship between weight status and dietary fat intake [103, 104]. Based on a comprehensive review of animal, epidemiological, and clinical studies exploring the relation between fat intake and obesity, Bray and Popkin [105] concluded that dietary fat is an important contributor to obesity in certain individuals. Likewise, data from a cross-sectional study [106] using food frequency questionnaires indicated a positive association between the consumption of red meats, fish, oil, poultry, eggs, fats, oils, and condiments and BMI and a negative association between the consumption of legumes, soy, tofu, fruit juice, cold cereals, and vegetables and BMI.

In summary, the consumption of certain foods and/or macronutrients (i.e., dietary fat) has been associated with increased weight status, however, this relationship may not hold true for all individuals. In general, it has been difficult to establish clear associations between weight status and intake of single foods or food groups [107].

vii. Influence of Fast-Food Consumption

Fast-food restaurants and the promotion of predominantly calorically dense, relatively inexpensive foods, are considered by some to be the cornerstones of the "toxic" environment. That is, increased consumption of fast foods has been associated with increased weight status. Among 891 adults enrolled in the "Pound of Prevention Study," greater frequency of eating fast foods was significantly associated with higher total energy intake, higher percent fat intake, more frequent consumption of hamburgers, french fries, and soft drinks, and less frequent consumption of fiber and fruit. Over 3 years, each additional fast-food meal/week was associated with an excess weight gain of 0.72 kg beyond the average weight gain observed during that period. In a prospective study of over 3,000 young adults enrolled in the Coronary Artery Risk Development in Young Adults (CARDIA) study, frequency of fast-food restaurant visits at baseline (visits/week) predicted excess 15-year weight gain and worsening of insulin resistance in Caucasian and African American respondents [108].

Visiting fast-food restaurants may promote obesity by promoting increased consumption of energy-dense foods. Prentice and Jebb [109] reviewed the nutritional content of the foods sold at three popular fast-food outlets. The average energy densities for the three menus were 1.7-fold greater than the average British diet. Other potential mechanisms by which fast-food restaurants may promote a positive energy balance include the increased portion sizes of foods (e.g., "super-sizing").

In conclusion, fast-food establishments may put certain individuals at an increased risk for the overconsumption of calories. As noted in Section 7, additional research is needed to test whether certain obesity-promoting genotypes moderate this association (i.e., a gene-environment interaction) or whether individuals with certain genotypes may be more likely to seek out such restaurants (i.e., a gene-environment correlation).

4b. Physical Activity and Sedentary Behavior

Living in the modern-day environment has decreased the need for individuals to be physically active. Decreased physical activity, and thus decreased energy expenditure, has been negatively associated with BMI [110-112] and maintenance of weight loss [68, 113]. The following two sections will highlight two activity-related activities in particular. One is television viewing and its association with weight status, the other is nonexercise activity thermogenesis (NEAT).

i. Television Viewing

Increased television viewing (TVV) has been associated with increased energy intake and body weight [114]. The mechanisms for this association are multifold. For one, increased TVV increases sedentary behavior which in turn is likely to displace time spent in physical activity.

Second, TVV also provides a setting during which food, especially energy-dense snack foods, can be consumed. A study conducted by Francis et al. [114] showed that TVV viewing was associated with increases snack food consumption in girls who were 5, 7, and 9 years old which in turn predicted girls' increase in BMI from age 5 to 9. Thus, TVV has been shown to be a risk factor for excessive snack consumption and in turn increased weight status, especially for those individuals who are predisposed for obesity. Another study [115] also demonstrated that physical activity (negatively associated) and TVV (positively associated) were the only significant predictors, beyond baseline BMI, of BMI in children between the ages of 3 and 4 years during a 3-year study phase.

A third mechanism by which increased TVV may lead to increased energy intake may involve the increased exposure to food advertising. A study conducted by Henderson and Kelly [116] was designed to analyze the content of food advertising appearing on either general market or African American TV programming. The results of the study showed that African American TV programs included more food advertisements, more advertisements for unhealthy foods such as fast food, candy, soda, or meat, and made more weight-related claims and those related to the fat content of foods compared to advertisements that appeared in general market television. Thus, food advertisements seem to be targeted at and tailored to specific populations to increase product sales.

In summary, television viewing has been associated with increased energy intake and weight status among individuals. It remains to be further investigated whether TVV is a behavior that, through its association with sedentarianism, may be fostered through an individual's biology, as the following section on NEAT alludes to.

ii. Nonexercise Activity Thermogenesis (NEAT)

NEAT has been defined as energy expenditure that is associated with daily activities such as sitting, standing, walking, and talking and as such is different from purposeful, planned physical activity [117]. NEAT can further be divided into thermogenesis that is associated with posture (standing, sitting, and lying) and that associated with movement (ambulation). Research conducted by Levine and colleagues [117] has shown that obese individuals, on average, were seated longer per day and spent less time in an upright position, compared to lean individuals. Overall this difference accounted for an additional energy expenditure of 352 calories per day, on average, for lean individuals. Interestingly, the difference in NEAT was not due to the differential body weights of the study participants per se, but seemed to be inherent in an individual's biological/genetic makeup. That is, even after the study team had obese subjects lose a considerable amount of weight (8 kg) and had lean individuals gain weight (4 kg), the two subject groups did not change their original posture allocation. These data suggest that interindividual differences in posture allocation (i.e., NEAT) may be genetically determined. The authors of the respective research [117] speculate that "(. . .) obese and lean individuals respond differently to the environmental cues that promote sedentary behavior" (p. 586). This type of research, again, provides a rich ground to further integrate more genetic-based research with studies on the social environment.

5. GENETIC INFLUENCES ON OBESITY AND OBESITY-PROMOTING BEHAVIORS

This section reviews evidence for genetic influences on obesity and obesity-promoting behaviors, corresponding to paths b and c in Figure C-1. The section is divided into two subsections. Subsection 5a examines evidence for genetic influences on BMI and body fat measure, while subsection 5b examines evidence for genetic influences on obesity-promoting behaviors related to food intake (i.e., pathway b in Figure C-1). Within each subsection, data are presented for studies that estimate heritability of the phenotype, followed by studies that tested the influence of specific genes or genomic regions.

5a. Genetic Influences on BMI and Fat Mass

i. Heritability of BMI and Fat Mass

"Heritability" refers to the extent to which variability in a trait is influenced by genetic variations within a population, and can be subdivided into "narrow-sense" or "broad-sense" heritabilities [118]. The former refers solely to additive genetic influences on the trait, whereas the latter refers to nonadditive interactions among genes. Beyond heritability, the remaining variance in weight status is due to environmental influences which can be partitioned into "shared environment" or the "nonshared environment" influences. Shared environment refers to aspects of the home environment that are perfectly shared by siblings from the same home (e.g., food in the home cupboards, the number of television sets at home). The nonshared environment refers to those aspects of the environment that are uncorrelated among siblings (e.g., differential interactions with parents or peers, or differential life experiences). Apropos to this report, specific aspects of the environment are rarely directly measured in behavioral genetics studies of obesity. Consequently, investigators generally could not test the influence of putative social-environmental factors when modeling genetic influences on fat mass.

Heritability of BMI has been estimated from a variety of designs, but most commonly from twin and adoption studies that compared phenotypic correlations for body fat between groups of individuals varying in genetic relatedness. To the extent that fat mass is genetically influenced, phenotypic correlations will be greater among individuals who are more genetically similar (e.g., monozygotic, MZ, twins) than less genetically similar (e.g., dizygotic, DZ, twins). Using biometric statistical models, heritability is estimated along with the magnitude of shared and nonshared environmental factors.

Maes et al. [118] put forward a most comprehensive review of this literature, the results of which provide indisputable support for a heritable component to BMI and fat mass. Heritability estimates fall in the range of 20 to 80 percent when estimated from family studies that compared parent-child and sibling correlations, 20 to 60 percent when estimated from adoption studies, and 50 to 90 percent when estimated from twin studies. Although the heritability estimates vary sizably, the most accurate estimates arguably come from twin studies, which have methodological strengths over other designs [118]. Also, studies of twins reared apart (i.e., twins separated during childhood and therefore not exposed to the same home environment) generally yield some of the higher heritability estimates. Stunkard et al. [119] conducted the first study of twins reared apart, the results of which estimated heritability at ~65 to 75 percent for BMI. Allison et al. [120] obtained comparable heritability estimates when pooling an international sample of twins reared apart from seven countries. Finally, the inconsistent results that have been reported across studies were likely due to small sample sizes.

Results of longitudinal behavior genetic studies suggest that there are age-specific genetic effects on BMI, such that different obesity-promoting genes may become active at different ages across the lifespan. This has been documented throughout the lifespan [121, 122]. Thus, although some genes exert a consistent influence over time and are partially responsible for the "tracking" of BMI, other genetic influences may appear at different stages of a child's development. Apropos to the theme of this paper, considering developmental milestones may be especially important for future studies testing the interplay of genetic and social-environmental influences on obesity.

Beyond heritability, unmeasured genotype studies have provided clues into the nature of environmental influences on obesity. First, most studies find no evidence for shared home environmental influences on fat mass variability in adulthood. Rather, most environmental influences on fat mass appear to be of the nonshared variety. In regards to this report, efforts to identify social-environmental influences on weight status might focus on unique life experiences that are unshared among family members. Second, among the few studies finding evidence for shared environmental influences on weight status, most examined pediatric samples. For example, in an analysis of over 3,500 twin pairs who were 4 years old, shared environmental factors accounted for 24 percent of variance in weight adjusted for height in boys and 25 percent of the variance in girls [123]. Jacobson and Rowe [124] reported shared environmental influences on BMI in white adolescent females, but not in African American adolescent females or adolescent males who were white or African American. Specific aspects of the shared environment were not tested in these studies.

ii. Specific Gene Associations with BMI and Fat Mass

The "Human Obesity Gene Map" is the most comprehensive annually updated compendium of specific genes that have been associated with obesity and obesity-related phenotypes (e.g., physiological and metabolic measures, peptides and hormones, and behavioral traits) [125]. Initially published in 1994, the Human Obesity Gene Map summarizes evidence from the following classes of human studies: (a) obesity due to single-gene or digenic mutations, (b) obesity associated with Mendelian disorders, such as Prader-Willi syndrome or Bardet-Biedl syndrome, (c) "association studies" that test whether candidate genes are associated with obesity phenotypes among samples of unrelated participants, and (d) "linkage studies" that test for causal associations between genomic regions and obesity phenotypes in cohorts of families. Within each of these categories, studies have been methodologically heterogeneous with respect to participant characteristics, sample sizes, phenotype measurements, and data analytic strategies.

From this voluminous literature, several broad conclusions about specific gene effects on weight status can be made. First, the number of genes shown to be statistically associated with fat mass and obesity-related traits increased dramatically over the past decade. This is true across each of the aforementioned study categories, as summarized in Table C-2. Thus, the platform of specific genes that might contribute to obesity is very large and involves loci throughout the genome. Similarly, the number of genes that might interact with the social environment could also be large.

Second, the failure to replicate positive associations has been a common occurrence. For example, although 204 genomic regions for obesity-related phenotypes were identified from 50 genome scans in the 2004 report, replication of positive findings was found for only 38 genomic regions. Similarly, although there were 358 significant associations with 113 candidate genes from association studies, only 18 positive associations were

TABLE C-2 Evolution in the Status of the Human Obesity Gene Map[a]

	1994	1995	1996	1997	1998	1999	2000	2001	2002	2003	2004
Single-gene mutations[b]				2	6	6	6	6	6	6/7	10
KO and Tg									38	55	166
Mendelian disorders with map location	8	12	13	16	16	20	24	25	33	41	49
Animal QTLs	7	9	24	55	67	98	115	165	168	183	221
Human QTLs from genome scans				3	8	14	21	33	68	139	204
Candidate genes with positive findings	9	10	13	21	29	40	48	58	71	90	113

[a]The growing number of genes or genomic regions associated with BMI, obesity, or obesity-related phenotypes during the past 10 years, as indexed in the Human Obesity Mene map publications. Source: [125].
[b]Number of genes, not number of mutations.

replicated across five studies. Probable reasons for nonreplication including population stratification (i.e., the combination of individuals from different racial/ethnic backgrounds), publication bias, Type 1 errors, and insufficient statistical power [126].

Third, among the phenotypes investigated that were not body composition measures, the vast majority were metabolic or physiological measures rather than measures of food intake, appetite, or food preferences. Thus, as described below, behavioral measures have been largely unrepresented in genotype studies and this may represent an opportunity for future research.

Fourth, single-gene mutations likely account for a small percent of the cases of human obesity in the general population. For most obese individuals, obesity likely results from the influence of multiple genes on different chromosomes that work additively and through gene-gene interactions [5]. Among the cases of monogenic obesity reported in the literature, most have been related to mutations in the melanocortin 4 receptor (MC4R) gene [125].

Finally, for most individuals in the population, the specific physiological mechanisms by which genes influence obesity probably involve both energy intake and expenditure pathways. A detailed discussion on this topic is beyond the scope of this report, which is more geared towards behavioral phenotypes, but is provided elsewhere [127-129]. Several genes have been implicated in the regulation of energy expenditure, including those related to mitochondrial uncoupling proteins (i.e., UCP1, UPC2, and UCP3 genes), the adrenergic systems (i.e., β_2-AR and β_3-AR genes), and the growth and development of the adipocyte (PPARγ gene) [125, 127]. Although these genes are involved in energy expenditure pathways, Loos and Bouchard [127] point out that most genetics studies examined these genes in relation to obesity-related phenotypes and not energy expenditure phenotypes per se.

The physiological pathways related to appetite are complicated and, as reviewed by Badman and Flier [130], involve the integration of short-term satiety signals from the gut to the brain along with longer-term homeostatic systems. Figure C-2 depicts an overview of these physiological pathways, which involve the integrated signaling of POMC, AGRP, MC4R, and NPY systems. As Loos and Bouchard note [127], there have been relatively inconsistent findings linking obesity phenotypes to genes for these proteins. Perhaps the most encouraging findings in the literature involve the MC4R gene, which, as discussed in the next section, has been associated with human food intake in a few preliminary studies.

5b. Genetic Influences on Food Intake

A relatively small number of studies have tested genetic influences on eating phenotypes, independent from body fat. They provide data pertinent

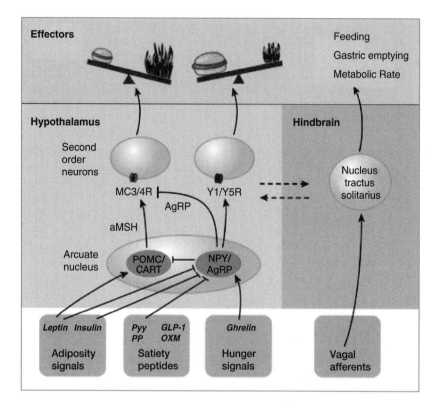

FIGURE C-2   Pictorial representation of potential action of gut peptides on the hypothalamus. Access circulating agents into the arcuate nucleus of the hypothalamus is facilitated by a relaxed blood-brain barrier. Primary neurons in the arcuate nucleus contain multiple peptide neuromodulators. Appetite-inhibiting neurons (red) contain pro-opiomelanocortin (POMC) peptides such as α–melanocyte-stimulating hormone (αMSH), which acts on melanocortin receptors (MC3 and MC4) and cocaine- and amphetamine-stimulated transcript peptide (CART), whose receptor is unknown. Appetite-stimulating neurons in the arcuate nucleus (green) contain neuropeptide Y (NPY), which acts on Y receptors (Y1 and Y5), and agouti-related peptide (AgRP), which is an antagonist of MC3/4 receptor activity. Integration of peripheral signals within the brain involves interplay between the hypothalamus and hindbrain structures including the NTS, which receives vagal afferent inputs. Inputs from the cortex, amygdala, and brainstem nuclei are integrated as well, with resultant effects on meal size and frequency, gut handling of ingested food, and energy expenditure. →, direct stimulatory; ⊣, direct inhibitory; ⋯→, indirect pathways. SOURCE: [130].

to pathway a in Figure C-1. These studies are summarized below, divided into those that estimated heritability and those that tested for specific genes or genomic regions.

i. Heritability of Eating Behaviors

A seminal investigation by de Castro studied adult twin pairs who recorded food and beverage intake continuously in the free-living environment for 7 days [131-134]. The initial cohort consisted of 109 identical and 86 fraternal twin pairs, who used diaries to record food and beverage intake, time of food consumption, amount of food consumed, food preparation methods, the number of other people present when eating, hunger and thirst levels, depression, anxiety, and perceived food attractiveness and palatability. These data lead to a series of publications on the broader genetic-environment architecture of adult food intake. In an initial report, significant heritability estimates were documented for reported total energy intake, weight of food intake, fat intake, carbohydrate intake, protein intake, and water intake, with the remaining variance due to nonshared environmental factors [132].

In a multivariate analysis that tested genetic and environmental influences on *meal-specific* energy intake, cumulative *daily* energy intake, and weight status, results supported the presence of independent genetic and nonshared environmental influences on meal-specific energy intake [133]. Genes accounted for 46 percent of the variance in the frequency with which meals were eaten and 56 percent of the variance in meal size, respectively. Thus, there may be genes and environmental influences on food intake at specific eating episodes that are different from those factors related to habitual dietary intake. Specific social-environmental influences on meal intake were not reported initially.

Gender differences in the heritability of food intake were reported in other twin cohorts (see below) and, apropos to this report, may be an important issue for studying the interplay of social-environmental and genetic influences on obesity.

There were genetic influences on the full range of appetite and eating phenotypes studied by de Castro [131-139] including premeal hunger levels, time of meals, premeal stomach content, energy intake of high-palatability foods, energy intake of low-palatability foods, and overconsumption of foods rated as being high versus low in palatability. Apropos to this report, the environmental conditions that promote social facilitation of eating may be genetically influenced, suggesting a gene-environment correlation. These data challenge the conceptual framework in Figure C-1, suggesting that covariation between genes and the social-environmental should be modeled. As de Castro concluded, "genes appear to affect the physiologic, psychologic, and social context in which eating occurs. In the past, heredity and environment were seen to operate separately on behavior. The present results indicated that there may not be such a clear separation [. . .] heredity ends up having a strong influence on the nature of the environment in which individuals immerse themselves" (p. 554). This issue is revisited in the final section of this report.

Building upon de Castro's work, other twin studies tested the heritability of dietary intake patterns. In a study of 4,640 adult twins who completed the National Cancer Institute food-frequency questionnaire, the heritability of consuming foods high in fat, sugar, and salt was 15 percent for women and 30 percent for men [140]. The heritability of consuming "healthy foods," including fruits and vegetables, was 15 percent for women and 30 percent for men. Thus, heritability estimates were larger for men than women and, for women only, there were significant shared environmental influences on both food categories.

A recent study of 5,250 male and female twin pairs, 16 years of age, examined the heritability of frequency of breakfast consumption [141]. Results indicated significant gender differences, in that heritability estimates were higher for boys while shared environmental influences were higher for girls. The authors concluded that "Breakfast eating is moderated differently in adolescent boys and girls. Unlike boys, girls are much influenced by the family and pair-specific environment. In girls, environmental influences may override genetically driven factors" (p. 512).

With one exception, all studies testing the heritability of eating traits used self-report measures of dietary intake. A recognized drawback to this method is the fact that respondents tend to underreport food intake, a finding that is more common among obese individuals [142]. To bypass this problem, Faith et al. [143] studied 36 MZ and 18 DZ twins who consumed a buffet lunch in a controlled feeding laboratory. The meal provided servings of 27 foods and beverages, including chicken nuggets, hot dog sandwiches, apples, grapes, carrots, chocolate cookies, and donuts. Results indicated that shared environmental factors had the biggest influence on total energy intake, accounting for 48 percent of the variance. Additive genetic factors accounted for 33 percent of the variance, and nonshared environmental factors accounted for 19 percent of the variance. Thus, the results suggest that both genes and the shared environment can influence total energy intake, although specific aspects of the environment were not measured. The study also suggests the potential value of laboratory-based protocols for study eating phenotypes, a point that is reviewed in the final section of this report for future research directions.

ii. Specific Genes Associated with Eating Behaviors

There is a dearth of information on specific genes that influence dietary patterns, at least for most individuals in the population [144]. There is a subset of individuals whose obesity resulted from single gene mutations and who were markedly hyperphagic [125]; however, these individuals are relatively uncommon in the population. Studies that used association or linkage designs to identify specific genes or genomic regions in larger cohorts have been relatively uncommon and represent an area for future research.

A series of reports documented associations between the serotonin (5-hydroxy-tryptamine, 5-HT) receptor gene and reported energy intake. Because serotonin has been shown to reduce food intake and may play a role in the etiology of eating disorders, it was deemed an appropriate candidate gene by some investigators. Aubert et al. [145] studied 276 unrelated overweight and obese adults who were genotyped for the –1438 GA polymorphism of the 5-HT_{2A} receptor gene. Results indicted that reported daily energy intake from 3-day food records was significantly associated with genotype. Specifically, individuals carrying the A allele of the gene (i.e., genotypes G/A or A/A) consumed less total energy per day than individuals not carrying the A allele (i.e., genotype G/G). Comparable findings were made by the same investigators in an analysis of 370 children and adolescents, 10 to 20 years old, who were participants in the Stanislas Family Study. Youth carrying the A allele consumed less daily energy intake than youth not carrying the allele, even when controlling for age, sex, weight, and height [146].

Associations between MC4R polymorphisms and eating traits have been reported, further implicating the role of this gene in obesity onset. In a study of 500 children with severe obesity, 5.8 percent of the sample was found to have mutations in the MC4R gene that were not found in nonobese controls [147]. The investigators then compared the ad libitum test meal intake of children who had fully inactive or partially inactive MC4R polymorphisms. When served a standardized breakfast, children with fully inactive mutations consumed more food than children with partially inactive mutations. In another study, MC4R genotype was associated with Binge Eating Disorder status in a cohort of severely obese Caucasians adults and nondieting controls [148].

Several linkage studies investigated genomic regions associated with reported Dietary Restraint, Eating Disinhition, and Hunger levels, as measured by the Eating Inventory (EI) [149]. Steinle et al. [150] studied 624 related individuals, from 28 Amish families, who completed the EI. Results of a genome-wide linkage analysis revealed five chromosomal regions that contained genes for EI subscales. Specifically, markers for Restraint were detected on chromosomes 3 (LOD score = 2.5) and 6 (LOD score = 2.3);

markers for Disinhibition were detected on chromosomes 7 (LOD score = 1.6) and 16 (LOD score = 1.4); and a marker for Hunger was detected on chromosome 3 (LOD score = 1.4). Specific genes on these regions were not identified. Thus, overall, it was dietary disinhibition which was identified as the behavior that had the highest heritability estimate (0.40 ± 0.10) and showed the strongest association with obesity phenotypes.

In a subsequent linkage analysis of the EI conducted in 660 adults from the Quebec Family study, fine-mapping strategies identified the Neuromedian β (NMB) gene on chromosome 15 as a possible gene contributing to eating behaviors and obesity [151]. A specific polymorphism of NMB gene (i.e., the *p.P73T* polymorphism) was associated with scores on the Disinhibition and Hunger subscales. Specifically, individuals with the *T/T* genotype had higher Disinhibition and Hunger scores than individuals with the *P/T* or *P/P* genotypes. Moreover, 6-year weight gain was significantly greater among individuals with the *T/T* genotype compared to the other genotypes. This study implicates a specific gene that appears to promote increased food intake and weight gain. Specific social-environmental factors were not identified in this report but remain an avenue for future research.

SECTION 6: EVIDENCE FOR INTERACTIONS AMONG SOCIAL ENVIRONMENTAL, GENETIC, AND BEHAVIORAL FACTORS AS THEY RELATE TO OBESITY

The conceptual framework for this paper (Figure C-1) does not explicitly posit interactions among social-environmental and genetic factors due to the paucity of research examining gene-environment interactions in the human obesity literature. This section reviews the handful of studies that have addressed this issue, which provides a potential framework for future investigations (see Section 7). The section is divided into subsections examining (a) the potential moderating effects of the social environment on the relationship between genetics and obesity, and (b) the potential moderating effects of genetic factors on the relationship between the social environment and obesity. The term "moderator" is used as is conventional for multiple regression analysis [152]. Specifically, a variable X is considered a "moderator" when the relationship between two other variables, Y and Z, depends on the level of X. Figures C-2 and C-3 depict a modified version of Figure C-1, allowing for interactions among the social environment and genetic factors. The social environment serves as the moderator in Figure C-2, while genetic factors serve as the moderator for Figure C-3.

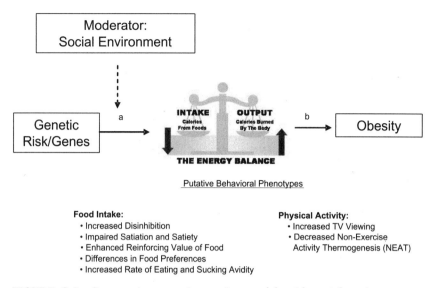

FIGURE C-3 Gene-environment interaction model, with social environment as moderator. In this model, the effects of genetic risk/specific genes on obesity and obesity-promoting behaviors depend on the level of exposure to a given social-environmental factor.

6a. Social Environment as a Potential Moderator Variable

Ravussin et al. [153] compared the weight status and diabetes-related comorbidities of Pima Indians living in remote rural regions of Mexico compared to those living in Arizona. The Pima Indians of Arizona have been extensively studied given their markedly high prevalences of obesity and type 2 diabetes. They are considered to be at genetically increased risk for these disorders. Compared to Pima Indians living in rural Mexico, those living in Arizona weighed significantly more (64.2 vs. 90.2 kg), had higher BMIs (24.9 vs. 33.4 kg/m^2), and had higher total cholesterol levels (146 vs. 174mg/dl). Among the Pima Indians from Mexico, 11 percent of the women and 6 percent of the men had type 2 diabetes; by contrast, among the Pima Indians living in Arizona, 37 percent of the women and 54 percent of the men had the diagnoses.

Bhatnagar et al. [154] compared 247 London residents who had migrated from the Indian subcontinent of Punjabi against 117 of their siblings who still lived in India. Compared to the siblings in India, those living in London had significantly higher BMI values, systolic blood pressure, serum cholesterol, apolipoprotein B, and fasting blood glucose.

These results collectively suggest that genetic influences on the development of obesity can be mitigated by environmental conditions. However, these data do not necessarily provide support for the presence of gene-by-environment interactions *within* the U.S. population at a single time period. As noted in Section 7, this represents an avenue for additional research, especially when considering the potential moderating effects of SES and environmental factors associated with lower income.

i. Genetic Factors as a Potential Moderator Variable

Bouchard and colleagues conducted a seminal "overfeeding" study in which 12 male MZ twin pairs were fed an additional 1,000 kcal/day beyond their baseline intake levels, for 6 days per week over 100 days [155]. The investigators tested whether changes in body composition in response to overfeeding differed as a function of twinship. Outcome measures included changes in body composition and metabolic parameters. Results provided clear evidence that response to overfeeding was related to twinship. Twins were significantly correlated with respect to changes in body weight, percent fat mass, fat mass, and estimated subcutaneous fat, and visceral adiposity. Table C-3 presents changes in study outcome measures associated with experimental overfeeding and the intraclass correlation coefficients representing the within-twin pair association for change scores.

In recent years, a series of candidate gene analyses evaluated whether specific genes were associated with response to overfeeding in this cohort. As reviewed by Ukkola and Bouchard [156], a number of candidate genes showed associations. For example, a polymorphism in the adipsin gene was associated with greater increases in body weight, total fat mass, and subcutaneous fat in response to overfeeding; the Gln27Glu polymorphism of the beta 2 adrenergic receptor gene was associated with greater gains in body weight and subcutaneous fat. Few associations were found for changes in visceral adiposity. Despite the limited sample size, these analyses have been critical to the field for demonstrating how specific genes might moderate the effects of a specific environmental manipulation that promotes weight gain (i.e., overfeeding).

Finally, Epstein et al. [157] recently reported that the association between the "reinforcing value of food" phenotype (see Section 5) and ad libitum energy intake in the laboratory was moderated by the dopamine transporter gene (SLC6A3) and the dopamine 2 receptor gene (DRD2). Participants were 88 smokers of European American ancestry who were evaluated before beginning a smoking cessation treatment. With respect to the SLC6A3 gene, subjects who scored high on the reinforcing value of food and who lacked the SLC6A3*9 allele consumed more total energy than participants with other SLC6A3 genotypes. With respect to the DRD2

TABLE C-3 Effect of 100d Overfeeding in 12 Pairs of Male Twins and Measures of the Similarity Within Pairs

Variable	Before Over-Feeding	After Over-Feeding	Similarity Within Pairs	
			F ratio	ICC
Body weight (kg)	60.3 (8.0)	68.4 (8.2)	3.4	0.55[a]
Fat mass (kg)	6.9 (3.5)	12.3 (4.5)	3.0	0.50[a]
Subcutaneous fat (mm)	75.9 (21.1)	129.4 (32.9)	2.8	0.47[a]
Abdominal fat[c]				
Total (cm^2)	106.0 (46.0)	199.0 (50.0)	4.1	0.58[b]
Subcutaneous (cm^2)	72.0 (40.0)	141.0 (46.0)	3.8	0.58[a]
Visceral (cm^2)	34.0 (9.0)	58.0 (15.0)	6.1	0.72[b]
Fasting insulin (pmol L^{-1})	45.0 (10.0)	67.0 (29.0)	5.9	0.71[b]
OGTT insulin area (pmol L^{-1} min^{-1} 10^{-3}	54.5 (20.3)	70.6 (33.5)	2.6	0.44
OGTT glucose area (mmol L^{-1} min^{-1} 10^{-3})	0.93 (0.13)	0.99 (0.11)	2.5	0.43
Total cholesterol (mmol L^{-1})	4.5 (0.8)	4.9 (1.2)	4.4	0.63[b]
HDL cholesterol (mmol L^{-1})	1.2 (0.2)	1.1 (0.2)	2.8	0.48[a]
Total triglycerides (mmol L^{-1})	1.1 (0.5)	1.8 (1.3)	12.1	0.85[b]

From: [156]. OGTT, oral glucose tolerance test; HDL, high-density lipoprotein. Values are expressed as means (SD). The statistical significance was determined by a two-way analysis of variance for repeated measures on one factor (time). The F ratio was the ratio of the variance between pairs to that within pairs. The intraclass correlation coefficient (ICC) was used to assess the similarity within pairs in the response to overfeeding.
[a]$p < 0.05$,
[b]$p < 0.01$.
[c]Similarity within pairs was adjusted for gain in fat mass.

gene, subjects who scored high on reinforcing-value-of-food and who had the A1 allele consumed more total calories compared to participants with any other DRD2 genotype. This study is unique in that it focused on the genetics of food reward as they relate to dopamine pathways, an area of research that has been understudied and may be promising for future research.

In sum, there have been very few studies of gene-environment interaction as they relate to obesity and obesity-related behaviors. As noted in the next and final section, this represents an avenue for additional research.

SECTION 7: OPPORTUNITIES FOR FUTURE RESEARCH THAT WOULD ENLIGHTEN RELATIONSHIPS BETWEEN GENETICS AND THE SOCIAL ENVIRONMENT

Building upon literature reviewed in this report, this final section reviews opportunities for additional research that would bridge two active, but so far separate, research areas: specifically, genetic and social-environmental influences on obesity. This list is not exhaustive and the ideas are not necessarily presented in order of importance. The overarching recommendation is that current knowledge on the causes of obesity may benefit from future research that explicitly tests the interactions between, or covariations among, genetic and social-environmental factors that promote obesity. This would require greater collaborations among social and behavioral scientists, physiologists, and molecular geneticists, with each discipline bringing its unique perspectives and methodological tools to a joint research effort. The potential benefits of integrative research need to be weighed against the potential drawbacks, including greater recruitment challenges, increased costs, and issues concerning adequate statistical power.

New insights into the joint influences of genetics and the social-environmental on obesity and obesity-promoting behaviors may be generated by:

• **Additional prospective studies that test genetic and environmental influences on obesity development during putative "critical growth periods."** Most genetic studies reviewed in Section 5 used cross-sectional designs to test for genetic effects at a single time point, whether it was a general heritability estimate or tests of a specific gene or genomic region. However, the onset of obesity is a developmental process that may be influenced by different genetic or environmental influences at different ages. Thus, research that uses prospective designs to identify genetic and environmental influences on the developmental trajectories of body fat stores would be informative. This would be especially useful for studying putative "critical growth periods" for obesity: growth in the intra-uterine environment, "adiposity rebound" in early childhood, and adolescence [158]. The extent to which body composition changes during these periods is influenced by life experiences *specific* to those periods, or age-specific genetic influences, warrants additional research. If there is evidence for genetic influences at specific ages, the role of individual genes needs to be elucidated. Very few studies have used twin designs to test critical growth periods for obesity onset. Indeed, the results of one such twin study suggested that the intra-uterine environment is a critical period for the development of adult height, but not for adult BMI [159].

Likewise, new research, so far only in the animal model, is emerging that suggests that maternal obesity during pre- and postnatal periods can have profound, genotype-specific effects on the development of obesity in offspring that is genetically predisposed to obesity (B. Levin, 2005: Oral presentation at the *Society for the Study of Ingestive Behavior*).

The first year of life may be an especially interesting period to study with respect to longer term obesity. Rapid weight gain during the first 4 months of life is a risk factor for obesity in childhood and adulthood. In one study of 300 full-term African American infants, rapid weight gain was defined as an increase in weight-for-age ≥ 1 SD between birth and 4 months [160]. After adjusting for confounding factors, infants who had experienced rapid weight gain by 4 months of age were 5.22 times more likely to be obese at 20 years of age compared to infants who did not experience rapid weight gain. In a separate analysis of 19,397 infants, results indicated that both birth weight and rate of weight gain were associated with an increased probability of childhood overweight at 7 years of age [161]; within each strata of birth weight, increased rate of weight gain was associated with increased childhood overweight prevalence. Potential genetic and home environmental influences on early life rate of weight gain are poorly understood and may be an important area for future research.

• **Additional studies that evaluate the heritability of, or specific genes associated with, refined behavioral phenotypes related to obesity.** Very little is known about the heritability of behavioral traits that are associated with obesity, particularly those reviewed in Section 4. Studies that clarify the genetic-environmental architecture of these traits would elucidate the extent to which those behavioral traits are genetically influenced, as well as the nature of environmental influences that influence those behaviors (i.e., shared vs. nonshared environmental effects). Such designs could also address important multivariate questions, including the extent to which the correlations between behaviors and body fat is influenced by the same genes (i.e., "genetic correlations") or the same environmental factors (i.e., "environmental correlations"). Especially interesting would be heritability studies of laboratory-based behavioral traits, such as the reinforcing value of food [90, 157], delayed satiation [78], disinhibition [64], or eating in the absence of hunger [48, 162], which have been linked to obesity status.

One of the difficulties in identifying obese phenotypes and associated eating behaviors lies in the existence of several subpopulations of overweight and obese populations. It is likely that individuals who are gaining or losing weight (i.e., reduced obese) exhibit different eating patterns and intake behaviors than do those individuals who are obese but weight-stable.

Behavioral measures evaluated during infancy would be uniquely informative because sucking rate at 3 months of age predicts subsequent weight gain during the first 2 years of life [95, 96]. It is possible that sucking behavior is genetically influenced, because infants born at high risk for obesity have been shown to suck at greater rates when studied in the laboratory compared to infants born at low risk for obesity [95]. The heritability of infant sucking rate is unknown. On the other hand, there is considerable evidence that infants learn flavor preferences during the first year of life through environmental exposure to specific foods [163-168], as well as data that restrictive feeding patterns during infancy are associated with excess infant weight gain [169]. Thus, the roles of learning and genetics, early life sucking, appetite, and food intake needs greater attention. In addition, the identification of genes for NEAT and other refined physical activity traits would advance the field.

In summary, the obese phenotype is likely to be characterized by a conglomerate of significant behaviors related to eating and physical activity which likely work in conjunction to affect energy balance. One of the goals could be to develop (a set of) tools that capture "obese" eating behaviors and physical activity behaviors in an unobtrusive way, if possible at an early age, to make predictions of an individual's weight development.

• **Additional research that incorporates specific measures of the environment into genetics studies.** Most genetic studies have not measured specific aspects of the environment. This includes aspects of the home environment, as well as components of the broader "macroenvironment" discussed in Section 3. Genetics studies provide clear evidence that obesity is influenced by the environment, with most studies suggesting that the non-shared environment is more influential. However, the identity of specific environmental influences has remained elusive, especially during child development. Adding specific measures of the environment might help address these issues.

It is noteworthy that valid measures of the home environment exist and, in principle, could be incorporated into genetic studies. One of the most extensively used instruments in the child development literature is the "Home Observation for Measurement of the Environment" (HOME) system [170]. Different versions of the HOME have been developed for different ages, specifically, infancy and toddlerhood, preschool and early childhood, school-age and middle childhood, and adolescence. The HOME has been used in at least one study of childhood obesity, finding that reduced levels of "cognitive stimulation" at home prospectively predicted increased obesity incidence [171].

In addition to measures of the home environment, measures of the broader environment would be informative for genetics research. In principle, genetic influences on food intake or physical activity may depend on the access to parks, playground, grocery shops, fast-food restaurants, or other environmental variables associated with SES strata. Recent studies have used Geographic Information Systems to "geocode" the physical distance between individual homes and these other components to the community [14]; however, there appear to be no studies to date that have used this technology in the context of genetics of obesity. A handful of studies in the child development literature used this approach to understand the interaction between genes and the broader social environment [172, 173]; these provide useful examples for obesity researchers. For example, in a study of 1,081 MZ twin pairs and 1,061 DZ twin pairs, Caspi et al. [172] found that 20 percent of the variability in 2-year-old children's "behaviors problems" were influenced by shared environmental factors. When a specific measure of "environmental deprivation" was added to the biometric model, however, it was found to account for 5 percent of the variance in the shared environment. Thus, geocoding and related tools that permit better measurement of the macroenvironment, or exposure to the "toxic environment," may advance the field of genetics research.

* **Additional observational and experimental research that evaluates gene-environment interactions.** As noted in Section 6, there are very few studies of gene-environment interaction in the literature. This could be an important area for research with respect to macroenvironmental variables, such as SES, ethnicity, and exposure to the "toxic environment." Thus, the effects of certain obesity-promoting genes may depend on the broader social environment in which a population lives; this is an avenue for additional research and is depicted in Figure C-4.

In addition, experimental studies that test for gene-environment interactions, in similar ways to the Quebec Overfeeding Study [155, 156], would be most informative. In principle, aspects of the "toxic environment" can be experimentally manipulated in a controlled feeding laboratory or metabolic ward, in a manner that cannot be done in the free-living environment. Examples include experimental manipulations of food portion size [74, 174], energy density [75], food deprivation status [175], and food variety [176]. These rigorous laboratory protocols, if used with genetics designs, could yield novel information regarding gene-environment interactions. Potential designs include: co-twin control designs, in which MZ twins are randomly assigned to different experimental conditions; classic twins de-

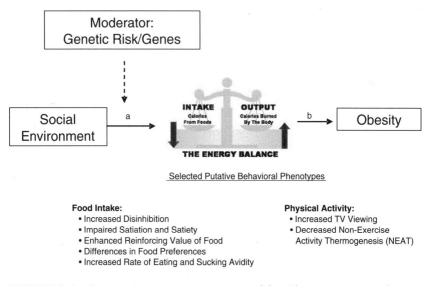

FIGURE C-4 Gene-environment interaction model, with genotype as moderator. In this model, the effects of social-environmental influences on obesity and obesity-promoting behaviors depend on genotype.

signs, in which MZ and DZ twins are used to estimate the heritability of response to an experimental manipulation; or candidate gene designs, in which participants are selected based on specific genotypes. In all cases, pertinent outcome variables could be behavioral and/or physiological measures, as well as changes in body weight if the manipulation is sustained over time.

• Additional research that evaluates gene-environment correlations. Genetic studies of obesity most commonly used BMI or body fat as the primary phenotype, followed by metabolic and physiological measures, and, least commonly, behavioral measures. However, in principle, obesity-promoting genes may operate by influencing the environments into which individuals place themselves. Such a scenario is depicted in Figure C-5. That is, social-environmental measures might be conceptualized as the phenotype in a genetics study, especially if genes influence whether certain individuals will seek out "obesity-promoting" environments (e.g., fast-food restaurants). As noted in Section 3, there is evidence that obese individuals may be more likely to attend restaurants than nonobese individuals on the days that buffets are served, which would be suggestive of a gene-environment correlation. Plomin et al. [177] provide a more detailed discussion of such "active" gene-environment correlations, in which genes influence people's tendencies to create their own environments.

The issue of gene-environment correlations is also relevant to the domain of child development and, in recent years, there has been increasing interest in the "genetics of parenting" [178-180]. Data suggest that certain parenting behaviors towards children are, in fact, elicited by child attributes and behavioral patterns that are probably genetically influenced. This may be a useful framework for studying parent-child feeding dynamics as they relate to obesity onset. As noted Section 3, there is evidence that parental restriction of child eating is elicited by child weight characteristics [43] and this in turn may exacerbate further weight gain by the child. Additional genetics studies could evaluate whether parental feeding restriction, or other parenting domains, are associated with specific candidate genes for obesity.

• **Additional research that builds upon existing conceptual models for "organism-environment interactions."** Conceptual models that explicitly address the integration of genetic and social-environmental influences on behavioral traits may help guide future studies. The field of developmental behavioral genetics has addressed this issue, although not in regards to obesity per se. Several pertinent books have been published [181-186]. In addition, several longitudinal behavioral genetics studies measured specific

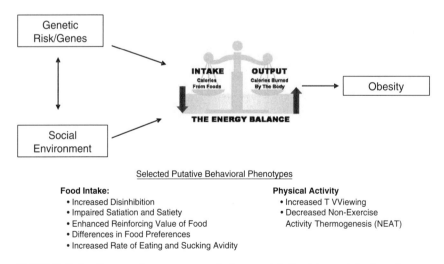

Selected Putative Behavioral Phenotypes

Food Intake:
• Increased Disinhibition
• Impaired Satiation and Satiety
• Enhanced Reinforcing Value of Food
• Differences in Food Preferences
• Increased Rate of Eating and Sucking Avidity

Physical Activity
• Increased T VViewing
• Decreased Non-Exercise
  Activity Thermogenesis (NEAT)

FIGURE C-5   Gene-environment correlation model. In this model, there is a correlation among genes and social-environmental factors that influence obesity and obesity-promoting behaviors.

aspects of the social environment and genetic factors and may provide useful models for obesity research. De Castro [187-189] has one of the few proposed models that integrates genetic and environmental influences on food intake.

• **Additional institutional and/or funding mechanisms to support integrative research projects or interdisciplinary training for scientists.** Interdisciplinary research of the sort reviewed in this report would likely require new collaborative relationships that bring together investigators from different "camps." Institutional and/or funding initiatives that encourage such collaborations may help advance such efforts, given the economic and logistical challenges of such research. Initial collaboration of this sort could be exemplars for other institutions and investigators.

SECTION 8: CONCLUSION

This report set out to highlight two distinct areas of research that share the common goal of identifying factors that contribute to weight gain and obesity in the population. The areas reviewed in this report included research on (social-) environmental factors, as well as the genetic factors, that may be associated with obesity or the onset thereof. Despite their unique focuses, the literature reviewed in this report shows that the two areas have the potential to complement each other and to stimulate future collaborations among investigators. The pathways that lead to obesity are complex and multivariate for most individuals in the population. Additional research that addresses how the genetics of obesity impacts on environmental choices made by certain individuals, and how certain environments moderate the expression of obesity-promoting genes, may advance the current state of knowledge and provide new insights for the prevention and the treatment of obesity.

LITERATURE CITED

1.　　Pi-Sunyer, F.X., *Health implications of obesity.* Am J Clin Nutr, 1991. **53**(6 Suppl): p. 1595S-1603S.
2.　　Zhu, S., et al., *Race-ethnicity-specific waist circumference cutoffs for identifying cardiovascular disease risk factors.* Am J Clin Nutr, 2005. **81**(2): p. 409-15.
3.　　Zhu, S., et al., *Combination of BMI and waist circumference for identifying cardiovascular risk factors in whites.* Obes Res, 2004. **12**(4): p. 633-45.
4.　　Park, Y.W., et al., *The metabolic syndrome: prevalence and associated risk factor findings in the US population from the Third National Health and Nutrition Examination Survey, 1988-1994.* Arch Intern Med, 2003. **163**(4): p. 427-36.
5.　　Dong, C., et al., *Interacting genetic loci on chromosomes 20 and 10 influence extreme human obesity.* Am J Hum Genet, 2003. **72**(1): p. 115-24.

6. Barnett, E. and M. Casper, *A definition of "social environment"*. Am J Public Health, 2001. **91**(3): p. 465.

7. Battle, E.K. and K.D. Brownell, *Confronting a rising tide of eating disorders and obesity: treatment vs. prevention and policy*. Addict Behav, 1996. **21**(6): p. 755-65.

8. Jacobson, M.F. and K.D. Brownell, *Small taxes on soft drinks and snack foods to promote health*. Am J Public Health, 2000. **90**(6): p. 854-7.

9. French, S.A., et al., *Fast food restaurant use among adolescents: associations with nutrient intake, food choices and behavioral and psychosocial variables*. Int J Obes Relat Metab Disord, 2001. **25**(12): p. 1823-33.

10. French, S.A., M. Story, and R.W. Jeffery, *Environmental influences on eating and physical activity*. Annu Rev Public Health, 2001. **22**: p. 309-35.

11. French, S.A., et al., *Pricing and promotion effects on low-fat vending snack purchases: the CHIPS Study*. Am J Public Health, 2001. **91**(1): p. 112-7.

12. Crawford, D.A., R.W. Jeffery, and S.A. French, *Television viewing, physical inactivity and obesity*. Int J Obes Relat Metab Disord, 1999. **23**(4): p. 437-40.

13. Jeffery, R.W. and J. Utter, *The changing environment and population obesity in the United States*. Obes Res, 2003. **11 Suppl**: p. 12S-22S.

14. Burdette, H.L. and R.C. Whitaker, *Neighborhood playgrounds, fast food restaurants, and crime: relationships to overweight in low-income preschool children*. Prev Med, 2004. **38**(1): p. 57-63.

15. Jeffery, R.W., et al., *The relationship between social status and body mass index in the Minnesota Heart Health Program*. Int J Obes, 1989. **13**(1): p. 59-67.

16. Sundquist, J., M. Malmstrom, and S.E. Johansson, *Cardiovascular risk factors and the neighbourhood environment: a multilevel analysis*. Int J Epidemiol, 1999. **28**(5): p. 841-5.

17. McMurray, R.G., et al., *The influence of physical activity, socioeconomic status, and ethnicity on the weight status of adolescents*. Obes Res, 2000. **8**(2): p. 130-9.

18. Ball, K. and D. Crawford, *Socioeconomic status and weight change in adults: a review*. Soc Sci Med, 2005. **60**(9): p. 1987-2010.

19. Goodman, E., *The role of socioeconomic status gradients in explaining differences in U.S. adolescents' health*. Am J Public Health, 1999. **89**(10): p. 1522-8.

20. Ellaway, A., A. Anderson, and S. Macintyre, *Does area of residence affect body size and shape?* Int J Obes Relat Metab Disord, 1997. **21**(4): p. 304-8.

21. Boslaugh, S.E., et al., *Perceptions of neighborhood environment for physical activity: is it "who you are" or "where you live"?* J Urban Health, 2004. **81**(4): p. 671-81.

22. Reidpath, D.D., et al., *An ecological study of the relationship between social and environmental determinants of obesity*. Health Place, 2002. **8**(2): p. 141-5.

23. Drewnowski, A. and N. Darmon, *Food choices and diet costs: an economic analysis*. J Nutr, 2005. **135**(4): p. 900-4.

24. Drewnowski, A. and S.E. Specter, *Poverty and obesity: the role of energy density and energy costs*. Am J Clin Nutr, 2004. **79**(1): p. 6-16.

25. Drewnowski, A., *Obesity and the food environment: dietary energy density and diet costs*. Am J Prev Med, 2004. **27**(3 Suppl): p. 154-62.

26. Neumark-Sztainer, D., et al., *Correlates of fruit and vegetable intake among adolescents. Findings from Project EAT*. Prev Med, 2003. **37**(3): p. 198-208.

27. Darmon, N., A. Briend, and A. Drewnowski, *Energy-dense diets are associated with lower diet costs: a community study of French adults*. Public Health Nutr, 2004. **7**(1): p. 21-7.

28. Burdette, H.L. and R.C. Whitaker, *A national study of neighborhood safety, outdoor play, television viewing, and obesity in preschool children*. Pediatrics, 2005. **116**(3): p. 657-62.

29. Herman, C.P., D.A. Roth, and J. Polivy, *Effects of the presence of others on food intake: a normative interpretation.* Psychol Bull, 2003. **129**(6): p. 873-86.

30. de Castro, J.M. and E.S. de Castro, *Spontaneous meal patterns of humans: influence of the presence of other people.* Am J Clin Nutr, 1989. **50**(2): p. 237-47.

31. de Castro, J.M., *Family and friends produce greater social facilitation of food intake than other companions.* Physiol Behav, 1994. **56**(3): p. 445-5.

32. Edelman, B., et al., *Environmental effects on the intake of overweight and normal-weight men.* Appetite, 1986. **7**(1): p. 71-83.

33. Kim, J.Y. and H.R. Kissileff, *The effect of social setting on response to a preloading manipulation in non-obese women and men.* Appetite, 1996. **27**(1): p. 25-40.

34. de Castro, J.M., et al., *Social facilitation of the spontaneous meal size of humans occurs regardless of time, place, alcohol or snacks.* Appetite, 1990. **15**(2): p. 89-101.

35. Kramer, M.S. and B. Moroz, *Do breast-feeding and delayed introduction of solid foods protect against subsequent atopic eczema?* J Pediatr, 1981. **98**(4): p. 546-50.

36. Armstrong, J. and J.J. Reilly, *Breastfeeding and lowering the risk of childhood obesity.* Lancet, 2002. **359**(9322): p. 2003-4.

37. Liese, A.D., et al., *Inverse association of overweight and breast feeding in 9 to 10-y-old children in Germany.* Int J Obes Relat Metab Disord, 2001. **25**(11): p. 1644-50.

38. Gillman, M.W., et al., *Risk of overweight among adolescents who were breastfed as infants.* JAMA, 2001. **285**(19): p. 2461-7.

39. Wadsworth, M., et al., *Breast feeding and obesity. Relation may be accounted for by social factors.* BMJ, 1999. **319**(7224): p. 1576.

40. Grummer-Strawn, L.M. and Z. Mei, *Does breastfeeding protect against pediatric overweight? Analysis of longitudinal data from the Centers for Disease Control and Prevention Pediatric Nutrition Surveillance System.* Pediatrics, 2004. **113**(2): p. e81-6.

41. Taveras, E.M., et al., *Association of breastfeeding with maternal control of infant feeding at age 1 year.* Pediatrics, 2004. **114**(5): p. e577-83.

42. Johnson, S.L. and L.L. Birch, *Parents' and children's adiposity and eating style.* Pediatrics, 1994. **94**(5): p. 653-61.

43. Faith, M.S., et al., *Parent-child feeding strategies and their relationships to child eating and weight status.* Obes Res, 2004. **12**(11): p. 1711-22.

44. Birch, L.L., et al., *Confirmatory factor analysis of the Child Feeding Questionnaire: a measure of parental attitudes, beliefs and practices about child feeding and obesity proneness.* Appetite, 2001. **36**(3): p. 201-10.

45. Wardle, J., et al., *Parental feeding style and the inter-generational transmission of obesity risk.* Obes Res, 2002. **10**(6): p. 453-62.

46. Baughcum, A.E., et al., *Maternal feeding practices and beliefs and their relationships to overweight in early childhood.* J Dev Behav Pediatr, 2001. **22**(6): p. 391-408.

47. Birch, L.L. and J.A. Fisher, *Appetite and eating behavior in children.* Pediatr Clin North Am, 1995. **42**(4): p. 931-53.

48. Fisher, J.O. and L.L. Birch, *Eating in the absence of hunger and overweight in girls from 5 to 7 y of age.* Am J Clin Nutr, 2002. **76**(1): p. 226-31.

49. Faith, M.S., et al., *Parental feeding attitudes and styles and child body mass index: prospective analysis of a gene-environment interaction.* Pediatrics, 2004. **114**(4): p. e429-36.

50. Hill, J.O., et al., *Obesity and the environment: where do we go from here?* Science, 2003. **299**(5608): p. 853-5.

51. Adams, N., et al., *The eating behavior of obese and nonobese women.* Behav Res Ther, 1978. **16**(4): p. 225-32.

52. Dodd, D.K., H.J. Birky, and R.B. Stalling, *Eating behavior of obese and normal-weight females in a natural setting.* Addict Behav, 1976. **1**(4): p. 321-5.

53. Hill, S.W. and N.B. McCutcheon, *Eating responses of obese and nonobese humans during dinner meals.* Psychosom Med, 1975. 37(5): p. 395-401.

54. Kaplan, D.L., *Eating style of obese and nonobese males.* Psychosom Med, 1980. 42(6): p. 529-38.

55. Mahoney, M.J., *The obese eating style: bites, beliefs, and behavior modification.* Addict Behav, 1975. 1(1): p. 47-53.

56. Marston, A.R., et al., *In vivo observation of the eating habits of obese and nonobese subjects,* in *Recent Advances in Obesity Research,* A. Howard, Editor. 1975, Newman: London. p. 207-210.

57. Stunkard, A., et al., *Obesity and eating style.* Arch Gen Psychiatry, 1980. 37(10): p. 1127-9.

58. Schachter, S., *Some extraordinary facts about obese humans and rats.* Am Psychol, 1971. 26(2): p. 129-44.

59. Schachter, S., R. Goldman, and A. Gordon, *Effects of fear, food deprivation, and obesity on eating.* J Pers Soc Psychol, 1968. 10(2): p. 91-7.

60. Schachter, S. and L.P. Gross, *Manipulated time and eating behavior.* J Pers Soc Psychol, 1968. 10(2): p. 98-106.

61. Nisbett, R.E., *Taste, deprivation, and weight determinants of eating behavior.* J Pers Soc Psychol, 1968. 10(2): p. 107-16.

62. Nisbett, R.E. and M.D. Storms, *Cognitive, social, psychological determinants of food intake,* in *Cognitive modification of emotional behavior,* H. London and R.E. Nisbett, Editors. 1975, Aldine: Chicago.

63. Nisbett, R.E. and L. Temoshok, *Is there an external cognitive style?* J Pers Soc Psychol, 1976. 33: p. 36-47.

64. Lindroos, A.K., et al., *Dietary intake in relation to restrained eating, disinhibition, and hunger in obese and nonobese Swedish women.* Obes Res, 1997. 5(3): p. 175-82.

65. Carmody, T.P., R.L. Brunner, and S.T. St Jeor, *Dietary helplessness and disinhibition in weight cyclers and maintainers.* Int J Eat Disord, 1995. 18(3): p. 247-56.

66. Westenhoefer, J., A.J. Stunkard, and V. Pudel, *Validation of the flexible and rigid control dimensions of dietary restraint.* Int J Eat Disord, 1999. 26(1): p. 53-64.

67. Lawson, O.J., et al., *The association of body weight, dietary intake, and energy expenditure with dietary restraint and disinhibition.* Obes Res, 1995. 3(2): p. 153-61.

68. McGuire, M.T., et al., *What predicts weight regain in a group of successful weight losers?* J Consult Clin Psychol, 1999. 67(2): p. 177-85.

69. Williamson, D.A., et al., *Association of body mass with dietary restraint and disinhibition.* Appetite, 1995. 25(1): p. 31-41.

70. Howard, C.E. and L.K. Porzelius, *The role of dieting in binge eating disorder: etiology and treatment implications.* Clin Psychol Rev, 1999. 19(1): p. 25-44.

71. Provencher, V., et al., *Eating behaviors and indexes of body composition in men and women from the Quebec family study.* Obes Res, 2003. 11(6): p. 783-92.

72. Bell, E.A. and B.J. Rolls, *Energy density of foods affects energy intake across multiple levels of fat content in lean and obese women.* Am J Clin Nutr, 2001. 73(6): p. 1010-8.

73. Rolls, B.J., E.L. Morris, and L.S. Roe, *Portion size of food affects energy intake in normal-weight and overweight men and women.* Am J Clin Nutr, 2002. 76(6): p. 1207-13.

74. Diliberti, N., et al., *Increased portion size leads to increased energy intake in a restaurant meal.* Obes Res, 2004. 12(3): p. 562-8.

75. Kral, T.V., L.S. Roe, and B.J. Rolls, *Combined effects of energy density and portion size on energy intake in women.* Am J Clin Nutr, 2004. 79(6): p. 962-8.

76. McConahy, K.L., et al., *Food portions are positively related to energy intake and body weight in early childhood.* J Pediatr, 2002. 140(3): p. 340-7.

77. Heitmann, B.L. and L. Lissner, *Dietary underreporting by obese individuals—is it specific or non-specific?* BMJ, 1995. **311**(7011): p. 986-9.
78. Westerterp-Plantenga, M.S., et al., *Energy intake adaptation of food intake to extreme energy densities of food by obese and non-obese women.* Eur J Clin Nutr, 1996. **50**(6): p. 401-7.
79. Meyer, J.E. and V. Pudel, *Experimental studies on food-intake in obese and normal weight subjects.* J Psychosom Res, 1972. **16**(4): p. 305-8.
80. Westerterp-Plantenga, M.S., L. Wouters, and F. ten Hoor, *Restrained eating, obesity, and cumulative food intake curves during four-course meals.* Appetite, 1991. **16**(2): p. 149-58.
81. Matsuda, M., et al., *Altered hypothalamic function in response to glucose ingestion in obese humans.* Diabetes, 1999. **48**(9): p. 1801-6.
82. Del Parigi, A., et al., *Neuroimaging and obesity: mapping the brain responses to hunger and satiation in humans using positron emission tomography.* Ann N Y Acad Sci, 2002. **967**: p. 389-97.
83. Kissileff, H.R., *Satiating efficiency and a strategy for conducting food loading experiments.* Neurosci Biobehav Rev, 1984. **8**(1): p. 129-35.
84. Rolls, B.J., et al., *Satiety after preloads with different amounts of fat and carbohydrate: implications for obesity.* Am J Clin Nutr, 1994. **60**(4): p. 476-87.
85. Fricker, J., et al., *Effect of a covert fat dilution on the spontaneous food intake by lean and obese subjects.* Appetite, 1995. **24**(2): p. 121-37.
86. Roe, L.S., et al., *A meta-analysis of factors predicting energy compensation in preloading studies.* FASEB Journal, 1999. **13**: p. A871.
87. Faith, M.S., et al., *Familial aggregation of energy intake in children.* Am J Clin Nutr, 2004. **79**(5): p. 844-50.
88. Epstein, L.H. and B.E. Saelens, *Behavioral economics of obesity: food intake and energy expenditure,* in *Reframing Health Behavior Change with Behavioral Economics,* W.K. Bickel and R.E. Vuchinich, Editors. 2000, Lawrence Erlbaum: Mahwah, NJ. p. 293 311.
89. Epstein, L.H.S., B.E, *Behavioral economics of obesity: Food intake and energy expenditure,* in *Reframing health behavior change with behavioral economics,* W.K.V. Bickel, R.E, Editor. 2000, Lawrence Erlbaum: Mahwah, NJ. p. 293-311.
90. Saelens, B.E. and L.H. Epstein, *Reinforcing value of food in obese and non-obese women.* Appetite, 1996. **27**(1): p. 41-50.
91. Lappalainen, R. and L.H. Epstein, *A behavioral economics analysis of food choice in humans.* Appetite, 1990. **14**(2): p. 81-93.
92. Ferster, C.B., J.I. Nurnberger, and E.B. Levitt, *The control of eating.* J Mathematics, 1962. **1**: p. 87-109.
93. Spiegel, T.A., E.E. Shrager, and E. Stellar, *Responses of lean and obese subjects to preloads, deprivation, and palatability.* Appetite, 1989. **13**(1): p. 45-69.
94. Kissileff, H.R. and J. Thornton, *Facilitation and inhibition in the cumulative food intake curve in man,* in *Changing Concepts of the Nervous System,* A.J. Morrison and P. Stick, Editors. 1982, Academic Press: New York. p. 585-605.
95. Stunkard, A.J., et al., *Energy intake, not energy output, is a determinant of body size in infants.* Am J Clin Nutr, 1999. **69**(3): p. 524-30.
96. Stunkard, A.J., et al., *Predictors of body size in the first 2 y of life: a high-risk study of human obesity.* Int J Obes Relat Metab Disord, 2004. **28**(4): p. 503-13.
97. Spiegel, T.A., et al., *Bite size, ingestion rate, and meal size in lean and obese women.* Appetite, 1993. **21**(2): p. 131-45.
98. Spiegel, T.A., *Rate of intake, bites, and chews—the interpretation of lean-obese differences.* Neurosci Biobehav Rev, 2000. **24**(2): p. 229-37.

99. Drewnowski, A., *Food perceptions and preferences of obese adults: a multidimensional approach.* Int J Obes, 1985. **9**(3): p. 201-12.

100. Mela, D.J. and D.A. Sacchetti, *Sensory preferences for fats: relationships with diet and body composition.* Am J Clin Nutr, 1991. **53**(4): p. 908-15.

101. Cox, D.N., et al., *Sensory and hedonic judgments of common foods by lean consumers and consumers with obesity.* Obes Res, 1998. **6**(6): p. 438-47.

102. Cox, D.N., et al., *Sensory and hedonic associations with macronutrient and energy intakes of lean and obese consumers.* Int J Obes Relat Metab Disord, 1999. **23**(4): p. 403-10.

103. Lissner, L., et al., *Dietary fat and the regulation of energy intake in human subjects.* Am J Clin Nutr, 1987. **46**(6): p. 886-92.

104. Tucker, L.A. and M.J. Kano, *Dietary fat and body fat: a multivariate study of 205 adult females.* Am J Clin Nutr, 1992. **56**(4): p. 616-22.

105. Bray, G.A., S. Paeratakul, and B.M. Popkin, *Dietary fat and obesity: a review of animal, clinical and epidemiological studies.* Physiol Behav, 2004. **83**(4): p. 549-55.

106. Maskarinec, G., R. Novotny, and K. Tasaki, *Dietary patterns are associated with body mass index in multiethnic women.* J Nutr, 2000. **130**(12): p. 3068-72.

107. Mela, D.J., *Determinants of food choice: relationships with obesity and weight control.* Obes Res, 2001. 9 Suppl 4: p. 249S-255S.

108. Pereira, M.A., et al., *Fast-food habits, weight gain, and insulin resistance (the CARDIA study): 15-year prospective analysis.* Lancet, 2005. **365**(9453): p. 36-42.

109. Prentice, A.M. and S.A. Jebb, *Fast foods, energy density and obesity: a possible mechanistic link.* Obes Rev, 2003. **4**(4): p. 187-94.

110. Parsons, T.J., C. Power, and O. Manor, *Physical activity, television viewing and body mass index: a cross-sectional analysis from childhood to adulthood in the 1958 British cohort.* Int J Obes (Lond), 2005. **29**(10): p. 1212-21.

111. Jakicic, J.M. and A.D. Otto, *Physical activity considerations for the treatment and prevention of obesity.* Am J Clin Nutr, 2005. **82**(1 Suppl): p. 226S-229S.

112. Wareham, N.J., E.M. van Sluijs, and U. Ekelund, *Physical activity and obesity prevention: a review of the current evidence.* Proc Nutr Soc, 2005. **64**(2): p. 229-47.

113. Wing, R.R. and S. Phelan, *Long-term weight loss maintenance.* Am J Clin Nutr, 2005. **82**(1 Suppl): p. 222S-225S.

114. Francis, L.A., Y. Lee, and L.L. Birch, *Parental weight status and girls' television viewing, snacking, and body mass indexes.* Obes Res, 2003. **11**(1): p. 143-51.

115. Jago, R., et al., *BMI from 3-6 y of age is predicted by TV viewing and physical activity, not diet.* Int J Obes Relat Metab Disord, 2005. **29**(6): p. 557-64.

116. Henderson, V.R. and B. Kelly, *Food advertising in the age of obesity: content analysis of food advertising on general market and African American television.* J Nutr Educ Behav, 2005. **37**(4): p. 191-6.

117. Levine, J.A., et al., *Interindividual variation in posture allocation: possible role in human obesity.* Science, 2005. **307**(5709): p. 584-6.

118. Maes, H.H., M.C. Neale, and L.J. Eaves, *Genetic and environmental factors in relative body weight and human adiposity.* Behav Genet, 1997. **27**(4): p. 325-51.

119. Stunkard, A.J., et al., *The body-mass index of twins who have been reared apart.* N Engl J Med, 1990. **322**(21): p. 1483-7.

120. Allison, D.B., et al., *The heritability of body mass index among an international sample of monozygotic twins reared apart.* Int J Obes Relat Metab Disord, 1996. **20**(6): p. 501-6.

121. Hewitt, J.K., *The genetics of obesity: what have genetic studies told us about the environment.* Behav Genet, 1997. **27**(4): p. 353-8.

122. Faith, M.S., S.L. Johnson, and D.B. Allison, *Putting the behavior into the behavior genetics of obesity.* Behav Genet, 1997. **27**(4): p. 423-39.

123. Koeppen-Schomerus, G., J. Wardle, and R. Plomin, *A genetic analysis of weight and overweight in 4-year-old twin pairs.* Int J Obes Relat Metab Disord, 2001. **25**(6): p. 838-44.

124. Jacobson, K.C. and D.C. Rowe, *Genetic and shared environmental influences on adolescent BMI: interactions with race and sex.* Behav Genet, 1998. **28**(4): p. 265-78.

125. Perusse, L., et al., *The human obesity gene map: the 2004 update.* Obes Res, 2005. **13**(3): p. 381-490.

126. Redden, D.T. and D.B. Allison, *Nonreplication in genetic association studies of obesity and diabetes research.* J Nutr, 2003. **133**(11): p. 3323-6.

127. Loos, R.J. and C. Bouchard, *Obesity—is it a genetic disorder?* J Intern Med, 2003. **254**(5): p. 401-25.

128. Korner, J. and R.L. Leibel, *To eat or not to eat—how the gut talks to the brain.* N Engl J Med, 2003. **349**(10): p. 926-8.

129. Rosenbaum, M. and R.L. Leibel, *The role of leptin in human physiology.* N Engl J Med, 1999. **341**(12): p. 913-5.

130. Badman, M.K. and J.S. Flier, *The gut and energy balance: visceral allies in the obesity wars.* Science, 2005. **307**(5717): p. 1909-14.

131. De Castro, J.M., *The effects of the spontaneous ingestion of particular foods or beverages on the meal pattern and overall nutrient intake of humans.* Physiol Behav, 1993. **53**(6): p. 1133-44.

132. de Castro, J.M., *Genetic influences on daily intake and meal patterns of humans.* Physiol Behav, 1993. **53**(4): p. 777-82.

133. de Castro, J.M., *Independence of genetic influences on body size, daily intake, and meal patterns of humans.* Physiol Behav, 1993. **54**(4): p. 633-9.

134. de Castro, J.M., *A twin study of genetic and environmental influences on the intake of fluids and beverages.* Physiol Behav, 1993. **54**(4): p. 677-87.

135. de Castro, J.M., *Stomach filling may mediate the influence of dietary energy density on the food intake of free-living humans.* Physiol Behav, 2005.

136. Pearcey, S.M. and J.M. de Castro, *Food intake and meal patterns of weight-stable and weight-gaining persons.* Am J Clin Nutr, 2002. **76**(1): p. 107-12.

137. de Castro, J.M., *Genes and environment have gender-independent influences on the eating and drinking of free-living humans.* Physiol Behav, 1998. **63**(3): p. 385-95.

138. De Castro, J.M., *Age-related changes in spontaneous food intake and hunger in humans.* Appetite, 1993. **21**(3): p. 255-72.

139. De Castro, J.M., *Social facilitation of duration and size but not rate of the spontaneous meal intake of humans.* Physiol Behav, 1990. **47**(6): p. 1129-35.

140. van den Bree, M.B., L.J. Eaves, and J.T. Dwyer, *Genetic and environmental influences on eating patterns of twins aged >/=50 y.* Am J Clin Nutr, 1999. **70**(4): p. 456-65.

141. Keski-Rahkonen, A.V., *Genetic and environmental factors in breakfast eating patterns.* Behav Genet, 2004. **34**: p. 503-514.

142. Lichtman, S.W., et al., *Discrepancy between self-reported and actual caloric intake and exercise in obese subjects.* N Engl J Med, 1992. **327**(27): p. 1893-8.

143. Faith, M.S., et al., *Evidence for genetic influences on human energy intake: results from a twin study using measured observations.* Behav Genet, 1999. **29**(3): p. 145-54.

144. Faith, M.S. and K.L. Keller, *Genetic architecture of ingestive behavior in humans.* Nutrition, 2004. **20**(1): p. 127-33.

145. Aubert, R., et al., *5-HT2A receptor gene polymorphism is associated with food and alcohol intake in obese people.* Int J Obes Relat Metab Disord, 2000. **24**(7): p. 920-4.

146. Herbeth, B., et al., *Polymorphism of the 5-HT2A receptor gene and food intakes in children and adolescents: the Stanislas Family Study.* Am J Clin Nutr, 2005. **82**(2): p. 467-70.

147. Farooqi, I.S., G.S. Yeo, and S. O'Rahilly, *Binge eating as a phenotype of melanocortin 4 receptor gene mutations.* N Engl J Med, 2003. **349**(6): p. 606-9; author reply 606-9.

148. Branson, R., et al., *Binge eating as a major phenotype of melanocortin 4 receptor gene mutations.* N Engl J Med, 2003. **348**(12): p. 1096-103.

149. Stunkard, A.J. and S. Messick, *The three-factor eating questionnaire to measure dietary restraint, disinhibition and hunger.* J Psychosom Res, 1985. **29**(1): p. 71-83.

150. Steinle, N.I., et al., *Eating behavior in the Old Order Amish: heritability analysis and a genome-wide linkage analysis.* Am J Clin Nutr, 2002. **75**(6): p. 1098-106.

151. Bouchard, L., et al., *Neuromedin beta: a strong candidate gene linking eating behaviors and susceptibility to obesity.* Am J Clin Nutr, 2004. **80**(6): p. 1478-86.

152. Kraemer, H.C., et al., *How do risk factors work together? Mediators, moderators, and independent, overlapping, and proxy risk factors.* Am J Psychiatry, 2001. **158**(6): p. 848-56.

153. Ravussin, E., et al., *Effects of a traditional lifestyle on obesity in Pima Indians.* Diabetes Care, 1994. **17**(9): p. 1067-74.

154. Bhatnagar, D., et al., *Coronary risk factors in people from the Indian subcontinent living in west London and their siblings in India.* Lancet, 1995. **345**(8947): p. 405-9.

155. Bouchard, C., et al., *The response to long-term overfeeding in identical twins.* N Engl J Med, 1990. **322**(21): p. 1477-82.

156. Ukkola, O. and C. Bouchard, *Role of candidate genes in the responses to long-term overfeeding: review of findings.* Obes Rev, 2004. **5**(1): p. 3-12.

157. Epstein, L.H., et al., *Relation between food reinforcement and dopamine genotypes and its effect on food intake in smokers.* Am J Clin Nutr, 2004. **80**(1): p. 82-8.

158. Dietz, W.H., *Critical periods in childhood for the development of obesity.* Am J Clin Nutr, 1994. **59**(5): p. 955-9.

159. Allison, D.B., et al., *Is the intra-uterine period really a critical period for the development of adiposity?* Int J Obes Relat Metab Disord, 1995. **19**(6): p. 397-402.

160. Stettler, N., et al., *Rapid weight gain during infancy and obesity in young adulthood in a cohort of African Americans.* Am J Clin Nutr, 2003. **77**(6): p. 1374-8.

161. Stettler, N., et al., *Infant weight gain and childhood overweight status in a multicenter, cohort study.* Pediatrics, 2002. **109**(2): p. 194-9.

162. Birch, L.L., J.O. Fisher, and K.K. Davison, *Learning to overeat: maternal use of restrictive feeding practices promotes girls' eating in the absence of hunger.* Am J Clin Nutr, 2003. **78**(2): p. 215-20.

163. Mennella, J.A., M.Y. Pepino, and D.R. Reed, *Genetic and environmental determinants of bitter perception and sweet preferences.* Pediatrics, 2005. **115**(2): p. e216-22.

164. Mennella, J.A., C.E. Griffin, and G.K. Beauchamp, *Flavor programming during infancy.* Pediatrics, 2004. **113**(4): p. 840-5.

165. Mennella, J.A., M.Y. Pepino, and G.K. Beauchamp, *Modification of bitter taste in children.* Dev Psychobiol, 2003. **43**(2): p. 120-7.

166. Gerrish, C.J. and J.A. Mennella, *Flavor variety enhances food acceptance in formula-fed infants.* Am J Clin Nutr, 2001. **73**(6): p. 1080-5.

167. Mennella, J.A. and G.K. Beauchamp, *Experience with a flavor in mother's milk modifies the infant's acceptance of flavored cereal.* Dev Psychobiol, 1999. **35**(3): p. 197-203.

168. Mennella, J.A. and G.K. Beauchamp, *Early flavor experiences: research update.* Nutr Rev, 1998. **56**(7): p. 205-11.

169. Fisher, J.O., et al., *Breast-feeding through the first year predicts maternal control in feeding and subsequent toddler energy intakes.* J Am Diet Assoc, 2000. **100**(6): p. 641-6.

170. Caldwell, B.M. and R.H. Bradley, *Home observation for measurement of the environment*. 1984, Little Rock: University of Arkansas.

171. Strauss, R.S. and J. Knight, *Influence of the home environment on the development of obesity in children*. Pediatrics, 1999. **103**(6): p. e85.

172. Caspi, A., et al., *Neighborhood deprivation affects children's mental health: environmental risks identified in a genetic design*. Psychol Sci, 2000. **11**(4): p. 338-42.

173. Kim-Cohen, J., et al., *Genetic and environmental processes in young children's resilience and vulnerability to socioeconomic deprivation*. Child Dev, 2004. **75**(3): p. 651-68.

174. Rolls, B.J., et al., *Increasing the portion size of a sandwich increases energy intake*. J Am Diet Assoc, 2004. **104**(3): p. 367-72.

175. Raynor, H.A. and L.H. Epstein, *The relative-reinforcing value of food under differing levels of food deprivation and restriction*. Appetite, 2003. **40**(1): p. 15-24.

176. Raynor, H.A. and L.H. Epstein, *Dietary variety, energy regulation, and obesity*. Psychol Bull, 2001. **127**(3): p. 325-41.

177. Plomin, R., J.C. DeFries, and J.C. Loehlin, *Genotype-environment interaction and correlation in the analysis of human behavior*. Psychol Bull, 1977. **84**(2): p. 309-22.

178. Loehlin, J.C., *Behavior genetics and parenting theory*. Am Psychol, 2001. **56**(2): p. 169-70.

179. Collins, W.A., et al., *Toward nature with nurture*. Am Psychol, 2001. **56**(2): p. 171-3.

180. Collins, W.A., et al., *Contemporary research on parenting. The case for nature and nurture*. Am Psychol, 2000. **55**(2): p. 218-32.

181. Wachs, T.D. and R. Plomin, *Conceptualization and Measurement of Organism-Environment Interaction*. 1991, Washington, DC: American Psychological Association.

182. Hetherington, E.M., D. Reiss, and R. Plomin, *Separate Social Worlds of Siblings*. 1994, Hillsdale, NJ: Lawrence Erlbaum Associates Publishers.

183. Lynch, M. and B. Walsh, *Genetics and Analysis of Quantitative Traits*. 1998, Sunderland, MA: Sinauer Associates, Inc.

184. Lytton, H., *Parent-Child Interaction. The Socialization Process Observed in Twin and Singleton Families*. 1980, New York: Plenum Press.

185. Emde, R.N. and J.K. Hewitt, *Infancy to Early Childhood. Genetic and Environmental Influences on Developmental Change*. 2001, New York: Oxford University Press.

186. Reiss, D., et al., *The Relationship Code. Deciphering Genetic and Social Influences on Adolescent Development*. 2000, Cambridge, MA. London: Harvard University Press.

187. de Castro, J.M. and S.S. Plunkett, *How genes control real world intake: palatability-intake relationships*. Nutrition, 2001. **17**(3): p. 266-8.

188. de Castro, J.M., *Eating behavior: lessons from the real world of humans*. Nutrition, 2000. **16**(10): p. 800-13.

D

The Interaction of Social, Behavioral, and Genetic Factors in Sickle Cell Disease

Robert J. Thompson, Jr., Ph.D. *
Professor of Psychology

INTRODUCTION

The genomics revolution has added powerful new potentialities and renewed impetus for understanding how biological, social, and behavioral processes act together in health and illness. More specifically, the genomics revolution is driving a paradigm shift from reductionistic approaches that focus on elements in isolation to systems approaches that focus on the interconnectedness of networks of elements acting as a whole. The challenge is "to connect the dots" and delineate patterns of transactions with regard to mechanisms of effect across scale. In particular, the genomics revolution has increased awareness of the role of promoters and enhancers in switching on and off specific genes as one mechanism of effect for health outcomes that can be triggered by social and behavioral factors as well as biological factors. In this way, genes are viewed as more than units of heredity but as mechanisms for extracting information from environmental experiences (Ridley, 2003).

The current paradigm shift, propelled by the genomics revolution, can be viewed as the most recent progression in conceptualization of health and illness. By the mid-1970s there was growing recognition of the limits of the biomedical model that explained illness in terms of single-factor biological malfunction with little attention to behavioral and social processes. George Engel (1977) traced the historical origins of the reductionistic biomedical model to assumptions of mind-body dualism and advo-

*Department of Psychology, Social and Health Sciences, Duke University.

cated a biopsychosocial model as a way to "broaden the approach to disease to include the psychosocial without sacrificing the enormous advantages of the biomedical approach" (p. 131). The biopsychosocial model maintains that health and illness are a function of multiple processes— biological, psychological, and social—and these processes must be considered simultaneously. In particular, the emergence of multifactorial approaches to the pathogenesis of disease enabled linkage between the behavioral and biomedical sciences (Weiss, 1987). Also important were systems theory perspectives and models of how biological and psychosocial processes act together in human development across the life span (Bronfenbrenner, 1977, 1979). A systems theory perspective focuses on the accommodations that occur through the life span between the developing organism and the changing environment.

The biopsychosocial model focuses on multiple factors in the etiology and progression of disease. Three primary mechanisms of effect have emerged: health behaviors, psychosocial processes, and genetics. Health behaviors include exercise, nutrition, smoking, and adherence to medical regimes. Psychosocial processes include a range of interpersonal and social processes that affect interpretation of environmental experiences and responses to stress. Risk-resiliency models are also prevalent and seek to identify factors and processes that enhance or decrease vulnerability to disease processes. A particular area of focus has been neuroendocrine and immune responses to stress. One mechanism of effect is through the impact of how individuals interpret and respond to the environment which influences the degree of stress experienced which in turn influences health behaviors and neuroendocrine and immune responses that in turn affect the etiology and progression of disease. Genetic mechanisms of effect involve the identification of internal and external factors that trigger the switching on or off of genes that modulate physiological processes.

The primary interest prompting this paper is enhanced understanding of the interaction of social, behavioral, and genetic factors on health. Sickle cell disease was selected as a good model for this investigation because it is a monogenetic event but the phenotype is multigenetic resulting in considerable individual differences in severity of the disease. More specifically, this paper addresses the following questions:

- What do we know about the influence of social and behavioral factors and the effects of other genes?
- What data do we have?
- What data do we need?
- What important questions remain to be answered about the influences of social and behavioral factors, including mechanisms, on sickle cell disease?

- Given the same genes, what is the evidence that social environment affects genes?
- What additional research on sickle cell would enlighten the broader relationship between single gene disorders and the social environment?

This review focused on the factors and processes associated with individual differences in clinical manifestations of sickle cell disease. Three lines of research are apparent that correspond to the three mechanisms of effect that have emerged from the biopsychosocial model. There are data about the effects of health behaviors on sickle cell disease, such as avoiding cold and maintaining hydration. Similarly, there are data regarding the role of psychosocial processes in the psychological adjustment of children, adolescents, and adults with sickle cell disease and with regard to the specific symptom of pain, and health services utilization. There are also data about the role of polymorphic genetic factors in the variability in the phenotypic expression of sickle cell disease as reflected in various indicators of pathophysiology. However, data do not yet exist regarding the interaction of psychosocial, behavioral, and genetic factors in the variability in the clinical manifestations and course of sickle cell disease. It is rare for markers of behavioral and psychosocial processes and genetic markers to be included in the same study. In contrast, the interaction of behavioral, psychosocial, and genetic factors in the variability in the physiological response to stress has been investigated. This suggests that the way to advance our understanding of this interaction of factors in sickle cell disease, as a model of a single gene disorder, is to focus on the interaction of behavioral, psychosocial, and genetic factors in the neuroendocrine and immune physiological response to stress and the subsequent impact on the pathophysiological processes of vasoocclusion, infection, and neurocognitive dysfunction that are central to sickle cell disease.

This paper is intended for a broad audience with varying degrees of background in the genetic, pathophysiological, and psychosocial aspects of sickle cell disease. The general, nontechnical level of this paper is a necessity given that the author's background is that of a pediatric psychologist and not a molecular biologist or physician. References are provided to facilitate fuller consideration and specific processes.

This report is organized in four parts. The first section reviews the etiology, pathophysiology, and clinical manifestations of sickle cell disease and considers what is known about the role of polymorphic genetic factors in the phenotypic expression of the disease. The second section reviews what is known about the impact of social and behavioral factors on the clinical manifestations of sickle cell disease, particularly on psychological adjustment, pain, and neurocognitive functioning. The third section considers stress as a common mechanism of effect through which behavioral, social, and genetic processes affect health outcomes. The paper concludes with a consideration of future research needs and directions.

SICKLE CELL DISEASE: ETIOLOGY, PATHOPHYSIOLOGY, AND CLINICAL MANIFESTATIONS

The adult hemoglobin molecule (Hb A) is compromised of a duplicated pair of alpha (α) and a pair of beta (β) chains. The α-globin gene cluster is located on chromosome 6 and the β-globin gene cluster is located on chromosome 11. The structure of hemoglobin changes during development. Embryonic hemoglobin is replaced by fetal hemoglobin (Hb F) shortly before birth which in turn is replaced by adult hemoglobin (Hb A) over the first year of life (Weatherall, 2001).

Sickle cell disease refers to a group of related autosomal recessive blood disorders caused by a variant of the β-globin gene called sickle hemoglobin (Hb S). A single nucleotide substitution (GTG → GAG) in the sixth codon of the β-globin gene results in the substitution of valine for glutomic acid which in turn allows Hb S to polymerase when deoxygenated. "A polymerization of deoxygenated Hb S is a primary indispensable event in the molecular pathogenesis of sickle cell disease" (Stuart and Nagel, 2004, p. 1343). Inherited autosomal recessively, either two copies of Hb S (Hb SS), referred to as sickle cell anemia, or one copy of Hb S plus another β-globin variant are required for sickle cell disease. In addition to sickle cell anemia, homozygotic Hb SS disease, there are several other compound heterozygote sickle genotypes of Hb S plus one copy of another β-globin gene variant, Hb C or Hb β-thalassemia. The carrier state, sickle cell trait, has one copy of the normal β-globin gene and one copy of the sickle variant (Hb AS) (Ashley-Koch et al., 2000).

Four major β-globin gene haplotypes have been identified. Three are named for regions in Africa in which the mutations first appeared: BEN (Benin), SEN (Senegal), and CAR (Central African Republic). The fourth haplotype, Arabic-India, occurs in India and the Arabic peninsula (Quinn and Miller, 2004).

Disease severity is associated with several genetic factors. "Genotype is the most important risk factor for disease severity" (Ashley-Koch et al., 2000, p. 842). The highest degree of severity is associated with Hb SS followed by Hb s/β0-thalassemia and Hb SC and Hb S/β+-thalassemia are associated with a more benign course of the disease (Ashley-Koch et al., 2000). Disease severity is also related to β-globin haplotypes, probably due to variations in hemoglobin level and fetal hemoglobin concentrations. The Senegal haplotype is most benign, followed by the Benin, and the Central African Republic haplotype is the most severe form (Ashley-Koch et al., 2000). Another genetic factor associated with disease severity is α-globin gene compliment. Thus, although sickle cell disease is a monogenetic disorder, its phenotypical expression is multigenetic. Epistatic or modifier genes include the co-presence of α-thalassemia, the .158 C → T mutation that enhances Hb F expression, particularly in the Senegal and the Arab-Indian globin cluster haplotypes, and the female population (Stuart and Nagel, 2004). Steinberg (2005) maintains that: "Understanding the vascular and inflammatory components of the disease pathophysiology provides many loci where the disease phenotype can be impacted by modifying genes" (p. 465).

Pathophysiology

There are two cardinal pathophysiologic features of sickle cell disease: chronic hemolytic anemia and vasoocclusion. The polymerization of the hemoglobin S molecule (Hb S) within the red blood cells upon deoxygenation causes the red blood cells to change from the usual biconcave disc to an irregular sickled or crescent shape. Upon reoxygenation, the red cell initially resumes a normal configuration but after repeated cycles, the erythrocyte is damaged permanently, resulting in red cell dehydration and erythrocyte destruction. Sickled red blood cells also have a propensity to adhere to the walls of blood vessels and are susceptible to hemolysis, causing chronic anemia (Ashley-Koch et al., 2000). The deformed red blood cells cause microcirculatory obstruction and prevent normal blood flow and decreased delivery of oxygen to organs and tissues resulting in the vasoocclusive crisis. However, information summarized by Stuart and Nagel (2004) indicates that the actual mechanism is more complicated.

One of the factors complicating the pathophysiology is cell heterogeneity. Sickle cells vary in their density and deformity because cation homeostasis is impaired in some cells. The amount of hemolysis is related to the number of irreversibly sickled cells and dense cells (Steinberg and Rodgers, 2001). Another factor that varies is fetal hemoglobin (Hb F) concentrations. Vasoocclusive events depend on the interaction of features intrinsic to the sickled erythrocyte, including degree of polymer formation and cellular damage, interacting with other factors in the cells environment such as endothelial cells and leukocytes (Steinberg and Rodgers, 2001). Other potentially contributing factors include neutrophil transmigration that "adds to the increased inflammation in the microvascularture" and "disregulation of vasomotor tone by perturbations in vasodilator mediators such as nitrous oxide (NO)" (Stuart and Nagel, 2004, p. 1345). The abnormal cation homeostasis contributes to dehydrated dense sickle cells which in turn contributes to anemia and hemolysis.

The recognition that the adherence of sickled erythrocytes to the endothelium correlated with disease severity focused attention on the mechanisms involved (Stuart and Nagel, 2004). As a barrier between blood and tissue, endothelial cells have a number of functions that may contribute to the vascular pathology of sickle cell disease and "genetic differences are likely to cause different responses among patients" (Steinberg and Rodgers, 2001, p. 300). One of the functions of endothelial cells is to control vascular tone by elaborating vasoconstrictors and vasodilators. Endothelial cells also express genes adhesion molecules for blood cells and proteins (Steinberg and Rodgers, 2001). Endothelial cell activators are generated by a number of factors such as hypoxia, thrombin, and infection (Steinberg and Rodgers, 2001). Other extra-erythrocyte related pathophysiological factors include leukocyte size, rigidity, and adhesive characteristics and coagulation activation, with thrombin hypothesized as potentially providing a crucial link between coagulation activation and adhesion (Stuart and Nagel, 2004). Of particular interest is the finding that laminin bonds strongly to sickle erythrocytes via the protein that carries Lutheran blood-group antigens (B-CAM/Lu) and epinephrine increases this adhesion. "Since stress is a potential initiation factor for vasoocclusion, epinephrine modulation of adhesion provides a powerful biological link between intraerythrocytic signaling pathways and the external milieu" (Stuart and Nagel, 2004, p. 1346).

Clinical Manifestations

Two primary consequences of hypoxia secondary to vasoocclusive crisis are pain and damage of organ systems. The organs at greatest risk are those where blood flow is slow, such as the spleen and bone marrow, or those with a limited terminal arterial blood supply, including the eye and the head of the femur and humerus, and lung as the recipient of deoxygenated sickle cells that escape the spleen or bone marrow. Major clinical manifestations of sickle cell disease include painful events, acute chest syndrome, splenic dysfunction, and cerebrovascular accidents.

Painful events occur as a result of ischemic tissue injury and can be precipitated by hypoxia, dehydration, and extreme cold. The frequency and severity of painful events are varied. Musculoskeletal pain is the most common, followed by abdominal pain, and low back pain. Painful events typically last 4-6 days. Transduction is the process whereby noxious inflammatory mediators that are generated by tissue damage in turn activate nociceptors to chemical or mechanical forms of energy to an electrochemical impulse, which is transmitted along the spirothalamic tract to the thalamus which in turn transmits the signal to the brain where it is perceived as pain (Ballas, 2001a). Descending fibers in the midbrain can inhibit the transmission of painful stimuli via endogenous endorphins and communications through the limbic system can modulate the emotional response to pain and thereby enhance or inhibit the intensity of the perception of pain (Ballas, 2001a).

Acute chest syndrome involves chest pain, fever, increased leukocytosis, hypoxemia, and pneumonia-like symptoms. Typical causes include infection and pulmonary infarction (Ballas, 2001b). This acute illness can be self-limiting or can rapidly progress and may be fatal. "Risk factors include HB SS genotype, low HB F concentrations and high steady state leukocyte and HB concentrations" (Stuart and Nagel, 2004, p. 1350).

Splenic dysfunction develops during infancy and predisposes the infant to overwhelming infection from encapsulated bacteria, particularly streptococcus pneumonia and haemophilus influenza. Between the ages of 5 months and 2 years, children with sickle cell anemia are at risk for sudden intrasplenic pooling of vast amounts of blood, known as splenic sequestration. The hemoglobin level can drop precipitously, causing hypovolemic shock and death. High concentrations of Hb F serve as a protection factor (Stuart and Nagel, 2004).

Stroke affects 6-12% of patients with sickle cell disease. In children, the most common cause of stroke is cerebral infarction; intracerebral hemorrhages become increasingly common with age. Recurrent stroke causes progressive impairment of cognitive functioning. "Risk factors include the HB SS phenotype, previous transient ischemic attacks, low steady state HB concentrations, high leukocyte counts, raised systolic blood pressure, and previous acute chest syndrome" (Stuart and Nagel, 2004, p. 1351). Silent brain lesions have been evidenced on magnetic resonance imaging (MRI) accompanied by neurocognitive deficits (Armstrong et al., 1996).

The efforts to enhance clinical care are focusing on increasing understanding of the pathophysiology of sickle cell disease to enable a precise prognosis and individualized treatment. What is required is knowledge about which genes are associated with the hemolytic and vascular complications of SCD and "how variants of these genes interact among themselves and with their environment" (Steinberg, 2005, p. 465).

Genetic Modulation of Disease Severity

Individual differences occur in part through differences in the order and pattern of gene expression (i.e., variations in the regulatory sequence of the genome, referred to as promoters). A promoter is a special sequence of bases usually found immediately upstream of the gene itself. A gene is expressed or transcribed into messenger RNA by the binding of a protein called a transcription factor to a promoter. The binding of a transcription factor and the expression of a gene can be altered by experience (Ridley, 2004). An example is the elevation of cortisol that occurs upon appraisal of a situation as stressful, which in turn alters gene expression in the immune system by reducing the expression of interleukin II and turning down the activity, number, and life span of lymphocytes (Ridley, 2004). Two broad molecular genetic strategies have been employed to identify the role of genes (de Gues, 2002). One strategy involves whole genome scans through linkage analysis. The advantage of this approach is that all relevant genes are examined but the disadvantage is that it requires large samples of genetically related subjects. A second approach is an allelic association or candidate gene studies. Associations with known functional candidate genes are investigated, for example "genes suspected to influence neurotransmission in the brain because they code for protein constituents of receptors, transporters, or enzymes involved in neurotransmitter synthesis and degradation (Plomin and Crabbe, 2000)" (de Gues, 2002, p. 4). The advantage of this approach is the ability to use smaller samples of unrelated subjects but the disadvantage is that some genetic influences are missed because they are not among the candidate genes studied.

It is easier to identify the effect of the gene on a more elementary trait than on a complex one. The strategy is to identify an endophenotype that is upstream of the more complex effect, determine the amount of variance that the gene explains in the endophenotype, and then determine the variance explained in the disease outcome by the endophenotype. Identifying allelic candidate genes is a matter of looking for genes that are part of a system known to influence the disease. The genes influence the disease by influencing the concentration of a protein or its functionality or efficiency or responsiveness to the environment.

Ridley (2004) maintains that "Diversity in the human population is starting to be explained at least as much by variations in the number of repeats of a genetic phrase in the regulatory region of the gene as by single-nucleotide polymorphisms" (p. 97). "Varying the number of repeats of a phase has a much subtler effect on gene function then does changing a single nucleotide in a codon, which tends to shut a gene down" (p. 97).

Steinberg (2005) views the use of Bayesian networks as a promising approach for the discovery of the genetic basis of complex traits in large association studies and describes a Bayesian network that was developed to analyze 235 single nucleotide polymorphisms (SNPs) in 80 candidate genes in 1398 unrelated patients with sickle cell anemia. The findings indicated that "SNP's on 11 genes and four clinical variables, including α-thalassemia and Hb F, interacted in a complex network of dependency to modulate the risk of stroke. This network of intersections included three genes, BMP6, TGFBR2, and TGFBR3 with a functional role in the TGF-β [transforming growth factor-β pathway and one gene (SELP) associated with stroke in the general population" (Steinberg, 2005, p. 472). Subsequently, this model was validated by predicting the occurrence of stroke in a different population with a true positive rate of 100%; a true negative rate of 98.14%; and an overall predictive accuracy of 98.2% (Sabastiani et al., 2005). In his comprehensive review of predictors of SCD complications, Steinberg (2005) considers both established predictors, including fetal hemoglobin and α-thalassemia, and potential predictors.

Fetal Hemoglobin

Fetal hemoglobin (Hb F) inhibits Hb S polymerization and higher levels are associated with a reduction of most vasoocclusive complications of sickle cell anemia (Steinberg, 2005). However, Hb F concentrations vary among patients with sickle cell anemia, ranging from 0.1% to 30%, and there is considerable variability in severity of complications among patients with similar concentration levels.

Typical levels of Hb F vary across the four major β-globin haplotypes. The highest Hb F level and mildest clinical course is found in carriers of the Hb S gene on the Senegal or Arab-India haplotype, intermediate levels and severity on the Benin haplotype, and the lowest levels and most severity on the Bantu (Central African Republic) haplotype (Steinberg, 2005).

Fetal hemoglobin expression is a quantitative trait and investigations are addressing complex interactions among transcription factors, genes modulating erythropoiesis, and elements linked to the β-globin cluster. In addition, similar genetic analyses are being undertaken in an effort to predict responsiveness to hydroxyurea, which is used to treat the complications of SCD and is thought to work by increasing Hb F levels (Steinberg, 2005).

α-Thalassemia

Alpha thalassemia is the result of the deletion of one of two α-globin genes from a chromosome (Nagel and Steinberg, 2001). Coincidental α-thalassemia occurs in approximately 30% of patients with sickle cell anemia and affects the phenotype of sickle cell anemia by reducing the concentration of Hb S polymerization (Steinberg, 2005). The presence of α-thalassemia with sickle cell anemia is also associated with less hemolysis, higher concentration of hemoglobin (Nagel and Steinberg, 2001) and higher packed cell volume (PCV), and lower mean corpuscular volume and reticulocyte counts (Steinberg, 2005). However, the clinical effects of coexisting α-thalassemia are mixed. Benificial effects are generally found with vasoocclusive events that are dependent on PCV, such as stroke and leg ulcer, whereas deleterious effects are associated with complications that are dependent on blood viscosity, such as painful episodes and acute chest syndrome (Steinberg, 2005).

Since the diversity of sickle cell anemia cannot be explained entirely by Hb F and α-globin gene-linked modulation, attention is being directed to epistatic or modifying genes that act independently of Hb S polymerization. The genes that potentially could modulate the phenotype of sickle cell anemia include: "mediators of inflammation, oxidant injury, NO biology, vasoregulation, cell-cell interaction, blood coagulation, haemostasis, growth factors, cytokine and receptors and transcriptional regulators" (Steinberg, 2005, p. 470). However, studies of candidate genes, seeking associations of SNP with phenotypes, are in the beginning stages and present many interpretative challenges (Steinberg, 2005).

ROLE OF BEHAVIORAL AND PSYCHOSOCIAL FACTORS IN SICKLE CELL DISEASE

Consistent with the biopsychosocial model, investigations of the role of behavioral and psychosocial factors in sickle cell disease have been bidirectional. One line of research has focused on the impact of sickle cell disease on psychological adjustment in children and adolescence with sickle cell disease and their parents, and adults with sickle cell disease. Another line of research has focused on the impact of behavioral and psychosocial processes on selected dimensions of disease outcome, particularly with regard to pain and neurocognitive functioning.

Psychological Adjustment

The findings with regard to the psychological adjustment of children with sickle cell disease are consistent with those for children with chronic illnesses in general (Thompson and Gustafson, 1996). The risk of psychological adjustment problems in children with chronic illness is 1.5 to 3 times as high as with their healthy peers (Thompson and Gustafson, 1996). In addition to determining the type and frequencies of adjustment problems, effort has been directed to identifying the mediating and moderating role of illness parameters, typically disease severity, and psychological and social processes to adjustment to the stress of chronic illness. The transactional stress and coping model (Thompson and Gustafson, 1996; Thompson et al., 1992) has proven to be a useful conceptual framework for these investigations and psychological adjustment was the target of a number of studies done through the Duke University of North Carolina Sickle Cell Center.

Psychological adjustment was assessed in a study of 50 children, age 7 to 17 years of age with sickle cell disease, (Hb SS 60%; Hb SC 12%; sickle β-thalassemia syndromes 16%). In terms of mother reported behavioral problems, 64% of the children were classified with poor adjustment, primarily of the internal behavior problem type. In terms of child self-report, as assessed through a semi-structured diagnostic interview, 50% reported symptoms that met the criteria for one or more DSM-III diagnosis. Internalizing problems reflected in anxiety, phobic, and obsessive-compulsive diagnoses were most frequent. In contrast, externalizing problems reflected in conduct disorder and oppositional disorder were relatively infrequent. Hierarchal multiple regression analysis was utilized to assess the increment in psychological adjustment accounted for by maternal psychological adjustment and children's cognitive processes and pain coping strategies over and above that accounted for by demographic parameters and illness severity parameters, including type of sickle cell disease, pain frequency, pain severity, and number of complications. In terms of the variance in mother-reported internalizing behavioral problems, the demographic variables of gender, socioeconomic status, and age accounted for 8% and the illness parameters of pain frequency and type of sickle cell disease accounted for another 9% and 8%, respectively. Maternal anxiety accounted for 16% of the variance in mother-reported internalizing behavioral problems and 33% in mother-reported externalizing behavioral problems. In terms of child-reported total symptom score, sickle cell type did not account for any of the variance, the number of illness complications accounted for 2%, and pain frequency accounted for 1%. The demographic variables of socioeconomic status and gender only accounted for 6% of the variance. However, children's pain coping strategies characterized by negative thinking accounted for a 21% increment in child reported total symptom score.

Psychological adjustment over time was assessed at 3 points across 2 years with a sample of 50 children with sickle cell disease (Hb SS, 54%; Hb SC, 34%; sickle β-thalassemia, 12%; males, 64%; females, 36%). In terms of child-reported symptoms, 12% met diagnostic criteria for a DSM-III diagnosis across all three time points whereas 17% consistently demonstrated good adjustment. The variability in report of symptoms meeting diagnostic criteria over time is also reflected by the percentage of children who had 1 (49%) or 2 (27%) changes in adjustment classification over the three-time periods. In terms of specific diagnoses, internalizing disorders were most frequent at each time but there was very little consistency in specific diagnoses across time. In terms of mother-reported behavioral problems, 47% met the criteria for poor adjustment and 19% for good adjustment across all three assessment points. One change in classification occurred for 25% and two-changes occurred for 4% (Thompson et al., 1999a).

Maternal psychological adjustment was assessed in a study of 78 mothers of children and adolescence, 7 to 17 years of age, with sickle cell disease (Hb SS, 62%; Hb SC 23%; and sickle β-thalassemia syndromes; 15%) (Thompson et al., 1993b). In terms of self-reported symptoms of psychological distress, 36% of mothers' met criteria for poor psychological adjustment. None of the illness or demographic parameters accounted for significant amounts of variance in mothers' symptoms scores. Over and above the 7% of variance in adjustment accounted for by illness parameters and 2% by demographic parameters, a 46% increment in variance was accounted for by three psychosocial processes: mother-reported use of palliative coping in relation to active coping (30%); stress associated with daily hassles (13%); and family functioning characterized by an emphasis on control (3%) (Thompson et al., 1993a).

In a study of maternal adjustment across three assessment points over 2 years, 43% of mothers of children with sickle cell disease consistently met the criteria for poor adjustment (Thompson et al., 1999b). Mothers with stable good adjustment differed significantly from those with stable poor adjustment in terms of lower levels of daily stress and use of palliative coping methods in relation to adaptive coping and lower levels of illness-related stress.

The psychological adjustment of adults with sickle cell disease was assessed in a sample of 109 patients (female, 55%; male, 45%;) ranging in age from 18-68 years (Hb SS, 77%; Hb SC, 12%; and sickle β-thalassemia syndromes, 11%). The criteria for poor adjustment in terms of self-reported symptoms of psychological distress was met by 56% of the patients with 40% demonstrating elevations into the clinical range of distress on depression and 32% on anxiety. In terms of illness parameters, type of sickle cell disease and number of complications accounted for no significant increment in reported psychological distress and pain frequency only accounted for a 2% increment. Similarly, the demographic parameters of socioeconomic status, gender, and age only accounted for an additional 9% of the variance. In contrast, with these variables controlled, daily stress accounted for an additional 35% of the variance in reported psychological distress and pain coping strategies characterized by negative thinking accounted for an additional 4% (Thompson et al., 1992).

The stability of psychological adjustment across three time periods spanning 20 months was assessed in a study of 59 African American adults with sickle cell anemia (Thompson et al., 1996). In terms of self-reported symptoms of psychological distress, consistently poor adjustment was demonstrated by 32% and consistently good adjustment by 25% of patients. Variability in adjustment at the individual level was also reflected in 26% of the patients changing classifications once and 17% changing twice over the three assessment points. With adjustment at the 20-month follow-up period as the outcome measure, the illness parameters of complications and pain frequency at baseline did not account for any significant increment in variance and the demographic parameter of socioeconomic status only accounted for 6%. With illness and demographic parameters controlled, baseline levels of daily stress accounted for a 29% increment in psychological distress at 20-month follow-up and illness-related stress accounted for another 8% increment.

A multisite Cooperative Study of Sickle Cell Disease (Farber et al., 1985; Gaston and Rosse, 1982) provided an opportunity to examine the independent and combined contributions of family functioning and neurocognitive functioning to behavioral problems in children with sickle cell disease. In an initial cross-sectional study of 289 children (Hb SS, 68%; Hb SC, 32%; males, 52%; females, 48%) 5.9 to 15.5 years of age completed a neuropsychological evaluation, brain MRI and mothers completed the child behavior checklist and family environment scale (Thompson et al., 1999a). Mother-reported behavior problems occurred with 30% of the patients. The subgroup with behavior problems had significantly lower verbal IQ, reading, and math scores and lower levels of family support and higher levels of family conflict. The rate of behavioral problems did not vary across the three subgroups formed on the basis of MRI status (normal, clinically apparent cerebral infarction, and silent infarction). Demographic parameters of child age and gender and mother age and education and the biomedical parameters of hematocrit level and type of SCD each accounted for only 2% of the variance in behavioral problems. However, family functioning characterized as conflicted, reflecting both high levels of conflict and a lack of organization and support, accounted for a 19% increment in variance in behavioral problems.

The relationship of behavioral problems, intellectual functioning, and family functioning was assessed longitudinally in a follow-up prospective study of 222 children with at least two complete sets of measures obtained across four assessment points over the study period of nine years. The findings indicated that overall 60% of the children were consistently classified in terms of behavioral problems (9%) or good adjustment (51%) based on at least three measures across four assessment points. The risk of consistent behavior problems was not related to MRI classification, gender, education level of the mother, or age of the child but significantly increased with higher baseline levels of family conflict and decreased with higher baseline full-scale IQ. More importantly, an increase in behavioral problems was associated with a reported increase in family conflict but was not related to change in intellectual functioning. There was a decline in neurocognitive functioning over time. On average, full-scale IQ decreased 1.2 points per year with age and compared with a child with a normal MRI, was 3.8 points lower for a child with silent infarction and 14.4 points lower for a child with stroke.

In summary, the findings across a number of studies indicate an increased risk for psychological adjustment problems in children and adolescents with sickle cell disease and their mothers and adults with sickle cell disease. However, there is considerable variability in adjustment over time and good adjustment is the norm. Biomedical indicators of disease severity, including type of sickle cell disease and number of complications and frequency of pain episodes, account for very little variance in psychological adjustment. Similarly, demographic factors of gender, age, and socioeconomic status also account for very little variance in adjustment. In contrast pain coping strategies characterized by negative thinking and passive adherence account for a significant portion of variance in children and adolescent psychological adjustment. Stress processing variables account for a significant portion of variance in the adjustment of adult patients with sickle cell disease and mothers of children and adolescents with sickle cell disease. More specifically appraisals of stress, especially daily stress, use of palliative coping strategies, and family functioning characterized by low levels of supportiveness and high levels of conflict account for significant increments in adjustment variance over and above that accounted for by illness and demographic factors.

Neurocognitive Functioning

The cooperative study of sickle cell disease included neuropsychological and MRI assessment of children 6-12 years of age (Armstrong et al., 1996). For children with Hb SS disease, 6.6% had a clinical CVA and 15.6% had a silent infarct. For children with Hb SC disease, none had evidence of CVA and 5.1% demonstrated a silent infarct. For children with Hb SS disease, those with a history of stroke had a significantly lower verbal, performance, and full-scale IQ scores and math achievement scores than children without MRI abnormalities and significantly lower performance and full-scale IQ scores than with children with silent infarcts. In turn, children with silent infarcts had significantly lower verbal scale IQ scores than children without MRI abnormalities.

The independent and combined contribution of biomedical risk and parenting risk to child neurocognitive functioning was assessed in a study of young children with sickle cell disease through 3 years of age (Thompson et al., 2002). The study sample included 89 African American children with sickle cell disease (Hb SS, N = 55; Hb SC, N = 27; and other, N = 7). Measures of cognitive and psychomotor development were obtained at 6, 12, 24, and 36 months of age. There was no significance decrease in psychomotor functioning (PDI) over time but cognitive functioning (MDI) declined, with a significant decrease occurring between the 12- and 24-month assessment points. By 24 months of age, 29% of the children have MDI scores and 24% had PDI scores more than one standard deviation below the mean for the normative group. There were no significant differences in MDI or PDI scores at any assessment time as a function of type of sickle cell disease. However, multiple regression analyses of developmental outcome at 24 months of age indicated that maternal learned helplessness attributional style accounted for 20% of the variance in MDI followed by type of sickle cell disease which accounted for another 22% increment in variance. The findings indicated that developmental functioning at 24 months of age was associated with both Hb SS phenotype and maternal learned helplessness attributional style, with parenting processes as the hypothesized mechanism of effect.

Sickle Cell Disease Pain

Painful episodes or crises are a cardinal aspect of sickle cell disease. The onset of pain is not predictable and the duration of pain is highly variable from a few hours to several days. Sickle cell disease pain has been associated with increased utilization of health care services, decreased social activities, and increased frequencies of psychological distress in children, adolescents, and adults with sickle cell disease (Gil et al., 1991). A number of studies have addressed the relationship of disease severity, demographic parameters, and pain coping strategies on pain associated with sickle cell disease (Gil et al., 1991).

In a study of 72 children and adolescents, ranging in age from 7 to 17 years, the relationship of reported pain coping strategies was assessed with three outcome measures: health care utilization, reflected in the medical record; psychological adjustment, as assessed through a structured diagnostic interview; and reduction in activities reported by parents (Gil et al., 1991). The Coping Strategies Questionnaire (Rosenstiel and Keefe, 1983) was used to assess pain coping strategies and three major patterns of coping were identified. Negative thinking is a pattern of coping in which children engage in catastrophizing and self-statements of fear and anger. Passive adherence is a pattern of coping in which children relied on concrete, passive strategies, such as resting. Coping attempts is a pattern in which the patient used multiple cognitive and behavioral strategies to deal with pain, such as diverting attention and calming self-statements. The sample included patients with Hb SS disease, Hb SC disease, and sickle β-thalassemia syndromes. Children and adolescents high on the negative thinking and passive adherence patterns were less active in school and social activities, had higher levels of psychological distress during painful episodes, and had higher levels of health care service utilization in comparison to those low on these patterns. Children and adolescents high on coping attempts were more active and required less frequent health care services. These coping strategy patterns accounted for significant portions of variance in household, school, and social activity reduction and emergency room (ER) visits and psychological distress even after controlling for the effects of age and frequency of painful episodes.

The extent to which pain coping strategies measured at baseline predict subsequent adjustment in children and adolescents with sickle cell disease was assessed with 70 patients ranging in age from 7 to 18 years (Hb SS, 58%; Hb SC, 13%; and sickle β-thalassemia syndromes, 29%). With age and pain frequency controlled, baseline levels of pain coping strategies characterized by coping attempts were associated with higher levels of school, household, and social activity during painful episodes. In contrast, baseline patterns of pain coping characterized by passive adherence were associated with more frequent health care contacts. Furthermore, increases in pain coping strategies characterized by negative thinking were associated with further increases in health care contacts and those with less negative thinking over time decreased their health care contacts (Gil et al., 1993).

In a study of 79 adults with sickle cell disease, pain coping strategies characterized by negative thinking and passive adherence were associated with more severe pain episodes, less activity during painful episodes, more frequent hospitalization and ER visits and higher levels of self-reported psychological distress (Gil et al., 1989). In a related study, the relationship between stress, coping, and psychological adjustment was assessed in 109 patients with sickle cell disease (Hb SS, 77%; Hb SC disease, 12%; and sickle β-thalassemia, 11%) ranging in age from 18 to 68 years (Thompson et al., 1992). With self-reported levels of psychological distress as the outcome variable, the demographic parameters of socioeconomic status, gender, and age accounted for 9% of the variance, and illness parameters of pain frequency, number of complications, and type of sickle cell disease accounted for 2% of the variance. Over and above the contribution of these variables, daily stress accounted for a 35% increment in psychological distress and pain coping strategies characterized by negative thinking accounted for an additional 4%.

The relationship between stress and pain was examined in a study of 53 adults ranging in age from 18 to 58 years with sickle cell disease (Hb SS, 85%; Hb SC, 8%; sickle β-thalassemia syndrome, 8%) (Porter et al., 1998). Patients completed daily pain ratings for a 2-week period. Stress was assessed in terms of daily hassles (Kanner et al., 1981). Activity reduction, pain occurrence and intensity, reported medication use, and health care use were not significantly related to type of sickle cell disease or the number of sickle cell disease-related complications. However, higher pain intensity ratings were associated with greater health care use in terms of ER visits, hospitalizations, physician visits, and phone calls. Intensity ratings of daily stress were significantly related to pain intensity levels, and reductions in housework and social activities, even after controlling for pain intensity. Thus higher levels of daily stress were related to greater pain and greater functional impairment (Porter et al., 1998).

This examination of the role of behavioral and social factors in sickle cell disease indicates the limitations of the approaches utilized and thus our state of knowledge. These studies were driven by biopsychosocial conceptual models. However, the specific studies predominantly focus on psychological adjustment as the outcome variable. Demographic and illness parameters, including type of sickle cell disease, pain frequency and severity, and number of complications have little effect on psychological adjustment but stress processing variables including appraisals, coping methods, and social support have a large effect. There are a significant number of studies that have addressed pain as an outcome measure and an emerging research literature on neurocognitive functioning. The findings with regard to pain are similar to those with psychological adjustment. Demographic and illness parameters account for relatively little variance in reported pain but stress appraisal and pain coping strategies account for significant amounts of variance. With regard to neurocognitive functioning, both phenotype and parenting process account for significant amounts of variance. There is a notable lack of studies in which physiological measures of illness severity or complications are included as outcome variables. Furthermore, genetic markers, other than type of sickle cell disease, are for the most part not included in behavioral and psychosocial studies. Although the field has not yet actualized the potential of a full biopsychosocial model, the pathway is discernable. Studies that assess the impact of candidate genes on multiple measures of pathophysiology need to be conjoined with studies that assess the impact of behavioral and social processes on stress and stress processing.

STRESS AS A MECHANISM OF EFFECT

A common pathway, that links genetic and environmental psychosocial variables with disease outcome, is through physiological response to perceived stress (Cruess et al., 2004). Stress is defined as the interpretation of an event as threatening that in turn elicits physiological and behavioral responses (McEwen, 2000). Stress hormones mediate both adaptive and maladaptive responses and are protective in the short term but deleterious in the long term if not shut off when no longer needed (McEwen, 2000). Psychosocial stressors can affect a number of disease processes through their impact on the autonomic nervous system, the hypothalamic-pituitary-adrenal (HPA) axis, and the immune system (Cruess et al., 2004). Physiological response to perceived stress can serve as an endophenotype, reflecting the interaction of genetic, behavioral, and psychosocial processes, that in turn affects the variability in manifestations of sickle cell disease.

Stress Activation of the Sympathetic Nervous System (SNS)

Activation of the SNS in response to stress results in increased secretions of catecholamines, epinephrine and norepinephrine, and higher levels of catecholamines lead to increase in blood pressure and heart rate and more oxygenated blood glucose is required (Cruess et al., 2004). There is evidence that increased SNS activity is a mechanism for atherogenesis, ventricular hypertrophy, and hypertension (Cruess et al., 2004). There is a large body of evidence that perceived stress, personality characteristics, and specific emotion states, including hostility and depression, are linked to decreases in the neurotransmitter serotonin in particular, and depression may have a link to coronary heart disease (CHD) through the serotonergic system (Cruess et al., 2004). Depressive symptoms are often associated with CHD and there are indications that proinflammatory cytokines mediate this relationship (Cruess et al., 2004).

Stress Activation of the HPA Axis

Stress can also have an effect through over activation of the HPA axis. Psychological stressors elicit a physiological response by activating specific cognitive and affective processes and their central nervous system underpinnings (Dickerson and Kemeny, 2004). Sensory information is integrated and the significance of environmental stimuli is appraised through the thalamus and frontal lobes. These cognitive appraisals can elicit emotional responses through the connections from the prefrontal cortex to structures of the limbic system including the amygdala and hippocampus which connect to the hypothalamus and serve as a pathway for activating the HPA axis (Dickerson and Kemeny, 2004). Activation of the HPA axis is initiated by the paraventricular nucleus of the hypothalamus releasing corticotropin releasing hormone, which in turn stimulates the anterior pituitary to secrete adrenocorticotropin hormone (ACTH) which in turn triggers the adrenal cortex to release the glucocorticoid (GC) cortisol into the bloodstream (Dickerson and Kemeny, 2004). GCs act to restore homeostasis. Cortisol affects metabolism by mobilizing energy resources by elevating blood glucose levels; surpresses the immune system by inhibiting proteins that play a central role in regulating inflammation; and affects the cardiovascular system through the catecholamines and other sympathetic products that induce vasoconstriction (Dickerson and Kemeny, 2004; Herman et al., 2003). "Although the effects of catecholamines are almost immediate and transient, cortisol is slower acting and more likely to influence blood flow and glucose production during prolonged stress responses" (Cruess et al., 2004, p. 43). Prolonged cortisol activation brought about by failure to shut down this response after stressor termination or by frequent exposure to stressors is associated with a number of negative health consequences including

immune system suppression, for example decreased lymphocyte proliferation and cytokine production, damage to the hippocampus, and hypertension (Dickerson and Kemeny, 2004). The vasoconstrictive and immunological impact of the activation of the HPA axis is of relevance for sickle cell
disease.

The effect of cortisol on tissues is mediated by the glucocorticoid receptor (GR) through direct binding to hormone-responsive elements in the
RNA or by interactions with, and modulation of, other transcription factors (Wüst et al., 2004b). The response of a cell to cortisol is a function of
the level of the steroid and its GC sensitivity. Variants of the GR gene
(located on chromosome 5, locus 5q31) affect sensitivity (Wüst et al.,
2004b). Support has been provided for the hypothesis that common polymorphisms in the GR gene may have modulating effects on the HPA response to psychological stress. In a recent study, the impact of three GR
gene polymorphisms (BclI RFLP, N363S, and ER22/23EK) on cortisol and
ACTH responses to psychological stress and pharmacological stimulation
was assessed (Wüst et al., 2004b). In comparison to subjects with two wild-
type alleles, 363S carriers showed a significant increased salivary cortisol
response to stress whereas the cortisol response of the BclI homozygotes
was diminished. This study provides evidence that common polymorphisms
of a single gene impact HPA regulation and contribute to the individual
variability in response to psychological stress. The impact of genetic factors
on HPA axis activity was reported from findings of twin studies and association studies with polymorphisms in the GR gene (Wüst et al., 2004a). In
addition, a number of polymorphisms were identified as good candidate
genes for future studies (Wüst et al., 2004a).

Evidence suggests that the GCs act through genetic mechanisms, to
modify transcription of key regulatory proteins, and by non-genetic mechanisms on cell signaling processes that have a more rapid impact on homeostatic regulation (Herman et al., 2003). The HPA mediated response to
stressful stimuli differ depending upon whether the threat to homeostasis is
"real" or "predicted." By real stressor is meant stimuli that are recognized
by somatic, visceral, or circumventricular sensory pathways as a challenge
to homeostasis. These stimuli include hormonal signals, such as renin-
angiotensin, visceral or somatic pain, or humoral inflammatory signals
such as blood-borne cytokines signaling infection (Herman et al., 2003). In
addition to these "reactive" responses, GC responses can occur in "anticipation" of homeostatic disruption under situations in which threat may be
predicted or associated with learned experience. The anticipatory responses
are under the control of limbic regions such as the hippocampus, amygdale,
and prefrontal cortex (Herman et al., 2003). These two systems act together in an integrated, hierarchal manner. The reactive pathway evokes
direct PVN activation whereas the anticipatory pathway involves forebrain

processing of polysensorial and associational input that also mediate reactive responses. "The resultant hierarchal organization of stress-responsive neurocircuitries is capable of comparing information from multiple limbic sources with internally generated and peripherally sensed information, thereby tuning the relative activity of the adrenal cortex" (Herman et al., 2003, p. 151). Both genetics and early life experiences can modulate response characteristics of the HPA axis (Herman et al., 2003). Changes in limbic system integration patterns as a function of experience are hypothesized to play a role in HPA axis dysfunction (Herman et al., 2003).

The importance of psychological stress processing for the understanding of the psychobiological stress response is becoming increasingly clear (Gaab et al., 2005). Conceptualizations of stress have moved from that of a stimulus or response to "A relationship between the person and the environment that is appraised by the person as taxing or exceeding his or her resources and endangering his or her well being" (Lazarus and Folkman, 1984, p. 19). A recent study provided support for the role of anticipatory cognitive appraisal, but not general personality factors or retrospective stress appraisal, in the salivary cortisol response to psychological stress (Gaab et al., 2005).

Whereas there is evidence that psychological stressors are capable of activating the HPA axis, the effects are highly variable. For example, several aspects of perceived chronic stress, more specifically worries, social stress, and lack of social recognition, were found to be significantly associated with increased cortisol awakening response (Wüst et al., 2000). To evaluate the characteristics of psychological stressors that evoke a cortisol response, a meta-analysis of 208 empirical studies was undertaken (Dickerson and Kemeny, 2004). The findings indicated that psychosocial stressors that involved social evaluative threat and uncontrollability were significantly associated with increased cortisol response. The findings were also similar for ACTH response. However, psychological distress in and of itself was not associated with increased cortisol response. The findings indicate that only those threats to central goals, such as physical self-preservation or preservation of the social self, and not having control over these situations, triggers cortisol activation. Sickle cell disease provides just such a situation of threat to self-preservation and social evaluative threat and the negative self-appraisals generated under these conditions rather than emotional stress in general could constitute psychological stressors that impact the HPA axis.

Stress and the Immune System

In understanding the relationship of psychosocial stressors to the immune system, Segerstrom and Miller (2004) maintain that it is useful to distinguish between natural and specific immunity. Natural immunity involves cells that do not provide a defense against a particular pathogen but operate broadly in a short time frame. These cells include the granulocytes, both neutrophil and macrophage, which releases cytokines such as interleukin, and natural killer (NK) cells. Specific immunity involves cellular response to intracellular pathogens and humoral responses to extracellular pathogens. Lymphocytes have receptor sites that respond to a specific antigen and when activated divide to create a population of cells in a process referred to as colonal proliferation (Segerstrom and Miller, 2004).

The immune system is of importance in sickle cell disease and one way of examining the impact of genetic, behavioral, and psychosocial processes on the immune system is through the impact of stress and stress processing. There are several ways that stress can affect the immune response (Segerstrom and Miller, 2004). The immune system is regulated both by neural inputs from the sensory, sympathetic, and parasympathetic system as well as by circulating catecholamines and GCs (McEwen, 2000). The substances released through the action of the nervous system bind to specific receptors on white blood cells and have a regulatory effect on their distribution and function (Segerstrom and Miller, 2004). More specifically, sympathetic fibers release substances that bind to receptors on lymphocytes, and "the hypothalamic-pituitary-adrenal axis, the sympathetic-adrenal-medullary axis, and the hypothalamic-pituitary-ovarian axis secrete the adrenal hormones, epinephrine, norepinephrine, and cortisol; the pituitary hormones prolactin and growth hormone; and the brain peptides melatonin, β-endorphin, and enkephalin" (Segerstrom and Miller, 2004, p. 604). Under acute stress, elevations of stress hormones (catecholamines and GCs) facilitate the movement of immune cells, lymphocytes, monocytes, and NK cells which are reduced in other tissues where other mediators of immune function activation become involved. For example, interferon gamma "is known to induce expression of antigen-presenting and cell-adhesion molecules on endothelia cells and macrophages and cell adhesion molecules on leukocytes" (McEwen, 2000, p. 175).

Stress also affects the immune system through behaviors, such as changes of sleep patterns, that could modify immune system processes (Segerstrom and Miller, 2004). Another association of the immune system with stress arises through the immunological activation of "sickness behavior" which refers to a constellation of behavioral changes that accompany infection that include a "reduction in activity, social interaction, and sexual activity, as well as increased responsiveness to pain, anorexia, and depressed mood" (Segerstrom and Miller, 2004, p. 604).

Support for the relationship of psychological stress and immune system response was provided through a meta-analysis of more than 300 empirical studies (Segerstrom and Miller, 2004). The findings across these studies indicated that acute stressors were associated with upregulation of natural immunity parameters and downregulation of specific immunity functions (Segerstrom and Miller, 2004). Acute stressors were associated with an increase in the number of NK cells, neutrophils, and large granular lymphocytes in peripheral blood, increased production of proinflammatory cytokines and cytokines, and decrease in colonal proliferation response (Segerstrom and Miller, 2004). Chronic stressors were associated with suppression of both cellular and humoral responses (Segerstrom and Miller, 2004). Furthermore, stress appraisal was found to be associated with a reduction in NK cell cytotoxicity (Segerstrom and Miller, 2004).

Chronic stress leading to sustained levels of stress hormones can also affect the immune system (Cruess et al., 2004). A proinflammatory cytokine, interleukin-6 (IL-6) is elevated under stress and stimulates SNS and HPA activation (Cruess et al., 2004). Furthermore, inflammation is critical in the development and progression of atherosclerosis which is associated with the rupture of plaque that can block blood flow (Cruess et al., 2004). Low-density lipoprotein cholesterol retained in the cell wall undergoes oxidative modification and the "resultant modified lipids can induce the expression of adhesion molecules and proinflammatory cytokines as mediators of inflammation in macrophages and vascular cell walls" (Cruess et al., 2004, p. 40). Psychological factors such as depression and stress have been associated with decrements in lymphocyte proliferative response and lower NK cell cytotoxicity (Cruess et al., 2004). Thus, alterations in neuroendocrine functioning affect the immune system and neurohormonal changes have been linked to a number of psychosocial factors including cognitive appraisals, coping responses, perceived loss of control, attributions of helplessness, and feelings of hopelessness, low self-efficacy, passive coping strategies, and lack of social support (Cruess et al., 2004).

Stress and Erythrocyte Adhesion

The vasoocclusive process in sickle cell disease is complex and increasing attention is focused on the role of the adhesion of sickle erythrocytes (SS RBCs) to endothelial cells (ECs). A direct relationship between the rating of vasoocclusive pain and biological markers of erythrocytes/EC adhesion has been reported (Dampier et al., 2004). In addition, there is evidence that the stress hormone epinephrine enhances adhesion of sickle erythrocytes (SS RBCs), but not normal RBCs, to ECs (Zennadi et al., 2004). Febrile episodes are frequently associated with vasoocclusive pain episodes in sickle cell anemia and are hypothesized to be viral in origin. Support was provided for the hypothesis that viruses, through double-stranded RNA, can induce sickle erythrocytes adherence to ECs through alpha4beta1-VCAM-1-mediated adhesion (Smolinski et al., 1995). A recent review summarizes the increasing knowledge about how membrane structures contribute to cell adhesion (Telen, 2005).

Stress and Neurocognitive Functioning

Chronic high levels of stress hormones and GCs contribute to impairment of cognitive function through effects on the hippocampus (McEwen, 2000). The hippocampus has two types of adrenal steroid receptors, type 1 (mineralocoiticoid), and type 2 (glucocorticoid), that mediate hormone effects on gene expression (McEwen, 2000). It is the combined action of circulating GCs and catecholamines interacting with local tissue mediators, such as cytokines, that affect the immune system and the excitatory amino acids, such as glutamate, and neurotransmitters, particularly serotonin, that affect the brain and cognitive functioning (McEwen, 2000). Brain atrophy has been shown to occur, particularly of the hippocampus, as a result of elevated GCs and severe stress and declines in hippocampally related cognitive functions such as episodic memory are correlated with increases in HPA activity (McEwen, 2000).

Adrenocortical stress responses to ordinary daily stress is sufficient to produce atrophy of hippocampal structures (McEwen, 2000). However, individual differences in stress responsiveness also play a role (McEwen, 2000). "Individuals with a more reactive stress hormone profile will expose themselves to more cortisol and experience more stress-related neural activity, than other people who can more easily habituate to psychosocial challenges" (McEwen, 2000, p. 183).

In assessing the impact of stress, it is useful to have multiple physiological measures within the same study. In a study of monozygotic and dizygotic female twin pairs, genetic and environmental effects on autonomic reactivity to a psychologically stressful situation was examined for both single physiological variables and functional combinations of seven of these variables (Lensvelt-Mulders and Hettema, 2001). The findings supported the hypothesis that autonomic response profiles would yield larger genetic effects than single autonomic measures and that the idiosyncratic relationship of a person and his/her environment is a heritable trait. Up to 80% of the variance in the functional profiles were accounted for by differences in individual genotypes. The authors comment, "there are at least two ways people physiologically respond to a situation: Directly, by making people more genetically liable to express a certain trait, and indirectly by influencing idiosyncratic interactions between a person and his environment" (Lensvelt-Mulders and Hettema, 2001, p. 38).

Stress and Cardiovascular and Renal Response

It has been hypothesized that exaggerated cardiovascular response to stress is a mechanism in the pathogenesis of essential hypertension and CHD. Snieder et al. (2002) have developed a biobehavioral model of stress-induced hypertension to explain how repeated exposure to stress, in combination with genetic susceptibility, could lead to the development of hypertension. This model is useful to consider, not only because of the cardiovascular problems in sickle cell disease, but because the biobehavioral model enables a systems perspective. The biobehavioral model focuses on the complex interrelationship of three underlying physiological systems that mediate the stress response of the heart, vasculature, and kidney: the SNS; the renin-angiotensin-aldosterone system (RAAS), and the endothelial system (ES). In support of this model, evidence is reviewed for a genetic influence on the two major intermediate phenotypes of the model: cardiovascular reactivity to psychological stress and the renal stress response in terms of stress-induced sodium retention. The data reviewed were from twin and family studies and a limited number of candidate gene association studies. The authors acknowledge that other biological systems, such as the HPA axis, parasympathetic autonomic reactivity, and serotonin functioning in the central nervous system may mediate the influence of stress on the development of the essential hypertension, and the importance of genetic variation of these systems has been demonstrated as well. The biobehavioral model of stress-induced essential hypertension proposes that in response to stress there is an increased central nervous system activity that in turn results in the release of catecholamines, norepinephine and epinephrine, which in turn increases heart rate. In addition, norepinephrine causes vasoconstriction and epinephrine causes vasoconstriction in some vessels and vasodilation in others (Snieder et al., 2002).

The ES influences the control of vascular smooth muscle function through the production of nitric oxide (NO), a vasodilator, and endothelin-1 (ET-1), a vasoconstrictor. SNS arousal potentiates the release of these vasoactive substances. Under stress there is evidence of increased release of ET-1 and decreased production of NO resulting in increased vasoconstictive tone (Snieder et al., 2002). The RAAS is activated by both the activity of the ES and SNS arousal. This results in further vasoconstriction and an increase in sodium retention enhances the vasoconstrictive effects of norepinephrine on peripheral vasculture (Snieder et al., 2002).

A complex interaction of these three systems contributes to increase total peripheral resistance in response to stress and repeated exposure leads to disregulation in appropriately activating, and/or turning off, cardiovascular function (Snieder et al., 2002). The responses to stress result in increases in cardiac and vascular wall tension and intravascular shear stress that leads to secondary renal damage and cardiovascular remodeling, including diminished endothelium-dependent arterial dilation to reactive hyperemia (Snieder et al., 2002). Another manifestation of vascular remodeling is increased arterial stiffness which in turn is associated with stroke, renal failure, and coronary artery disease and left ventricular hypertrophy, which is a strong predictor of cardiovascular morbidity and mortality (Snieder et al., 2002).

Snieder et al. (2002) also examined the evidence for the role of specific candidate genes on cardiovascular response to stress. Since the β2-adrenergic receptor mediates peripheral vasodilation, polymorphic variation in this gene may influence response to stress. Evidence has been provided for an association between Arg16Gly polymorphism β2-adrenergic receptor gene (ADRB2) and the Arg389Gly and Arg16Gly polymorphisms in the β1-adrenergic receptor gene (ADRB1) were associated with blood pressure at rest and reactivity to stress. The Gln27Glu polymorphism of the β2-adrenergic receptor gene also showed significantly higher levels of blood pressure at rest and stress but interestingly, no associations were found between these polymorphisms and cardiovascular reactivity for African Americans (Snieder et al., 2002). It should also be noted that an increase in cardiovascular response to stress has also been associated with a promoter polymorphism of the serotonin transporter gene (5HTTLPR) through higher levels of serotonin (Williams et al., 2001).

Snieder et al. (2002) suggested that future studies investigating genetic influences on cardiovascular and renal stress should employ measures of polymorphic variation in candidate genes that underlie the SNS, the ES, and the RAAS. They argue that rather than studying the effects of candidate genes in isolation that the biobehavioral model provides a framework for describing the interrelated physiological network underlying blood pressure regulation in response to stress. More specifically, Snieder et al. (2002) suggest the following candidate genes for the respective systems. SNS: "the α1- and α2-adrenergic receptor gene (ADRA1, ADRA2) and the β1- and β2-adrenergic receptor genes (ADRB1, ADRB2)"; RAAS: "the genes for angiotensin converting enzyme, and the angiotensin II type-1 receptor (AGTR1), aldosterone synthase (CYP11B2) and angiotensinogen"; ES: "the ET-1 gene (EDN1), the gene for ET-1 receptor A (EDNRA) and the genes for the three types of nitric oxide synthase (NOS1, NOS2, NOS3)" (Snieder et al., 2002, p. 87).

SUMMARY AND RECOMMENDATIONS

This review provides some information with regard to the specific questions of interest but may have its most significant contribution in terms of guidance of future research. In terms of what knowledge/data we have, the following findings are most salient:

• Among social and behavioral factors, stress—primary related to daily hassles, and stress processing—primarily in relation to cognitive appraisals and attributions, coping methods, and family support, are associated with variability in the manifestation of sickle cell disease—primarily psychological adjustment, pain, and neurocognitive functioning.

• Stress and stress processing are related to an array of neuroendocrine-mediated physiological responses, that in turn are associated with variability in vascular and inflammatory processes of importance in sickle cell disease.

• Pain management is related to variability in health care utilization and activity level.

• A number of candidate genes have been identified as mediators/modulators of the physiological response to stress and of the vascular and inflammatory manifestations of sickle cell disease.

The data that we do not yet have and the questions remaining to be answered are at the systems level of analysis, to which the biopsychosocial model aspires but has not yet reached. The current stage of research can most appropriately be described as multiple dimensions—biological, psychological, and social—considered concurrently but not transactionally. That is, current studies examine the contribution of biological and psychosocial factors in terms of their independent and combined contributions to variability in some aspect of sickle cell disease manifestations. This is one level of consideration of how multiple processes "act together." The next level is considering "acting together" in terms of mutual influence through continuous transactions over time. In addition, the studies of the contribution of behavioral and social factors have been limited in terms of outcome measures to primarily psychological adjustment, pain and health care utilization, and neurocognitive impairment but not other physiological manifestations of sickle cell disease. Finally, there are very few studies that include an examination of behavioral and psychosocial factors and candidate genes in the same study. This review suggests that the next research step is to develop requests for proposals for studies that are longitudinal, evaluate the role of stress appraisal, stress processing, and candidate genes on physiological stress responses as the endophenotype and on vascular and immunological physiological measures and cell adhesion as the endpoints.

ACKNOWLEDGMENT

I want to thank Meghan Von Isenburg, Information and Education Services Librarian, Duke University Medical Center Library, for her assistance with the literature search.

REFERENCES

Armstrong, D.F., Thompson, R.J., Jr., Wang, W., Zimmerman, R., Pegelow, C., Miller, S., Moser, F., Bello, J., Hurtig, A., and Vass, K. (1996). Cognitive functioning and brain magnetic resonance in children with SCD. *Pediatrics*, 97, 864-870.

Ashley-Koch, A., Yang, Q., and Olney, R.S. (2000). Sickle hemoglobin *(Hb S)* allele and sickle cell disease: A HuGE review. *American Journal of Epidemiology*, 151(9), 839-845.

Ballas, S.K. (2001a). Effect of α-globin genotype on the pathophysiology of sickle cell disease. *Pediatric Pathology and Molecular Medicine*, 20, 107-121.

Ballas, S.K. (2001b). Sickle cell disease: Current clinical management. *Seminars in Hematology*, 38(4), 307-314.

Bronfenbrenner, U. (1977). Toward an experimental ecology of human development. *American Psychologist*, 32, 513-531.

Bronfenbrenner, U. (1979). *The ecology of human development*. Cambridge, MA: Harvard University Press.

Cruess, O.G., Schneiderman, N., Antoni, M.H., and Penedo, F. (2004). Biobehavioral bases of disease processes (pp. 31-79). In T.J. Boll (Series Ed.) and R.G. Frank, A. Baum, and J.L. Wallander (Vol Eds). *Handbook of Clinical Health Psychology. Volume 3: Models and Perspectives in Health Psychology*. Washington, DC: American Psychological Association.

Dampier, C., Setty, B.N., Eggleston, B., Brodecki, D., O'Neal, P., and Stuart, M., (2004). Vaso-occlusion in children with sickle cell disease: Clinical characteristics and biologic correlates. *Journal of Pediatric Hematology/Oncology*, 26(12), 785-90.

de Gues, E.J.C. (2002). Introducing genetic psychophysiology. *Biological Psychology* 61, 1-10.

Dickerson, S.S., and Kemeny, M.E. (2004). Acute stressors and cortisol responses: A theoretical integration and synthesis of laboratory research. *Psychological Bulletin*, 130(3), 355-391.

Engel, G.L. (1977). The need for a new medical model: A challenge for biomedicine. *Science*, 196, 129-136.

Farber, M.D., Koshy, M., and Kinney, T.R. (1985). Cooperative study of sickle cell disease: Demographic and socioeconomic characteristics of patients and families with sickle cell disease. *Journal of Chronic Diseases*, 38, 495-505.

Gaab, J., Rohleder, N., Nater, U.M., and Ehlert, U. (2005). Psychological determinants of the cortisol stress response: The role of anticipatory cognitive appraisal. *Psychoneuroendocrinology*, 30, 599-610.

Gaston, H.H., and Rosse, W. (1982). The cooperative study of sickle cell disease: Review of study design and objectives. *American Journal of Pediatric Hematology and Oncology*, 4, 197-200.

Gil, K.M., Abrams, M.R., Phillips, G., and Keefe, F.J. (1989). Sickle cell disease pain: Relation of coping strategies to adjustment. *Journal of Consulting and Clinical Psychology*, 57, 725-731.

Gil, K.M., Thompson, R.J., Jr., Keith, B.R., Tota-Faucette, M., Noll, S., and Kinney, T.R. (1993). Sickle cell disease pain in children and adolescents: Change in pain frequency and coping strategies over time. *Journal of Pediatric Psychology*, 18, 621-637.

Gil, K.M., Williams, D.A., Thompson, R.J., Jr., and Kinney, T.R. (1991). Sickle cell disease in children and adolescents: The relation of child and parent pain coping strategies to adjustment. *Journal of Pediatric Psychology*, 16, 643-663.

Herman, J.P., Figueredo, H., Mueller, N.K., Ulrich-Lai, Y., Ostrander, M.M., Choi, D.C., and Cullinan, W.E. (2003). Central mechanisms of stress intergration: Hierarchical circuitry controlling hypothalamo-pituitary-adrenocortical responsiveness. *Frontiers in Neuroendocrinology*, 24, 151-180.

Kanner, A.D., Coyne, J.C., Schaefer, C., and Lazarus, R.S. (1981). Comparison of two modes of stress measurements: Daily hassles and uplifts versus major life events. *Journal of Behavioral Medicine*, 4, 1-39.

Lazarus, R.S., and Folkman, S. (1984). *Stress, appraisal, and coping.* New York: Springer Publishing Company.

Lensvelt-Mulder, G., and Hettema, J. (2001). Genetic analysis of autonomic reactivity to psychologically stressful situations. *Biological Psychology*, 58, 25-40.

McEwen, B.S. (2000). The neurobiology of stress: From serendipity to clinical relevance. *Brain Research Bulletin*, 886, 172-189.

Nagel, R.L., and Steinberg, M.H. (2001). Role of epistatic (modifier) genes in the modulation of the phenotypic diversity of sickle cell anemia. *Pediatric Pathology and Molecular Medicine*, 20, 123-136.

Plomin, R., and Crabbe, J. (2000). DNA. *Psychological Bulletin*, 126(6), 806-828.

Porter, L.S., Gil, K.M., Sedway, J.A., Ready, J., Workman, E., and Thompson, R.J., Jr. (1998). Pain and stress in sickle cell disease: An analysis of daily pain records. *International Journal of Behavioral Medicine*, 5, 185-203.

Quinn, C.T., and Miller, S.T. (2004). Risk factors and prediction of outcomes in children and adolescents who have sickle cell anemia. *Hematology/Oncology Clinics of North America*, 18, 1339-1354.

Ridley, M. (2003). *Nature via nurture: Genes, experience, and what makes us human.* New York: Harpers Collins Publishers.

Ridley, M. (2004). The biology of human nature. *Daedalus*, 89-98.

Rosenstiel, A.K., and Keefe, F.J. (1983). The use of coping strategies in low back pain patients: Relationship to patient characteristics and current adjustment. *Pain*, 17, 33-40.

Sabastiani, P., Ramoni, M.F., Nolan, V.G., Baldwin, C.T., and Steinberg, M.H. (2005). Genetic dissection and prognostic modeling of overt stroke in sickle cell anemia. *Nature Genetics*, 37(4), 435-440.

Segerstrom, S.C., and Miller, G.E. (2004). Psychological stress and the human immune system: A meta-analytic study of 30 years of inquiry. *Psychological Bulletin*, 130, 601-630.

Smolinski, P.A., Offermann, M.K., Eckman, J.R., and Wick, T.M. (1995). Double-stranded RNA induces sickle erythrocyte adherence to endothelium: A potential role for viral infection in vaso-occlusive pain episodes in sickle cell anemia. *Blood*, 85(10), 2945-2950.

Snieder, H., Harshfield, G.A., Barbeau, P., Pollock, D.M., Pollock, J.S., and Treiber, F.A. (2002). Dissecting the genetic architecture of the cardiovascular and renal stress response. *Biological Psychology*, 61, 73-95.

Steinberg, M.H. (2005). Predicting clinical severity in sickle cell anaemia. *British Journal of Haematology*, 129, 465-481.

Steinberg, M.H., and Rodgers, G.P. (2001). Pathophysiology of sickle cell disease: Role of cellular and genetic modifiers. *Seminars in Hematology*, 38(4), 299-306.

Stuart, M.J., and Nagel, R.L. (2004). Sickle-cell disease. *The Lancet*, 364, 1343-1360.

Telen, M.J. (2005). Erythrocyte adhesion receptors: Blood group antigens and related molecules. *Transfusion Medicine Reviews*, 19, 32-44.

Thompson, R.J., Jr., and Gustafson, K.E. (1996). *Adaptation to chronic childhood illness.* Washington, DC: American Psychological Association Press.

Thompson, R.J., Jr., Gil, K.M., Abrams, M.R., and Phillips, G. (1992). Stress, coping and psychological adjustment of adults with sickle cell disease. *Journal of Consulting and Clinical Psychology*, 60, 433-440.

Thompson, R.J., Jr., Gil, K.M., Burbach, D.J., Keith, B.R., and Kinney, T.R. (1993a). Psychological adjustment of mothers of children and adolescents with sickle cell disease: The role of stress, coping methods and family functioning. *Journal of Pediatric Psychology*, 18, 621-637.

Thompson, R.J., Jr., Gil, K.M., Burbach, D.J., Keith, B.R., and Kinney, T.R. (1993b). Role of child and maternal processes in the psychological adjustment of children with sickle cell disease. *Journal of Consulting and Clinical Psychology*, 61, 468-474.

Thompson, R.J., Jr., Gil, K.M., Abrams, M.R., and Phillips, G. (1996). Psychological adjustment of adults with sickle cell anemia: Stability over 20 months, correlates, and predictors. *Journal of Clinical Psychology*, 52, 253-266.

Thompson, R.J., Jr., Armstrong, F.D., Kronenberger, W.G., Scott, D., McCabe, M.A., Smith, B., Radcliffe, J., Colangelo, L., Gallagher, D., Islam, S., and White, E. (1999a). Family functioning, neurocognitive functioning, and behavior problems in children with sickle cell disease. *Journal of Pediatric Psychology*, 24, 491-498.

Thompson, R.J., Jr., Gustafson, K.E., Gil, K.M., Kinney, T.R., and Spock, A. (1999b). Change in the psychological adjustment of children with cystic fibrosis or sickle cell disease and their mothers. *Journal of Clinical Psychology in Medical Settings*, 6, 373-391.

Thompson, R.J., Jr., Gustafson, K.E., Bonner, M.J., and Ware, R.E. (2002). Neurocognitive development of young children with sickle cell disease through three years of age. *Journal of Pediatric Psychology*, 27, 235-244.

Thompson, R.J., Jr., Armstrong, F.D., Link, C.L., Pegelow, C.H., Moser, F., and Wang W.C. (2003). A prospective study of the relationship over time of behavior problems, intellectual functions, and family functioning in children with sickle cell disease: A report from the Cooperative Study of Sickle Cell Disease. *Journal of Pediatric Psychology*, 28, 59-65.

Weatherall, D.J. (2001). Phenotype-genotype relationships in monogenic disease: Lessons from the thalassaemias. *Nature Reviews Genetics*, 2, 245-255.

Weiss, S.M. (1987). Behavioral medicine in the trenches. In J. Blumenthal and D. McKee (Eds.), *Applications in Behavioral Medicine snd Health Psychology: A Clinician's Source Book* (pp. xvii-xxiii). Sarasota, FL: Professional Resource Exchange.

Williams, R.B., Marchuk, D.A., Gadde, K.M., Barefoot, J.C., Grichnik, K., Helms, M.J., et al. (2001). Central nervous system serotonin function and cardiovascular responses to stress. *Psychosomatic Medicine,* 63(2), 300-305.

Wüst, S., Federenko, I.S., Hellhammer, and D.H., Kirschbaum, C. (2000). Genetic factors, perceived chronic stress, and the free cortisol response to awakening. *Psychoneuroendocrinology*, 25(7), 707-720.

Wüst, S., Federenko, I.S., Van Rossum, E.F.C., Koper, J.W., Kumsta, R., Entringer, S., and Hellhammer, D.H. (2004a). A psychobiological perspective on genetic determinants of hypothalamus-pituitary-adrenal axis activity. *New York Academy of Sciences,* 1032, 52-62.

Wüst, S., Van Rossum, E.F.C., Federenko, I.S., Koper, J.W., Kumsta, R., and Hellhammer, D.H. (2004b). Common polymorphisms in the glucocorticoid receptor gene are associated with adrenocortical responses to psychosocial stress. *Journal of Clinical Endocrinology and Metabolism,* 89, 565-673.

Zennadi, R., Hines, P.C., DeCastro, L.M., Cartron, J.P., Parise, L.V., and Telen, M.J. (2004). Epinephrine acts through erythroid signaling pathways to activate sickle cell adhesion to endothelium via LW-$\alpha v\beta 3$ interactions. *Blood*, 104, 3774-3781.

E

Modern Epidemiologic Approaches to Interaction: Applications to the Study of Genetic Interactions

Sharon Schwartz, Ph.D. *

INTRODUCTION

Epidemiology attempts to discern the causes of disease through an analysis of the patterns of exposure/disease relationships that are brought into view by our study designs. The types of designs and methods that are developed are largely influenced by the health challenges that the population faces as well as any methodologic and technological constraints.

Current epidemiologic methods were sparked by the rise of chronic diseases that did not fit well within the causal models underlying infectious disease epidemiology. Infectious disease models, based on the Henle-Koch principles, reserved the term "cause" for factors that were both necessary and sufficient for disease occurrence. Although this assumption did not apply strictly to the identified causes of many infectious diseases, this model worked well enough to provide utility over time.

A crisis arose, however, over the study of the relationship between smoking and lung cancer. Although the association between smoking and lung cancer was strong and seemed persuasive, smoking clearly was neither necessary nor sufficient for the development of lung cancer. This led to a paradigmatic crisis that over time resulted in the development of a new framework for the identification of causes, which crystallized as "risk factor epidemiology." This framework is rooted in the notion that there are

*Associate Professor of Clinical Epidemiology, Mailman School of Public Health, Columbia University, 722 West 168th Street, Room 720 b New York, NY 10032, sbs5@columbia. edu.

multiple pathways to the same disease and that within each pathway there are multiple causes that work in tandem to lead to the disease. These types of causes are often referred to as "risk factors."

The risk factor framework generally is "egalitarian" in its assumptions about causation; all types of factors that contribute to disease occurrence can be called a cause. There may be some factors that are necessary causes in the sense that the disease never occurs in their absence, but other causes may not be necessary at all. In addition, even necessary causes require the presence of causal partners to lead to disease occurrence. These causal partners also are considered to be causes of the disease.

The necessity of a causal partner for disease occurrence is what we mean by "biologic interaction." Thus, the very definition of a cause in risk factor epidemiology places the issue of interaction front and center. It is assumed that virtually all diseases arise from the interaction of two or more causes.

Despite the centrality of interaction to this causal framework, methodologic advances have focused mainly on the isolation of single causes and the identification of individual risk factors that contribute to disease occurrence in a population. New designs were developed to allow us to see the relationships between exposures[1] and disease in our data that would provide clues to the identification of these causes. Statistical methods were developed to aid in causal inference.

The identification of the causal partners of particular risk factors, the assessment of interaction, was a more complex notion that awaited conceptual clarification and methodological advances. Considerable progress has been made; however, often a lag occurs between the development of new methods and approaches and their application and appearance in the literature. Thus, the way in which interaction is assessed in epidemiologic studies is only now beginning to reflect these newer methods.

What follows is a discussion of this newer way of thinking about how to identify "biologic interaction." I prefer the use of the term "synergy" in this discussion because it is more neutral to the level of organization at which interaction is being described. Although these methods have developed separately from those in the field of genetics, they are fully applicable to the field, and while genes have characteristics that are distinct from many of the risk factors studied in epidemiology, an epidemiologic approach to causation easily and naturally encompasses genes as causes. However, this application requires a shift in perspective. From a genetic point of view there is a hierarchy of causes, with "the gene" having centrality as the defining cause and all other factors being ancillary to it. Factors that are

---

[1]This paper uses the term "exposure" to mean any factor that is being examined to see if it is a cause of disease. The term applies to any factor under consideration—genetic or environmental.

considered equal causes from an epidemiologic frame are sometimes labeled in genetics in a way that gives them secondary status. One example is the use of the term "phenocopy" to distinguish a case of disease caused in the absence of a putative genetic cause. Another example is the concept of "reduced penetrance." This term refers to the inexact relationship between a genotype and a phenotype and implies that this slippage is a characteristic of a gene; the gene evidences "reduced penetrance," or the gene is "fully penetrant." From an epidemiologic perspective, reduced penetrance is simply a normal characteristic of all causes—the lack of a one-to-one relationship between causes and diseases due to interaction. From an epidemiologic perspective, reduced penetrance is not a characteristic of the gene, but rather a characteristic of the distribution of the causal partners with which the gene works to cause disease. It is the natural state of most causal relationships.

Thus epidemiologic approaches to interaction provide an exciting perspective on genetic concepts that may shed new light on genetic issues. Likewise, the integration of genetic thinking into epidemiology can advance methodology.

I begin this task with a discussion of why the assessment of interaction is so problematic, and then I will discuss the current epidemiologic resolution to the problem. However, to fully understand the solution and its applicability to a genetic context, we need to probe the concept of causation in epidemiology more fully. Although this may seem a bit off topic, it is central to understanding the elements of the new ways of thinking about synergy. Finally, more specific problems of application and design will be addressed.

CURRENT EPIDEMIOLOGIC FRAMEWORK FOR ASSESSING INTERACTION

The Problem

Because the testing of our hypotheses and the assessment of our data rely on statistical tools, we already are most familiar with the concept of statistical interaction. From a statistical perspective, we can say that there is interaction when in the presence of two factors the outcome occurs more frequently than would be expected based on the independent effects of each factor. By independent effect, we mean the effect of one factor in the absence of the other factor. To make this more concrete, we would say that interaction can be identified when among people with both a genetic variant and an environmental exposure the disease rate is higher than would be expected if the genetic factor and environmental exposure each worked independently.

Proportion of Respondents Developing Depression

		Intimacy Problems	
		Yes	No
Severe life event or major difficulty	Yes	32%	10%
	No	3%	1%

FIGURE E-1 Assessment of interaction: example from Brown and Harris (1978).

Although this definition is clear in statistical terms, it begs the question of "what would be expected." As it turns out, what would be expected depends on the effect measure or statistical model used to express the relationship between exposures and disease. This can be seen in the data from a study in psychiatric epidemiology that proved to be very enlightening in this regard. Brown and Harris (1978) wanted to test the theory that stressful life events and problems with intimacy interacted in causing depression. They hypothesized that, while both stressful life events and intimacy problems each may confer a risk of depression, when they are both present they confer a greater risk than would be expected if each worked through a separate causal pathway. The data derived from a study to test this hypothesis are depicted in Figure E-1.

Brown and Harris interpreted these data as supporting their claim for an interaction between intimacy problems and stressful life events. The risk of depression in those with neither stressful life events nor intimacy problems was 1 percent, while among those with only stressful life events was 10 percent, and among those with only intimacy problems was 3 percent. The difference in the risk conferred by stressful life events alone was therefore 9 percent (10 percent – 1 percent), and the risk difference conferred by intimacy problems alone was 2 percent (3 percent – 1 percent). If there were no interaction, one would expect that when both factors were present the risk conferred would be 11 percent (9 percent + 2 percent). However, the data show that the risk conferred when both were present was 32 percent, which is substantially greater than would be expected based on the independent effects of each risk factor. Brown and Harris therefore concluded that these data supported their theory of an interaction between stressful life events and intimacy problems in causing depression.

Tennant and Bebbington (1978) challenged this conclusion. They reanalyzed these data using log linear modeling. This analysis calculated the effects on a different scale by calculating risk ratios. Using this model, life events acting alone increase the risk of depression by a factor of 10 (10 percent = 1 percent * 10). Intimacy problems alone increased the risk by a factor of 3 (3 percent = 1 percent * 3). Therefore, based on this calculus one would expect the co-presence of these risk factors to increase the effect by a factor of 30 (1 percent * 3 * 10) if they were acting independently of each other. This is very close to the 32 percent risk actually found. Thus, Tennant and Bebbington concluded from these same data that there was no support for Brown and Harris's conclusion.

What was not fully appreciated at the time was that both Brown and Harris and Tennant and Bebbington provided absolutely correct interpretations of the data based on the unarticulated statistical assumptions of their approaches. Brown and Harris, using risk differences to express the effects of risk factors, used a model that implicitly assumed that, absent interaction, risks add in their effects. They used an additive model. Tennant and Bebbington, on the other hand, analyzed the data using a log linear model that implicitly assumed that absent interaction, risks multiply in their effects. They used a multiplicative model. Thus, based on statistical definitions of interaction the same data both did and did not support a theory of interaction.

This state of affairs is disconcerting, to say the least. We depend on our data and statistical tools to give us a rough estimate of the state of affairs in the real world, and it is problematic when the answers to our questions differ depending on the statistical model we use to assess our data. To make matters worse, the choice of statistical model often is based on statistical considerations. For example, we usually employ additive models, such as linear regression, when our outcome variables are continuous. When our outcomes are dichotomous, as they frequently are in genetic and epidemiologic contexts, we employ logistic regression, because such outcomes violate the statistical assumptions of linear regression models. Although this choice meets statistical requirements, it shifts us to a multiplicative model. Linear regression assumes that risks add in their effects, and thus interaction is indicated by an appreciable deviation from additivity (i.e., sub- or superadditivity). Logistic regression assumes that risks multiply in their effects, and thus interaction is indicated by an appreciable deviation from multiplicativity (i.e., sub- or supermultiplicativity).

The problem is that if both risk factors have an effect, there always will be interaction on at least one of these scales. As illustrated in Figure E-2, additivity implies submultiplicativity, and multiplicativity implies superadditivity. Thus, except in instances of supermultiplicativity (in which both models will index positive interaction) and subadditivity (in which both models index negative interaction), the answer to the question of whether or not there is interaction will depend on the statistical model that we choose.

This is very unsettling, because we want our statistical models to repre-

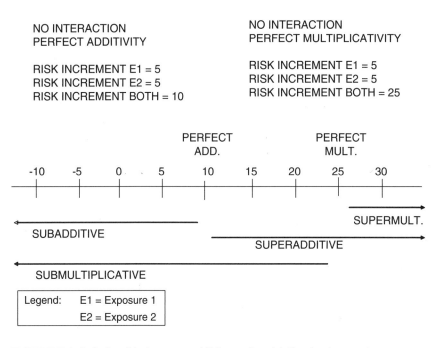

FIGURE E-2 Relationship between additive and multiplicative interaction.

sent our concepts rather than having them define our concepts. So, the question would be one of what model best represents the "true" relationship between risk factors. That is, do risk factors really add or multiply in their effects? Darroch (1997), Rothman and Greenland (1998), and others have grappled with this problem. It appears that the additive model with a twist best represents what we mean by interaction. The twist is due to redundancy in causes, as we shall see.

To appreciate this argument, and to assess its applicability to the context of assessing interactions that include genetic factors, a fuller discussion of the causal model on which this assessment is based is necessary. This casual model—the counterfactual or potential outcomes model—developed in philosophy and statistics (Mackie, 1974; Maldonado and Greenland, 2002; Rubin, 2004; Shadish et al., 2002) underlies much of the causal thinking today in epidemiology and allied fields such as history, sociology, and economics.

The solution to the interaction problem derives from the application of this causal model to synergy. The advantage of this approach is obvious. It provides a way to assess what we mean by interaction conceptually and asks what mathematic representations support our concepts, rather than providing a statistical model and then contorting our concepts to fit the requirements of that model. Whether or not you agree that Darroch's solution is correct, this approach toward the solution seems reasonable.

The Underlying Causal Model

The counterfactual or potential outcomes model underlies many current developments in epidemiologic methods. Although at first blush it sounds intimidating, this way of thinking about causes echoes simple notions that we apply in everyday circumstances. Nonetheless, its articulation has many interesting implications for causal thinking and is an immensely useful tool for grappling with difficult design decisions, and, as we shall see, for assessing the relationship between our conceptual and statistical tools.

From a counterfactual perspective, a cause is any factor without which the disease event would not have occurred, at least not when it did, given that all other conditions are fixed (Greenland and Robins, 1986; Maldonado and Greenland, 2002; Rothman and Greenland, 1998). Note that this defines causation at the level of the individual, with the definition applying to individual disease events.

The counterfactual way of thinking is familiar to all of us when we second guess our actions and think about what would have happened had we taken a different action. We compare what happened to what would have happened had we made a different choice. Similarly, when we try to make a decision about how to act in the future, we often imagine the outcome under alternative sets of actions. We compare what we think would happen under one action with what we think would happen under a different action.

We also use this type of thought experiment to conceptually separate co-occurrences that are coincidental from those that are causal. So, for example, if a teakettle whistles and then the doorbell rings, we do not assign causality to the teakettle's whistle, because we think that the doorbell would have rung even without the teakettle whistling, assuming all else remained the same. This is the essence of causation from a counterfactual perspective.

Rothman (1976) has developed a heuristic based on this definition of a cause—referred to as causal pies—that provides a useful framework for understanding the implications of this approach.

In this heuristic, the causes of each disease event are depicted by a causal pie (a circle), cut into its constituent pieces. Each piece of the pie represents an exposure that contributes to the occurrence of the disease event. When all of the pieces of the pie are present, the disease occurs. Thus, each pie represents a sufficient cause of disease that is comprised of component causes each of which are necessary for the completion of this sufficient cause of disease. For example, as depicted in Figure E-3, there are three

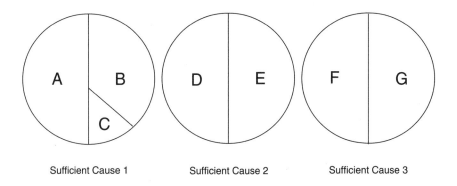

FIGURE E-3 Rothman's causal pies.

posited causal pathways to this disease outcome; individuals can get this disease from sufficient causes 1, 2, or 3. For sufficient cause 1 to occur, an individual must be exposed to components A, B, and C. If any one of the components is missing, the pie will not be complete and disease will not occur, at least not through this mechanism.

Thus, each component in the pie is a cause according to the counterfactual definition, because given that all else is fixed (i.e., all of the causal partners are in place), if we remove component A, for example, the outcome would not have occurred.

Thus, from this perspective, biologic interaction is the relationship between two factors in the same causal pie. In more technical language, biologic interaction occurs when one risk factor allows the other to be expressed in a disease outcome. I prefer to refer to this process as synergy (a term also favored in the epidemiologic literature), because two factors may have causal effects when they influence each other on some level of organization other than the biologic. In the Brown and Harris example above, the interaction between stressful life events and intimacy problems in causing depression might be considered "psychologic interaction." Of course, these psychological factors need to have biologic consequences to cause disease, but the joint effects occur at the psychological level. The counterfactual perspective and Rothman's causal pies are neutral to the level of organization under discussion.[2]

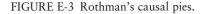

[2]The caveat to this is that an antecedent and a mediator cannot be considered simultaneously, because under that circumstance each component would not be necessary for the pie to form. The pies cannot contain redundant "slices." There is also an affinity for individual-level variables from the causal pie schema, but it can accommodate levels below and above the individual.

$$\text{Causal effect} = \frac{\text{Proportion of exposed people with the disease}}{\text{Proportion of these same people who would have gotten disease without the exposure}}$$

FIGURE E-4 Causal effect (causal contrast).

Thus, for example, A could be a genetic mutation and B one of the environmental factors that stimulates synthesis of a detrimental gene product, or A and B could be two genes that interact to cause disease. Therefore, several concepts that are distinguished from one another in genetics (e.g., epistasis, gene-environment interaction) would be considered to be the same phenomenon in epidemiology. Note that because all of the components in the same casual pie interact in this way, when we ask about interaction, we always must specify the particular components for which we are assessing interaction.

What becomes apparent from this model is that the effect of an exposure depends on the presence of its causal partners. Thus, A will have an effect if and only if its causal partners B and C are present. In contexts in which the causal partners are ubiquitous, the exposure will have a huge effect, since the conditions that activate it always will be present. In contexts in which the causal partners are absent, the exposure will have no effect. In genetics, the classic example used to illustrate this point is phenylketonuria (PKU). The genetic variant that causes PKU has a huge effect in societies in which phenylalanine is a ubiquitous part of the human diet, but a small effect in those in which it is not. Thus, the effect of the "PKU gene" depends on the prevalence of its causal partners. Indeed, intervention on the causal partner is the way in which we largely prevent the deleterious effects of this genetic variant.

As noted above, causes are defined for the individual who gets the disease, which makes sense because the disease occurs in the body of the individual. However, although we use individuals as the units of our analysis, we cannot draw conclusions about the units, but only about the average of the units. Thus, the causal effect (also called the causal contrast) is indexed by the difference between the proportion of exposed people who got the disease at a particular moment in time and the proportion of these same people who would have gotten the disease at that particular moment in time had the exposure not occurred, all things being equal (Mackie, 1974; Rothman and Greenland, 1998), as illustrated in Figure E-4.

Estimating Causal Effects from a Counterfactual Perspective

It is apparent that although we can observe the amount of disease that exposed people experience, we cannot observe the amount of disease that they would have experienced during that same period had they not been exposed. We cannot see both the "fact" (the exposure and disease state of a person) and the "counterfactual" (the disease state under the condition of nonexposure). The counterfactual is, by definition, counter to the facts and therefore not visible. This is a reiteration of the central problem in disease etiology—that causation is not observable. We can see the co-occurrence of exposures and disease, but causation itself cannot be observed, it can only be inferred.

Since we cannot observe the counterfactual state, we select a group of unexposed people as a substitute, or proxy, for the unobservable counterfactual. This substitute gives us the "correct answer" (i.e., represents the true casual effect) to the extent that it is a good proxy. What we mean by a good proxy is that the disease proportion (i.e., disease risk) in this group of unexposed people represents the disease risk the exposed would have had had they not been exposed (i.e., the counterfactual risk).

For the unexposed to be a good proxy, the exposed and the unexposed should be equal on all causes of disease other than the exposure of interest. When this occurs, the exposed and unexposed are said to be "exchangeable." A lack of exchangeability—that is, when the disease risk in the unexposed does not equal that of the exposed had they not been exposed—is what we mean by confounding. When there is confounding, we cannot see whether the exposure had an effect or not. However, assuming exchangeability, or assuming that the unexposed are a good proxy for the counterfactual, the difference in the disease risk between the exposed and unexposed provides an index of the effect of the exposure.

We will discuss this issue of confounding in a bit more detail in order to more fully understand the implications of the counterfactual approach for interaction. This simpler scenario, in which we are attempting to identify the causal effect of a single exposure, will ease the discussion of the application to the more complex scenario of synergy.

Suppose we have a disease such as depression, whose sufficient causes are depicted in Figure E-5. Our hypothesis is that A (perhaps some genetic variant) is a cause of depression. We assume that A has causal partners, which are unidentified but indicated in this model by B. Note that B is simply a stand-in for all of the factors that must be present for A to have an effect. We also assume that there are other pathways to the disease that do not include A. We will note all these other causal pathways by a causal pie with X. X is neither a single exposure nor a single causal pathway. Rather, X is a stand-in for all combinations of exposures that lead to disease that do not include A. Another complication is that it is possible for A to prevent

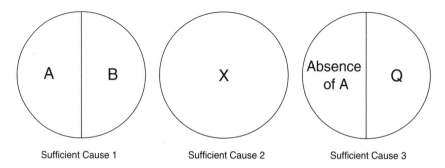

FIGURE E-5 Hypothetical example—causes of depression.

disease in some situations. If so, this means that some people have a combination of exposures (depicted by Q) that require the absence of A to get the disease.

If we consider causation under the counterfactual model, we can imagine what would happen to people with different causal partners if they were exposed to the risk factor under investigation—A in this instance. These potential outcomes are depicted in Figure E-6.

People exposed to X will get the disease if they are exposed to A or not exposed to A (i.e., under the counterfactual they will get disease as well). The exposure does not cause the disease for these people, since even without the exposure they would have gotten it. We label these people Type 1, Doomed.[3] The word is a little stronger than the meaning implied. It simply means that during the period under consideration these people will get the disease under study with or without the exposure of interest. Types are also not inherent characteristics of people; rather, they are a categorization of people by the causal partners (i.e., all risk factors other than those under study) to which they have been exposed by the end of the study period.

People with B will get the disease if they are exposed but not if they are not exposed (i.e., under the counterfactual they will not get the disease). We call these people Type 2, Causal Types (i.e., the exposure under investigation is causal for them). When we ask the question, "Is A a cause of disease?" what we really want to know is whether there are any Causal Types in the population.

People with exposure Q will not get the disease if they are exposed, but under the counterfactual, if they were unexposed, they would get the dis-

---

[3]In this paper I will, in general, use terminology from the original sources to allow easy translation when consulting the original texts. Sometimes the terminology is confusing or can be misinterpreted. In those instances, I will try to clarify the terms, but not invent new ones.

			DISEASE EXPERIENCE IF	
"Type" Person Is:	CAUSAL PARTNERS PERSON HAS	CAUSE PERSON IS SUSCEPTIBLE TO	Exposed to A	Not exposed to A
DOOMED	X	X	DISEASE	DISEASE
A CAUSAL	B	A | B	DISEASE	NO DISEASE
A PREVENTIVE	Q	NOT A | Q	NO DISEASE	DISEASE
IMMUNE			NO DISEASE	NO DISEASE

FIGURE E-6 Exposure of interest A: Potential outcomes of people with different causal partners.

ease. For these people the exposure has an effect, but it is preventive. They are called Type 3, Preventive Types. People who do not have B, X, or Q are labeled Type 4, Immune Types. Regardless of their exposure to A, they will not get the disease.

Thus, the true causal effect of the disease is indicated by the proportion of exposed people who get the disease compared with the proportion of exposed people who would have gotten disease without the exposure. This is the proportion of Doomed and Causal Types in the population compared with the proportion of Doomed and Preventive Types.

But types are unobservable. They represent the potential disease outcomes under the exposed and unexposed circumstances. If we could discern a person's type, we would know how he or she got the disease and know whether or not the exposure under study is, in fact, a cause for him or her. What can be observed is the disease experience of a cohort of people under one of the two conditions, either exposed or unexposed, but not both. Thus, if we take a cohort of people who are exposed, we can assess their actual disease experience, but not their counterfactual disease experience (i.e., what the disease proportion would have been among them had they not been exposed).

We use a comparison of the proportion of disease in an exposed and unexposed group as our best representation of the causal contrast. For the purposes of this discussion, I will hereafter assume that the unexposed are a good proxy for the counterfactual (i.e., the exposed and unexposed are exchangeable, and there is no confounding or bias of any type). Exchangeability also can be understood in terms of the types: it means that the distribution of types is the same in the exposed and unexposed cohorts (or, more specifically, the proportions of Doomed and Protective Types are the same in the two cohorts).

It should be noted again that the causal contrast of a group of people indexes the average effect of the exposure. If the exposure can have both causal and preventive effects, then our measures tell us only whether or not there are more people for whom the exposure is causal in the population than people for whom the exposure is preventive. For example, the risk difference is the difference in the proportion of Causal (Type 2) and Preventive Types (Type 3) in the population, and the risk ratio is the ratio of Types 1 and 2 (Doomed and Causal) to Types 1 and 3 (Doomed and Protective). If we can assume that the exposure can have only a causal effect, and never a preventive effect, then the difference in these proportions tells us the proportion of people for whom the exposure is, in fact, causal.

In sum, any time that we calculate a risk ratio or a risk difference, we are using the data we have on patterns of exposure/disease relationships to infer something about the types of people in the population. In particular, we make inferences about the presence and proportion of Causal Types (Type 2's) in the population. This is the basis for causal inference in epidemiology.

Darroch (1997) and Rothman and Greenland (1998) built on this insight to assess the particular mathematical representations that would arise in our data if, in fact, there were people in the population who got the disease because of the biological interaction of two particular exposures. From our definition of a cause, this has a particular meaning: There is biologic interaction (synergy) if and only if there are some people in the population who got the disease because they were exposed to both the exposures under consideration and who would not have gotten the disease otherwise. By "otherwise," I mean they would not have gotten the disease if they were exposed to only one of the two exposures or if they were exposed to neither.

Estimating Synergy Under a Counterfactual Perspective

For purposes of exposition, I will assume that we are interested in the hypothesis that a particular genetic polymorphism (Gene A) interacts with a particular toxin (Environmental Exposure B) to cause a particular disease

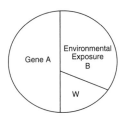

Gene A and Environmental Exposure B as causal partners (other casual partners represented by W).

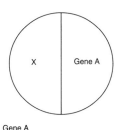

Gene A (causal partners represented by X)

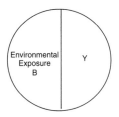

Environmental Exposure B (causal partners represented by Y)

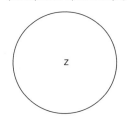

All other causes of depression that do not include Gene A or Environmental Exposure B.

FIGURE E-7 Sufficient causes of an outcome: interest in the interaction of Gene A and Environmental Exposure B.

outcome. We can think about this as a genetic variant that makes the body's cells unable to clear some toxin from the system. We do not assume that this is the only route to the disease.

One limitation to the detection of synergy is that we must assume that Gene A and Environmental Toxin B are not causal in some people and protective in others. In other words, to make the detection of synergy possible at all, we must assume that this particular genetic polymorphism and this particular environmental toxin can only cause damage and are never protective (although they may be neutral).

We can then conceptualize the causal pies that would depict the causes of disease from the point of view of interest in both Gene A and Environmental Exposure B and, in particular, in their interaction. There would then be four sufficient causes for this disease; one sufficient cause requires both A and B and their causal partners, one requires A and its causal partners, another B and its causal partners, and, finally, one requires neither A nor B (see Figure E-7). Our interest in synergy means that we want to know if there are any people, and, if so, how many, who got the disease from the first sufficient cause.

Based on this model, there are different response types to our two exposures of interest. People who are exposed to W will get the disease if and only if they are exposed to both Gene A and Environmental Exposure

B. These are the Synergistic Types whose presence or absence in the population we want to detect. People with X will get the disease if they are exposed to A, regardless of their exposure to B. They are the Gene A Susceptible Types. Likewise, people exposed to Y will get the disease if they are exposed to B, regardless of exposure to A. They are the Environmental Exposure B Susceptible Types. People exposed to Z, the Doomed, will get the disease whether or not they are exposed to A or B; they will get the disease from a sufficient cause that does not include A or B in the causal pathway. The Immune Type will not get disease by the end of the study, regardless of their exposure to A or B. Finally, there is another type, more recently discovered, that provides the "twist" alluded to above—people who have both X and Y and who will get the disease if they are exposed to either A or B. Such people are called Parallel Types.

One can think of the four right-hand columns of the figure as different, exchangeable cohorts of people. By exchangeable, we mean that the distribution of types across the cohorts is the same. That is, if 20 percent of the cohort exposed to A is Doomed, then 20 percent of the cohort exposed to B also is Doomed. This is simply an expansion of the "no confounding" assumption described earlier in the context of the detection of single causes.

One cohort is exposed to A only, one to B only, one to both A and B, and one to neither. The notation, diseased or not disease, in each column indicates the disease outcome for the type described in each row under each exposure condition. For example, row 2, the Gene A Susceptible Type, will get the disease if exposed to A or if exposed to A and B, but not otherwise. It should be noted that if exposed to A and B, the causal effect for this type is still only A.

What we want to know are the types of people (or the proportion of each type) in the population. But types are not visible. The only thing that is visible is the pattern of exposure for the disease experience; we can see the proportion of each exposure group that gets the disease. We want to use these patterns of exposure disease associations to identify the proportion of each type (and in particular the proportion of synergistic types) in the population.

By looking down the four right-hand columns in Figure E-8, we can easily identify the proportion of two types in the population—the Immune and the Doomed. The Doomed are represented by the proportion of people who get the disease among the cohort that is exposed to neither A nor B. Similarly, the Immune are the proportion of people exposed to both A and B who do not get the disease.

We also can see from Figure E-8 the types that contribute to disease under each exposure condition, which are summarized in Figure E-9. Among those exposed to both A and B, the Synergistic, Doomed, A Susceptible, B Susceptible, and Parallel types all get disease and thus contribute to

"Type" Person Is:	CAUSAL PARTNERS PERSON HAS	CAUSE PERSON IS SUSCEPTIBLE TO	DISEASE EXPERIENCE IF EXPOSED TO:			
			GENE A	ENVIRON-MENTAL EXPOSURE B	BOTH	NEITHER
SYNERGISTIC	W	W (GENE A / ENVIRON. B)	No disease	No disease	**Disease**	No disease
GENE A susceptible	X	X (GENE A)	**Disease**	No disease	**Disease**	No disease
ENVIRONMENTAL EXPOSURE B susceptible	Y	Y (ENVIRON B)	No disease	**Disease**	**Disease**	No disease
DOOMED	Z	Z	**Disease**	**Disease**	**Disease**	**Disease**
IMMUNE		NONE	No disease	No disease	No disease	No disease
PARALLEL	X / Y	X / Y (GENE A or ENVIRON B)	**Disease**	**Disease**	**Disease**	No disease

FIGURE E-8 Assessing interaction between Gene A and Environmental Exposure.

Disease risk:	Among those exposed to both Gene A and Environmental Exposure B	Among those exposed only to Gene A	Among those exposed only to Environmental Exposure B	Among those exposed to neither Gene A nor Environmental Exposure B
Observation:	Proportion diseased among those exposed to both	Proportion diseased in those exposed only to Gene A	Proportion diseased in those exposed to only Environmental Exposure B	Proportion diseased among those exposed to neither
Numeric representation:	R_{12}	R_1	R_2	R
Types contributing to this risk:	Synergistic, doomed, Gene A susceptible, Environmental Exposure B susceptible, parallel	Doomed, Gene A susceptible, parallel	Doomed, Environmental Exposure B susceptible, parallel	Doomed

$R_{12} - R_1 - R_2 + R =$ [Synergistic, doomed, Gene A susceptible, Environmental Exposure B susceptible, parallel] $-$ [Doomed, Gene A susceptible, parallel] $-$ [Doomed, Environmental Exposure B susceptible, parallel] $+$ [Doomed]

$R_{12} - R_1 - R_2 + R =$ Synergistic $-$ parallel

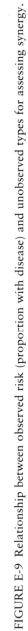

FIGURE E-9 Relationship between observed risk (proportion with disease) and unobserved types for assessing synergy.

the proportion of diseased people (i.e., the risk) in this exposure cohort. Among those exposed only to Gene A, the Doomed, A Susceptible Types, and Parallel Types contribute to the risk; among those exposed to B only, the Doomed, B Susceptible, and Parallel Types contribute to the risk; and among those exposed to neither A nor B, only the Doomed contribute to the risk.

We can see the risk (the proportion diseased) in each exposure group for which we provide specific labels. R_{12} is the risk (the proportion diseased) among those exposed to both A and B; R_1 is the risk for those exposed only to A; R_2 the risk for those exposed only to B; and R the baseline risk (i.e., the risk among those exposed to neither A nor B). We can now translate the proportion diseased (the risk) we observe under each exposure category into the underlying types that contribute to the risk in each exposure category.

Using basic mathematical tools, we attempt to isolate Synergistic types from the others. The closest we can come is the isolation of the balance between Synergistic and Parallel Types. The proportion of (synergistic – parallel) types in the population = $R_{12} - R_1 - R_2 + R$.[4] This is the additive model $(R_{12} - R) - (R_1 - R) - (R_2 - R)$[5] that assumes risks add in their effects, with the twist that parallelism makes the relationships somewhat less than additive. Thus, if the risk of disease among those exposed to both factors is more than the sum of the risk differences for each factor alone, there is evidence of Synergistic Types in the population. This is evidence that Gene A and Environmental Factor B work in a synergistic way to cause disease for at least some people.

Note, however, that we cannot definitively state what proportion of the disease is due to synergy. We can only say that the proportion of Synergistic Types is greater than the proportion of Parallel Types. In addition, perfect additivity is compatible with either no Synergistic Types in the population or a perfect balance of Synergistic and Parallel Types. Just as in the simple case of identifying single causes, we only identify the average risk—that is, the preponderance of causal over protective effects of an exposure—so too

---

[4]If we take the types that contribute to disease among those exposed to both A and B (the first box in Figure E-9), subtract from them those that contribute in the second box, subtract from them those that contribute in the third box, and then add those in the fourth box, we are left with (synergy – parallel). The Synergistic Types appear in only one box, so they cannot be canceled out, and the Parallel Types occur in three boxes, so their cancellation leaves the Parallel Type. All other types cancel out in this formula.

[5]$R_{12} - R_1 - R_2 + R = (R_{12} - R) - (R_1 - R) - (R_2 - R)$. In the absence of synergy, the risk difference for those with both factors $(R_{12} - R)$ will simply equal the risk difference for factor A $(R_1 - R)$ + the risk difference for factor B $(R_2 - R)$. Thus in the absence of synergy and parallelism, or a balance of synergy and parallelism, $(R_{12} - R) - (R_1 - R) - (R_2 - R) = 0$.

in the face of parallelism, we cannot rule out synergy if we find less than superadditivity, but we do find support if there is superadditivity.

It is important to note the constraints on this conclusion. First, this analysis makes all of the usual assumptions that apply in the way we currently conduct research; it assumes such things as independence of outcomes between units and no feedback loops. Second, it makes the important assumption that the exposures under consideration express either synergy or antagonism, but not both; it is assumed that a risk factor has only a casual effect or a preventive effect, but not both. How realistic this assumption is depends on the exposures under consideration. In psychology, this assumption is often unrealistic. For example, there may be parenting practices (such as strict discipline) that would be beneficial for children with one type of temperament, but detrimental for children with another. In genetics, the "norm of reaction," where a genetic factor has positive or negative effects depending on the context (Levins and Lewontin, 1985), could violate this assumption. However, this is simply a recognition that under these circumstances there are too many unknowns for any of our traditional mathematical models to handle. These caveats notwithstanding, Darroch's argument begins with the conceptual model and then brings us to the mathematical model that represents synergy most closely.

Applications in Practice

The conclusion drawn from these analyses is that synergy is indexed by deviations from additivity. In practice then, how do we estimate synergy using this approach? One method is to calculate an "interaction contrast" (Rothman and Greenland, 1998). To illustrate how this is done, I will use an example based on the interaction between a serotonin transporter gene polymorphism and life stress in causing depression, as reported from the Dunedin birth cohort (Caspi et al., 2003). The hypothesis was that there is a synergistic relationship between a short "s" allele and multiple stressful life events in causing depression.

As illustrated in Figure E-10, the disease prevalence among those with neither the susceptible genotype nor life events was 10 percent; among those with only the susceptible genotype, 10 percent; among those with only life events, 17 percent; and among those with both life events and the susceptible genotype, 33 percent. In this instance the interaction contrast would be $.33 - .17 - .10 + .10 = .16$. The interaction contrast thus equals the risk among those with both factors (.33), minus the risk among those with one (.17), minus the risk among those with the other (.10), plus the baseline risk (.10). Since the interaction contrast here is greater than zero (.16), it indicates the presence of synergy in this population.

In this example, the risks required for the computation were directly

Percentage of Individuals in Each Category Meeting Criteria for Depression

		"S" Genotype	
		Yes	No
4+ Life events	Yes	33%	17%
	No	10%	10%

FIGURE E-10 Estimation of the interaction contrast.

provided by the report. However, in a cohort study we can compute the interaction contrast, regardless of the form in which the results are analyzed and presented. Suppose we analyzed the data under a logistic regression model. The baseline odds of disease would be derived from the intercept. The odds ratios from the logistic regression then would be used to obtain the odds of disease under the other conditions. Finally, the odds would be converted to risks (odds = $p/1 - p$).

When we cannot estimate the baseline risk of disease, as in a case-control study, we can calculate an interaction contrast ratio using the odds ratios computed from a logistic regression analysis. The interaction contrast ratio is the odds ratio for those with both factors, minus the odds ratio for those with one factor, minus the odds ratio for those with the other factor, plus one. For illustration, I computed the odds ratios for the Dunedin study from the prevalence estimates given in Figure E-10. The baseline odds of depression among those with neither the "*s*" allele nor stressful life events are .11 (.10/1.10). The odds for those with both factors are .49 (.33/1.33); for those with only the "*s*" allele they are .11 (.10/1.10), and for those with only life event they are .20 (.17/1.17). Therefore the odds ratios would be 4.4 for those with both factors, 1.8 for life events alone, and 1 for the "*s*" allele alone. The interaction contrast ratio in this context would be 4.4 − 1.8 − 1 + 1 = 2.6. Since the interaction contrast ratio is greater than 0, this indicates the presence of Synergistic Types in the population. Several methods have been developed to calculate p values and confidence intervals around these estimates (see, e.g., Assmann et al., 1996; Hosmer and Lemeshow, 1992; Rothman and Greenland, 1998).

Although this is the understanding of synergy that is accepted in the methodologic literature, it has begun to filter down into actual research articles only recently (e.g., Li et al., 2005; Olshan et al., 2001; Rauscher et al., 2003; Shen et al., 2005). It is interesting to note that many of these articles assess gene-environment interactions. However, this model of assessing synergy is applicable to genetic interactions only to the extent that the underlying counterfactual causal model is applicable.

APPLICABILITY TO THE STUDY OF GENETIC INTERACTIONS

Applicability of a Counterfactual Approach to a Genetic Context

The counterfactual approach requires a thought experiment in which we hold everything constant and manipulate the exposure to see what the outcome would be under this new condition. The causal contrast—the index of the true effect of the exposure—is the difference between what was, given the exposure, and what would have been had the exposure been altered but everything else remained constant. Because this thought experiment requires the consideration of an alteration in the exposure and nothing else, the applicability of a counterfactual approach to nonmanipulable exposures has been questioned (e.g., Kaufman and Cooper, 1999). Since, currently, genes are not easily manipulable, this might open the question of the applicability of this approach to the consideration of genetic effects. In a similar vein, some have argued that personal characteristics, such as age, gender, ethnicity, and social class, should not be considered as causes because they are not manipulable.

However, others (Shadish et al., 2002; Susser and Schwartz, 2005) argue that the counterfactual can apply to nonmanipulable causes, although their detection is more difficult. Nonmanipulable causes cannot be randomly assigned to rule out the many potential sources of nonexchangeability between the exposed and unexposed group that cause confounding. Nonetheless, at the least, one can conduct the thought experiment and search for, or design, studies that approximate the thought experiment as closely as possible.

In addition, what is nonmanipulable today may, in the future, become manipulable. The use of animal "knock-out models" clearly indicates the possibility of genetic manipulation and, with increasing knowledge, even when the gene itself is not manipulable the active ingredients of the gene vis-à-vis the disease, the gene product, may be manipulable.

In the final analysis, it seems that in genetic studies in which people are compared who do and do not have a particular gene variant, or who do or do not have a proxy for a genetic predisposition (e.g., family history), the comparison only makes sense if there is some underlying notion of a causal contrast underlying it. The association may not reflect causation due to the nonexchangeability of the exposed and unexposed, but the logic of the methods assumes that barring such methodological problems, the contrast would imply a causal contrast. Otherwise, why do we use such methods to try to detect causes? The counterfactual approach is merely the clear articulation of the framework that supports the logic that underlies all of our study designs.

Primacy of the Genetic Effect

The egalitarian assumptions regarding causation constitute another possible objection to the application of this approach to a genetic context. As discussed above, from a counterfactual perspective, genes, behaviors, and the external environment share equally in the appellation "cause." There is no hierarchy of enabling factors and triggers versus the "real cause." This view is in contrast to genetic approaches that see the gene as the central actor, with all other "causes" playing a supporting role. However, this should not be a significant impediment to the application of epidemiologic approaches to interaction. There are many possible approaches to its resolution. First, one can impose a hierarchy on this approach by declaring a genetic factor to be a necessary cause of the outcome and by defining the phenotype based on the genetic component. As discussed above, there is nothing in this approach that precludes a cause that is found in every causal pie (i.e., in every causal pathway to disease). Of course, if the genetic factor is known to be a necessary cause, the detection of interaction is simplified. In such an instance, one would look for the main effects of a hypothesized causal partner among those with the genetic factor. But even if the genetic factor is not necessary, one could give it prominence by referring to the causal pies that do not contain the genetic effects as phenocopies. Similarly, in discussing the interaction between a genetic and an environmental cause, one can refer to the environmental factor as triggering a genetic effect. These interpretational preferences would not be inconsistent with a counterfactual approach.

On the other hand, the counterfactual approach also suggests that there may be benefits, under some circumstances, to dismantling the hierarchy. That is, one can often describe a gene-environment interaction equally well as the interaction between an environmental factor and a genetic vulnerability that allows the environmental factor to be expressed or as an interaction between a genetic factor and an environmental context that allows the gene to be expressed. The counterfactual approach points out the symmetry of interaction.

Application to Study Designs Used to Detect
Gene-Environment Interactions

Many of the study designs used to detect gene-environment interactions are indistinguishable from those used to detect interactions between environmental or other nongenetic factors. Cohort studies and case-control studies and their variants are prominent designs in general and genetic epidemiology (Hunter, 2005). Thus, the statistical models used to analyze the data—linear regression, logistic regression, Cox proportional hazards models, and Poisson regression—are used in both fields. The problems and arguments discussed above therefore apply directly.

There are other study designs, however, that have been developed specifically for the assessment of genetic exposures—for example, familial aggregation studies, twin studies, and the case-only design. The problem of the model dependence of interaction applies to these situations as well. In each instance, the data are analyzed using a model that makes some assumption about how independent effects influence risk and therefore about how interaction is indicated. Even case-only studies, which assess gene-environment interactions without the use of controls, make such an assumption. This design is predicated on a multiplicative model. Thus, case-only studies are also conservative if we think that synergy is best indicated by deviations from additivity (Gatto et al., 2004). Twin studies are perhaps the most problematic for assessing interaction, since the genetic and environmental factors are not measured. Their effects are derived from the pattern of results, which often have to assume the absence of interaction to be interpretable. To the best of my knowledge, the basic problem of the model dependence of measures of synergy is not solved by the use of specific genetic designs.

THE MESSINESS OF REAL-WORLD APPLICATIONS

One of the advantages of the counterfactual approach is that it illuminates a central problem of causal inference: it is uncertain. Causality is an unobservable construct that leaves footprints in the real world that are open to misinterpretation. It is important to note that the counterfactual approach does not cause these problems, but rather articulates them and thus forces us to confront them. But forewarned is forearmed. Once we recognize the reality of the uncertainty and subjectivity of causal inference, we can think about the factors that exacerbate and mitigate these uncertainties and design our studies and analyses accordingly. This approach also should warn us against demanding more of our data than they can provide and against interpreting our data beyond their inherent limitations.

The assessment of synergy is no exception. Our data can provide us with evidence that is consistent or inconsistent with synergy, but they can never provide definitive evidence for or against it. Each study has its own strengths and weaknesses. The most productive approach is to consider all of the extant evidence, consider our uncertainties about the data, and then design new studies that confront those uncertainties directly. No one study will provide us with an answer, but carefully designed studies that directly confront alternative hypotheses will move us toward greater clarity.

All of the threats to validity that apply to the detection of single causes apply to the detection of synergy. Some become even more salient. I will, therefore, only briefly touch on some of the issues that were specifically raised in the mandate for this paper.

Power

Power, the ability to detect an association of a designated magnitude when it exists in the population, is a problem in all studies, but it is one that is particularly problematic for detecting synergy. Power is based on three factors: how well variables are measured, how large the true effect is in the population, and the sample size. It follows, therefore, that we can increase power by measuring our variables well, looking for effects that are large (or looking for them where they are large), and conducting studies with sufficient numbers of people.

The genomic revolution should improve power because the genetic effect is more clearly and closely measured. When family history of a disease, for example, is used as a proxy measure for a genetic effect, the bias toward that null that derives from measurement error is enormous. A true genetic effect of 50 can look like a genetic effect of 2 or less, depending on the prevalence of the outcome and other factors (Zimmerman, 2003). Thus, measuring actual genetic markers decreases measurement error and increases power.

In detecting gene-environment interactions, accurate measurement of the environmental factor is equally important. The more clearly articulated the hypothesis, the more carefully the measures can be chosen, and the more power there will be to detect an effect. Vague theories about gene-environment interactions will be more likely to lead to poor construction of measurable variables and therefore decreased ability to detect synergistic effects. However, it should be noted that measurement error also can masquerade as interaction as well as mask it. Thus, false positive as well as false negative results can be produced by poor measurement.

Power also can be enhanced by looking for situations or populations in which the interaction is strong. As discussed in earlier sections of this paper, the effect of an exposure depends not only on its biologic effects, but on the prevalence of its causal partners and the number of sufficient causes in which it is not a partner. Therefore, the same biologic effect will be easier to detect in situations in which the other sufficient causes are rare and the causal partners are common. Along these lines, one suggestion for enhancing power regarding main effects is to look for the effect of an exposure in a group in which the outcome is rare (Rothman and Poole, 1988). This would enhance power because the base rate of disease in the unexposed group would be low. Similarly, to detect specific gene-environment interactions for a particular outcome, looking for populations in which the outcome is less common may help. In these situations, the same biologic effects will produce a larger risk ratio.

Power also should be a consideration in the choice of study design. For the same number of people, case-control studies will in general provide more power when the outcome is rare, and cohort studies will provide more power when the exposures are rare.

But whatever the choice, sample sizes need to be sufficient. Articulating hypotheses in advance has the added advantage of providing the basis for more accurate power estimates. However, methods for estimating power are less developed for synergy than they are for main effects, although some work has been done on proper power analyses for both additive and multiplicative interaction (e.g., De Gonzalez and Cox, 2005; Greenland, 1983). What is clear is that the detection of interaction requires considerably larger sample sizes than the detection of the exposures' main effects.

Multiple Comparisons

Multiple comparisons may be a particular problem regarding interaction because researchers are less likely to hypothesize them in advance. This raises the concern that we will increase the number of Type I errors in our studies; we will frequently reject the null in error. Some have suggested an adjustment to our alpha levels (e.g., Bonferroni adjustments) to take multiple comparisons into account. To fully address this issue requires a detailed discussion of the meaning of p values and confidence intervals, which is beyond the scope of this paper. I will, however, touch on some issues to consider.

The use of adjustments to the alpha level to correct for multiple comparisons reifies the p value and potentially contributes to a misuse of null hypothesis testing. Null hypothesis testing tells us the probability of our data if the null is true. What we really want to know is the probability that the null is true, given our data. Unfortunately, these two probabilities are not the same. There is a tendency, however, to treat significant results as though they told us the latter rather than the former. In addition, p values do not strictly apply in the context of observational studies because the statistical premises on which they are constructed are often violated in nonexperimental settings. For both reasons, the use of confidence intervals rather than p values is preferred. Confidence intervals provide a rough estimate of the precision of our data. Wide confidence intervals tell us that our data do not provide much information about the effect. Narrow confidence intervals suggest that our data are more precise. Of course, there may be confounding and other biases reflected in our estimates, but the association is more trustworthy. The use of confidence intervals, with a statement of the number of comparisons made, provides more information for the reader to decide how seriously to take the results of a study. But nothing solves the problem of multiple comparisons. Data that are consistent with well-formulated hypotheses that are developed in advance of the study provide better evidence than data that result from studies for which the hypotheses are developed after the fact.

Population Stratification

From an epidemiologic perspective, population stratification is simply confounding; the exposed and unexposed may differ for reasons other than the exposure under study. One advantage of genetic epidemiology is that the confounders of genetic associations are limited, and the more carefully specified and measured the genetic factor, the more limited the potential sources of nonexchangeability. For example, if you measure a genetic effect by a family history of the outcome, the exposed and unexposed may differ on a large number of factors other than the exposure of interest, which is a genetic effect. However, if you measure the exposure as a particular genetic variant or marker, it becomes less likely that there will be nonexchangeability of the exposed and unexposed on other causes of disease beyond what would occur by chance.

Population stratification is simply confounding that arises because the groups with unequal distributions of a particular genetic variant also have unequal distributions of other risk factors for the disease. The problem is exacerbated in case-control studies for which the selection of cases and controls can create population stratification even when it does not exist in the naturally occurring populations that gave rise to the cases, as is true with most problems of confounding. In a cohort study, population stratification would be more easily detected and controlled.

To the extent that population stratification is a problem in studies of single exposures, it will be a problem in studies of synergy. Just as single studies require the exposed and unexposed cohorts to be exchangeable regarding all causes of the disease other than the exposure of interest, so too the assessment of synergy requires that all four exposure cohorts (exposed to both factors, each of the two alone, and neither) be exchangeable regarding all causes of disease other than the two under investigation.

CONCLUSION

Epidemiologic approaches to biologic interaction have benefited from a full articulation of the underlying causal assumptions of risk factor epidemiology. The counterfactual or potential outcomes approach clarifies methodologic principles and provides a guide for methodologic choices. This starting point suggests that risks add in their effects. Therefore, synergy is best indicated by deviations from an additive rather than a multiplicative model, with a twist. I think that this model applies as well to genetic causes as it does to the environmental and behavioral causes that are more frequently examined in traditional epidemiologic contexts. It has the added advantage of clarifying and unifying other genetic constructs, providing a basis for understanding confounding in general and population stratification in particular, and providing a bridge between genetic and risk factor epidemiology. Although this approach has limitations, the transparency of its conceptual basis makes the limitations transparent as well. It brings the limitations inherent in all forms of causal inference to light, making them more amenable to amelioration.

ACKNOWLEDGMENTS

Ann Madsen reviewed and provided helpful comments on this paper. Many of the ideas and examples in this paper derive from Schwartz and Susser (in press) "Causal Explanation Within a Risk Factor Framework," Chapter 35, in: Susser, Schwartz, Morabia, and Bromet, *Psychiatric Epidemiology: the Search for Causes of Mental Disorders*, Oxford University Press.

REFERENCES

Assmann SF, Hosmer DW, Lemeshow S, Mundt KA (1996). Confidence intervals for measures of interaction. *Epidemiology* 7:286-290.

Brown GW, Harris T (1978). Social origins of depression: a reply. *Psychological Medicine* 8:577-588.

Caspi A, Sugden K, Moffitt TE, et al. (2003). Influence of life stress on depression: moderation by a polymorphism in the 5-HTT gene. *Science* 301:386-389.

Darroch J (1997). Biologic synergism and parallelism. *American Journal of Epidemiology* 145:661-668.

De Gonzalez AB, Cox DR (2005). Additive and multiplicative models for the joint effect of two risk factors. *Biostatistics* 6:1-9.

Gatto NM, Campbell UB, Rundle AG, Ahsan H (2004). Further development of the case-only design for assessing gene-environment interaction: evaluation of and adjustment for bias. *International Journal of Epidemiology* 33(5):1014-1024.

Greenland S (1983). Test for interaction in epidemiologic studies: a review and a study of power. *Statistics in Medicine* 2:243-251.

Greenland S, Robins JM (1986). Identifiability, exchangeability, and epidemiological confounding. *International Journal of Epidemiology* 15:413-419.

Hosmer DW, Lemeshow S (1992). Confidence interval estimation of interaction. *Epidemiology* 3:452-456.

Hunter DJ (2005). Gene-environment interactions in human diseases. *Nature Reviews* 6:287-298.

Kaufman JS, Cooper RS (1999). Seeking causal explanations in social epidemiology. *American Journal of Epidemiology* 150:113-120.

Levins R, Lewontin R (1985). *The Dialectical Biologist.* Cambridge, MA: Harvard University Press.

Li Y, Millikan RC, Bell DA, Cui L, Tse CJ, Newman B, Conway K (2005). Polychlorinated biphenyls, cytochrome P450 1A1 (CYP1A1) polymorphisms, and breast cancer risk among African American women and white women in North Carolina: a population-based case-control study. *Breast Cancer Research* 7:R12-R18.

Mackie JL (1974). *Cement of the Universe.* Oxford, England: Clarendon Press.

Maldonado G, Greenland S (2002). Estimating causal effects. *International Journal of Epidemiology* 31:422-429.

Olshan AF, Weissler MC, Watson MA, Bell DA (2001). Risk of head and neck cancer and the alcohol dehydrogenase 3 genotype. *Carcinogenesis* 22:57-61.

Rauscher G, Sandler DP, Poole C, Pankow J, Shore D, Bloomfield CD, Olshan AF (2003). Is family history of breast cancer a marker of susceptibility to exposures in the incidence of de novo adult acute leukemia? *Cancer Epidemiology, Biomarkers and Prevention* 12:289-294.

Rothman KJ (1976). Reviews and commentary: causes. *American Journal of Epidemiology* 104:587-592.

Rothman KJ, Greenland S (1998). *Modern Epidemiology,* 2nd ed. Philadelphia, PA: Lippincott-Raven.

Rothman KJ, Poole C (1988). A strengthening programme for weak associations. *International Journal of Epidemiology* 17:955-959.

Rubin DB (2004). Direct and indirect causal effects via potential outcomes. *Scandinavian Journal of Statistics* 31:161-170.

Shadish WR, Cook TD, Campbell DT (2002). *Experimental and Quasi-Experimental Designs for Generalized Causal Inference.* Boston, MA: Houghton Mifflin.

Shen J, Gammon MD, Terry MB, Wang L, Wang Q, Zhang F, Teitelbaum SL, Eng SM, Sagive SK, Gaudet MM, Neugut AI, Santella RM (2005). Polymorphisms in XRCC1 modify the association between polycyclic aromatic hydrocarbon-DNA adducts, cigarette smoking, dietary antioxidants, and breast cancer risk. *Cancer Epidemiology, Biomarkers and Prevention* 14:336-342.

Susser E, Schwartz S (2005). Are social causes so different from all other causes? A comment on Sander Greenland. *Emerging Themes in Epidemiology* 2:4.

Tennant C, Bebbington P (1978). The social causation of depression: a critique of the work of Brown and his colleagues. *Psychological Medicine* 8:565-575.

Zimmerman R (2003). Familial aggregation study designs: causes of discrepancies in case-control and reconstructed cohort effect estimates. Ph.D. Dissertation: Columbia University, AAT 3071403.

F

Acronyms

5-HT T	serotonin transporter 5-hydroxytryptamine
A1AT	alpha-1-antitrypsin
ACTH	adrenocorticotropin hormone
ADA	Americans with Disabilities Act
ADHD	Attention Deficit Hyperactivity Disorder
ADR	adverse drug reaction
ANS	autonomic nervous system
Apoe	apolipoprotein E
ATM	ataxia telangiectasia mutated
AVP	arginine vasopressin
BEN	"Benin" β-globin gene haplotype
BIND	Biomolecular Interactions Newtwork Database
BMI	body mass index
bp	base pairs
BRCA1	Breast Cancer Gene 1
BRCA2	Breast Cancer Gene 2
CAR	"Central African Republic" β-globin gene haplotype
CDC	Centers for Disease Control and Prevention
CF	cystic fibrosis
CGAP	Cancer Genome Anatomy Project
CLIA	Clinical Laboratory Improvement Amendments
CNS	central nervous system

CPS	Current Population Study
CRF	corticotropin releasing factor
CRH	corticotropin-releasing hormone
CRISP	Computer Retrieval on Information on Scientific Projects
CRP	C-reactive protein
CVD	cardiovascular disease
CYP	cytochrome P450
CYP2D6	cytochrome P450, family 2, subfamily D, polypeptide 6
CYP3A4	cytochrome P450, family 3, subfamily A, polypeptide 4
dbSNP	Single Nucleotide Polymorphism Database
DES	diethylstilbestrol
DNA	deoxyribonucleic acid
DRD4	dopamine receptor D4
DVT	deep vein thrombosis
EAE	experimental allergic encephalomyelitis
EC	Enzyme Commission
EcoCyc	Encyclopedia of Escherichia Coli K-12 Genes and Metabolism
EGP	Environmental Genome Project
ENRICHD	Enhancing Recovery in Coronary Heart Disease Patients
ERα	estrogen receptor a
ERCC2	excision repair cross-complementing group 2
F344	Fischer 344 rat strain
FDR	false discovery rate
FEW	family-wise type I error rates
GAI	Genetic Annotation Initiative
GSF	German National Center for Environment and Health
GSTM1	glutathione S-transferase M1
GSTP1	glutathione S-transferase pi
GSTT1	glutathione S-transferase theta 1
HaploChIP	haplotype-specific chromatin immunoprecipitation
HapMap	International Haplotype Mapping Project
Hb S	sickle hemoglobin
HER2	human epidermal growth factor receptor 2
HIPAA	Health Insurance Portability and Accountability Act of 1996
HIV	human immunodeficiency virus

HPA	hypothalamic-pituitary-adrenal
HUD	Department of Housing and Urban Development
IDR	interdisciplinary research
IL-1α	interleukin-1α
IOM	Institute of Medicine
IRB	Institutional Review Board
KEGG	Kyoto Encyclopedia of Genes and Genomes
LD	linkage disequilibrium
LDL	low-density lipoprotein
LDL-R	low-density lipoprotein receptor
LEW	Lewis rat strain
MALDI-TOF	matrix assisted laser desorption/ionization-time of flight
MAOA	monoamine oxidase A
MC4R	melanocortin 4 receptor
MeSH	medical subject headings
MIAME	Minimum Information About a Microarray Experiment
MIPS	Munich Information Center for Protein Sequences
mRNA	messenger ribonucleic acid
MTO	Moving To Opportunity
NAS	National Academy of Sciences
NAT2	N-acetyltransferase 2
NCBI	National Center for Biotechnology Information
NCHS	National Center for Health Statistics
NCI	National Cancer Institute
NHGRI	National Human Genome Research Institute
NIEHS	National Institute of Environmental Health Sciences
NIGMS	National Institute of General Medical Sciences
NIH	National Institutes of Health
NMR	nuclear magnetic resonance
NRC	National Research Council
NRSA	National Research Service Act
NSF	National Science Foundation
OBSSR	Office of Behavioral and Social Sciences Research
OC	oral contraceptive
OMB	Office of Management and Budget
OMIM	Online Mendelian Inheritance in Man
OR	odds ratio

p53	tumor suppressor gene p53
p450	cytochrome P450
PBMC	peripheral blood mononuclear cells
PI	Principle Investigator
PNI	psychoneuroimmunology
PKU	phenylketonuria
P/T	promotion and tenure
RD	risk difference
RFA	Request for Applications
RNA	ribonucleic acid
RR	risk ratio
SAGE	serial analysis of gene expression
SCPD	Promoter Database of *Saccharomyces cerevisiae*
SCW	streptococcal cell wall
SEI	Socioeconomic Index
SEN	"Senegal" β-globin gene haplotype
SEPA	Science Education Partnership Award
SES	socioeconomic status
SIV	simian immunodeficiency virus
SNP	single nucleotide polymorphism
SNS	sympathetic nervous system
SSRI	selective serotonin re-uptake inhibitor
TPMT	thiopurine S-methyltransferase
U.S. DHHS	U.S. Department of Health and Human Services
V1aR	vasopressin-1a receptor
WHO	World Health Organization
WIT	What Is There?
XRCC1	x-ray cross complementing group 1

G

Biographical Sketches

Dan G. Blazer, M.D., Ph.D. *(Chair)*, is J.P. Gibbons Professor of Psychiatry and Behavioral Sciences at Duke University Medical Center. Dr. Blazer is the author or editor of more than 30 books and the author or co-author of more than 300 peer-reviewed articles on topics including depression, epidemiology, and consultation liaison psychiatry. He is a fellow of the American College of Psychiatry and the American Psychiatric Association and is a member of the Institute of Medicine (IOM), with expertise in medical education (both undergraduate and graduate), religion and medicine, and preventive medicine and public health. Much of Dr. Blazer's research has focused on the prevalence of physical and mental illness in the elderly, such as the Epidemiologic Catchment Area Project and the Established Populations for Epidemiologic Studies of the Elderly (EPESE). He has served as the Principal Investigator of the Duke University EPESE, the Piedmont Health Survey of the Elderly. He served as a member of the Board of Directors of Retired Persons' Services (AARP pharmacy service) and currently is a Board member of the American Association of Geriatric Psychiatry. He served as Chair of the Committee for the IOM review efforts of the Department of Defense to provide adequate medical care to Persian Gulf War Veterans and the IOM Committee on Testosterone Replacement Therapy in the Elderly. He also served as the President of the American Association of Geriatric Psychiatry.

Melissa A. Austin, Ph.D., is Professor of Epidemiology in the School of Public Health and Community Medicine, Director of the Institute for Public Health Genetics (IPHG) at the University of Washington (UW), and

Associate Dean for Academic Programs in the UW Graduate School. Dr. Austin's National Institutes of Health (NIH)-funded research program focuses on the genetic epidemiology of cardiovascular disease, diabetes, and pancreatic cancer. She is currently investigating candidate genes for pancreatic cancer in two case-control studies funded by the National Cancer Institute. She also holds an adjunct position in the Department of Medical History and Ethics and is a co-investigator of the UW Center for Ecogenetics and Environmental Health. In her role as Director of the IPHG, Dr. Austin leads an interdisciplinary team of faculty members from seven different schools and colleges (Public Health, Law, Medicine, Pharmacy, Nursing, Arts and Sciences, and Public Affairs) that offer M.P.H. and Ph.D. degrees in Public Health Genetics. As Associate Dean of the Graduate School, Dr. Austin oversees academic review of all graduate programs at the UW, approval of new graduate degree programs, and coordination of the 17 interdisciplinary degree programs administered by the Graduate School.

Wendy Baldwin, Ph.D., is the Executive Vice President for Research at the University of Kentucky. Previously she served as the Deputy Director for Extramural Research at the NIH, and prior to that as the Deputy Director of the National Institute of Child Health and Human Development. With a background in social demography, her research areas are adolescent pregnancy and childbearing, AIDS risk behavior, child care and low birth weight, and international aspects of reproductive health. She received the 1997 National Public Service Award for her accomplishments in science administration and reinvention at NIH. Her work at NIH also addressed issues of data sharing and bioethics. She has served on two National Academy of Sciences committees, most recently the Committee on the Assessment of Behavioral and Social Science Research on Aging.

Ellen Wright Clayton, M.D., J.D., the Rosalind E. Franklin Professor of Genetics and Health Policy, Professor of Pediatrics, Professor of Law, and Co-Director of the Center for Biomedical Ethics and Society at Vanderbilt University, has been studying and teaching the ethical, legal, and social implications of developments in genetics for more than a quarter of a century and has published 2 books and more than 75 peer-reviewed articles and book chapters. She has been an active participant in policy debates advising the National Human Genome Research Institute as well as numerous other federal and international bodies on an array of topics, ranging from issues in children's health, including newborn screening, to the ethical conduct of research involving human subjects. In these roles, she has helped develop policy for numerous national and international organizations. She is a member of the Health Sciences Policy Board of the IOM and recently served on the Committee on the Use of Third Party Toxicity Research with

Human Research Participants of the Science, Technology, and Law Program and the Committee on Genomics and the Public's Health in the 21st Century.

Firdaus S. Dhabhar, Ph.D., received his Ph.D. in Biomedical Sciences from The Rockefeller University in 1996. He is currently Associate Professor of Psychiatry and Behavioral Sciences at the Stanford University School of Medicine. Dr. Dhabhar's laboratory has elucidated the psychophysiological, cellular, and molecular mechanisms by which acute versus chronic stressors respectively enhance or suppress in vivo immune responses. A large part of his research is focused on examining the newly appreciated immuno-enhancing and potentially health-promoting effects of acute or short-term stress. Dr. Dhabhar has received the Council of Graduate Schools Distinguished Dissertation Award and the PsychoNeuroImmunology Research Society Young Investigator Award. He has served on three National Academies committees, been a grant reviewer for the NIH, and is an elected member of the Scientific Council of the PsychoNeuroImmunology Research Society.

Guang Guo, Ph.D., combines research expertise in the sociological analysis of adolescents' well-being, statistical methods, and genetic analysis of complex traits in humans. Dr. Guo's primary interest is in gene-environment interactions in human behavior. He is leading efforts in the field of sociology to incorporate recent advances in genetic research. In addition, he is editing two special issues for two major sociology journals (*Social Forces* and *Sociological Methods & Research*) on sociology and biology/genetics. In addition to his publications in the traditional sociological literature, he has published in the *American Journal of Medical Genetics, Behavior Genetics, Twin Research and Human Genetics*, and *Human Mutation*.

Sharon L.R. Kardia, Ph.D., is an Associate Professor of Epidemiology at the University of Michigan. She is Director of the Public Health Genetics Program, Co-Director of the Michigan Center for Genomics and Public Health, and Co-Director of the Life Sciences & Society Program housed in the University of Michigan School of Public Health. Dr. Kardia received her Ph.D. in human genetics from the University of Michigan, was a postdoctoral fellow in the Department of Microbiology and Immunology, and continued postdoctoral work in the Department of Human Genetics. She joined the faculty of the University of Michigan School of Public Health in 1998. Dr. Kardia's main research interests are in the genomic epidemiology of cardiovascular disease and its risk factors. She is particularly interested in gene-environment and gene-gene interactions, and in modeling complex relationships among genetic variation, environmental variation, and risk of common chronic diseases. Her work also includes using gene expression and proteomic profiles for molecular classification of tumors and survival analysis in lung and ovarian cancers. As a part of her center activity, Dr. Kardia is also actively working on moving genetics into chronic disease programs in state departments of health. Dr. Kardia has also been a member of two other National Academy of Sciences committees (Genomics and the Public's Health in the 21st Century and Applications of Toxicogenomics Technologies to Predictive Toxicology).

Ichiro Kawachi, M.D., Ph.D., is Professor of Social Epidemiology and the Director of the Harvard Center for Society and Health, both at the Harvard School of Public Health. Dr. Kawachi received his M.D. and Ph.D. both from the University of Otago, New Zealand. His research is focused on uncovering the social and economic determinants of population health. He was the co-editor (with Lisa Berkman) of the first textbook on *Social Epidemiology*, published by Oxford University Press in 2000. Dr. Kawachi is the Senior Editor (social epidemiology) of the international journal *Social Science & Medicine*, as well as an Editor of the *American Journal of Epidemiology*. Recently, he served on the IOM's Committee to Review and Assess NIH's Strategic Plan to Reduce Health Disparities.

Caryn Lerman, Ph.D., is Mary W. Calkins Professor in the Department of Psychiatry and Annenberg Public Policy Center at the University of Pennsylvania. She is also Associate Director for Cancer Control and Population Science at the Abramson Cancer Center. As the Principal Investigator of a National Cancer Institute (NCI)/National Institute on Drug Abuse Transdisciplinary Tobacco Use Research Center (P50) Grant, her research focuses on genetic influences on nicotine dependence and response to pharmacotherapy. She has published more than 210 peer-reviewed articles on genetics, cancer, and tobacco use, including work on health policy and ethical issues. Dr. Lerman has been a recipient of the Society of Behavioral Medicine New Investigator Award, the American Psychological Association Award for Outstanding Contributions to Health Psychology, and the Cullen Award for tobacco research from the American Society of Preventive Oncology. She has served on the NCI Board of Scientific Advisors (1998-2003) and the National Human Genome Research (ELSI) Evaluation and Planning Review Board (1997-2000).

Martha K. McClintock, Ph.D., is the Director of the Institute for Mind and Biology, Co-Director of the Center for Interdisciplinary Health Disparities Research, and a David Lee Shillinglaw Distinguished Service Professor of Psychology at the University of Chicago. Dr. McClintock's research interests focus on the interactions among social behavior, neuroendocrinology, and gene expression, particularly those that affect reproduction and health throughout the life span. Working with both animal and parallel clinical processes in humans, she concentrates on the social and pheromonal control of fertility and reproductive hormones, as well as the effect of social isolation on mammary tumors. She is also interested in the evolutionary function of hormone-behavior interactions, particularly their role in sexual selection. Dr. McClintock is an IOM member and has served on the Board on Behavioral, Cognitive, and Sensory Sciences.

Ruth Ottman, Ph.D., is Professor of Epidemiology, Mailman School of Public Health, and Deputy Director for Research, Gertrude H. Sergievsky Center, Columbia University. She is also a research scientist in the Epidemiology of Brain Disorders Research Department, New York State Psychiatric Institute, the biological sciences core leader for the Robert Wood Johnson Health & Society Scholars Program, and the Pre-doctoral Training Director of the Genetics of Complex Diseases Training Program, both at the Mailman School of Public Health, Columbia University. She is a member of numerous advisory committees, including the Board of Directors and the Professional Advisory Board of the Epilepsy Foundation, the Genetics Task Force of the American Epilepsy Society, and the Genetics Commission of the International League Against Epilepsy. Dr. Ottman has published extensively and is recognized internationally for her work in epidemiology, genetic epidemiology, human genetics, and neurology. Her research addresses the role of inherited factors in susceptibility to neurologic disorders, primarily focusing on seizure disorders. She is also interested in methodological issues in genetic epidemiology, including research designs for testing gene-environment interaction, methods for collection of valid family history data, and approaches to assessing familial aggregation.

David Rimoin, M.D., Ph.D., was the founding president of the American College of Medical Genetics and the American Board of Medical Genetics. He has also served as the president of the American Society of Human Genetics, the Western Society for Pediatric Research, and the Western Society for Clinical Research. Dr. Rimoin is currently the Director of the Medical Genetics Institute at Cedars-Sinai Medical Center and holder of the Steven Spielberg Chair. He also serves as the Program Director of the Intercampus Medical Genetics Training Program at the University of California, Los Angeles, School of Medicine, where is he Professor of Pediatrics, Medicine, and Human Genetics. His primary research interests have focused on medical genetics and include the areas of genetic causes of birth defects, prevention of common genetic diseases, policy on the provision of genetic services, dwarfism and growth disorders, and heritable disorders of connective tissue. Dr. Rimoin is the author of more than 385 peer-reviewed publications, as well as the book *Emery and Rimoin's Principles and Practices of Medical Genetics*. Dr. Rimoin is an IOM member and has served on several committees, most recently the IOM committee on Genomics and the Public's Health in the 21st Century and the Clinical Research Roundtable.

Keith E. Whitfield, Ph.D., is a Research Professor in the Department of Psychology and Neuroscience at Duke University. Dr. Whitfield received his bachelor's degree from the College of Santa Fe and his master's and doctoral degrees from Texas Tech University. He also completed postdoctoral work at the Institute for Behavioral Genetics at the University of Colorado. His research focuses on the interplay that genetic, environmental, and cultural factors play in the health and aging process, particularly among African Americans. Overall, he has directed or been associated with projects that have received more than $17 million in funding, and he has authored or co-authored more than 60 journal articles and 18 books and chapters on his research. Dr. Whitfield is Chair of the Gerontological Society of America's Task Force on Minority Issues and is past Chair of the organization's Emerging Scholars Program. He is also a member of the American Psychological Association and past Chair of its Minority Aging Networks in Psychology Program, and he serves on the advisory boards of the Center for Urban African American Aging Research (University of Michigan) and the Export Center to Reduce Health Disparities in Rural South Carolina (Clemson University). Dr. Whitfield has also served on two National Academies committees, the most recent dealing with research opportunities in social psychology, personality, and adult developmental psychology.

IOM Staff

Lyla M. Hernandez is a Senior Program Officer with the IOM and Study Director for the IOM Committee on Assessing Interactions Among Social, Behavioral, and Genetic Factors in Health. During her tenure with the IOM, Ms. Hernandez has directed numerous studies within the Board on Health Sciences Policy and the Board on Population Health and Public Health Practice on topics including the health of Gulf War veterans, the evaluation of complementary and alternative medicine, the education of public health practitioners, and the implications of genomics for public health. Prior to joining IOM, Ms. Hernandez coordinated policy development and health information for the American Pharmaceutical Association, served as Executive Vice President of the American Medical Peer Review Association, and was Program Coordinator for the California Quality of Care Program.

Andrew M. Pope, Ph.D., is Director of the Board on Health Sciences Policy at the IOM. With expertise in physiology and biochemistry, his primary interests focus on environmental and occupational influences on human health. Dr. Pope's previous research activities focused on the neuroendo-crine and reproductive effects of various environmental substances on food-producing animals. During his tenure at the National Academy of Sciences and since 1989 at the IOM, Dr. Pope has directed numerous reports on topics that include injury control, disability prevention, biologic markers, neurotoxicology, indoor allergens, and the enhancement of environmental and occupational health content in medical and nursing school curricula. Most recently, Dr. Pope directed studies on NIH priority-setting processes, fluid resuscitation practices in combat casualties, and organ procurement and transplantation.

Andrea M. Schultz is a Research Assistant for the Board on Health Sciences Policy at the IOM. Since joining the Board on Health Sciences Policy in December of 2004, she has worked with a number of committees, including the committees on Establishing of a National Stem Cell Bank Program, Increasing Rates of Organ Donation, Assessing Interactions Among Social, Behavioral, and Genetic Factors in Health and most recently the Develop-ment of a Reusable Facemask During an Influenza Pandemic. Ms. Schultz is a 2004 graduate of the University of Michigan where she earned a B.S. in cellular and molecular biology. While at the University of Michigan, she worked as a research assistant on a political priming study in the Depart-ment of Psychology. Ms. Schultz is currently continuing her education as a part-time graduate student at George Washington University's School of Public Health and Health Services, where she is pursuing an M.P.H. in health policy.

DBASSE Staff

Christine R. Hartel, Ph.D., is the Director of the Center for Studies of Behavior and Development at the National Research Council, where she also directs the Board on Behavioral, Cognitive, and Sensory Sciences. Pre-viously, she served as Associate Executive Director for Science at the Ameri-can Psychological Association and as Deputy Director for Basic Research at the National Institute on Drug Abuse. She also was a consultant to the World Health Organization on the effects of marijuana. She has written many scientific articles and edited or co-edited four books. As a research psychologist at the U.S. Army Research Institute for the Behavioral and Social Sciences, she earned the Army Research and Development Award, the Army's highest civilian award for technical excellence. She is a Fellow of the American Psychological Association and a member of the American

Psychological Society, the Society for Neuroscience, and the Gerontological Society of America. She has a Ph.D. in biopsychology from the University of Chicago.

Commissioned Paper Authors

Steve W. Cole, Ph.D., is an Assistant Professor for the Division of Hematology-Oncology at the University of California, Los Angeles (UCLA), School of Medicine. He is also an Associate Member of the UCLA/Department of Energy Molecular Biology Institute and the Cousins Center for Psychoneuroimmunology at the UCLA Neuropsychiatric Institute. Dr. Cole was awarded a Ph.D. in psychology from Stanford University in 1993, followed by postdoctorate work in neuroimmunology at the UCLA School of Medicine. Dr. Cole now studies the molecular mechanisms by which the nervous system controls viral and human gene expression. Much of his research focuses on HIV infection, including epidemiological studies of psychological risk factors, clinical studies on the autonomic nervous system's effect on disease pathogenesis, and in vitro studies on the molecular signaling pathways by which neurotransmitters accelerate HIV replication. Other projects analyze neural control of Human Herpesvirus 8 (Kaposi's Sarcoma herpesvirus) and cytokine production by dendritic cells (IL-12 and Interferon-alpha). He recently developed a new vaccine vector that capitalizes on psychoneuroimmunology signaling principles to enhance cytotoxic T-cell responses to viral infections and tumors. His present studies use novel bioinformatic and statistical genetics tools to enhance the effectiveness of these vectors in the search for an AIDS vaccine.

Myles S. Faith, Ph.D., is an Assistant Professor of Psychology in Psychiatry at the University of Pennsylvania School of Medicine. Dr. Faith received his Ph.D. in clinical psychology from Hofstra University in 1995. He then completed a postdoctoral fellowship at the NIH-funded New York Obesity Research Center at St. Luke's-Roosevelt Hospital, Columbia University. Dr. Faith's research focuses on the development of child food preferences, eating styles, and body weight. With his colleagues, Dr. Faith studies the interplay of genetic and environmental influences on child eating patterns, parent-child feeding dynamics, and the measurement of child appetite and satiety. Dr. Faith and colleagues test interventions to help treat and/or prevent obesity in children. He holds multiple grants from the NIH to study these issues.

Tanja V.E. Kral, Ph.D., is a Research Associate at the University of Pennsylvania School of Medicine. She received her B.S. from the University of Applied Sciences at Muenster in 1998 and her M.S. and Ph.D. in nutritional sciences from Pennsylvania State University in 2003. Dr. Kral's research interests include the study of human ingestive behavior in children/adolescents and adults in order to develop strategies for the prevention and the treatment of obesity. In particular, she is interested in characterizing individual differences in eating behavior among individuals of different weight status and in identifying factors (e.g., metabolic, environmental, psychological) that may predispose them to differential food and energy intakes.

Sharon Schwartz, Ph.D., is an Associate Professor of Clinical Epidemiology at the Mailman School of Public Health. She specializes in psychiatric epidemiology and received her Ph.D. from the Department of Sociology at the Graduate School of Arts and Sciences of Columbia University in 1985. She then did postdoctoral work at the Psychiatric Epidemiology Training program at Columbia, receiving an M.S. in epidemiology in 1988. She is currently the Training Coordinator of the Psychiatric Epidemiology Training Program. Her research focuses on methodological issues, particularly in psychiatric research, and the integration of methods from sociology, genetics, and epidemiology.

Robert J. Thompson, Ph.D., is Dean of Trinity College and Vice Provost for Undergraduate Education at Duke University. He also holds appointments in the Department of Psychology: Social and Health Sciences, Department of Psychiatry and Behavioral Sciences, and Department of Pediatrics at Duke. Dr. Thompson completed his graduate work at the University of North Dakota, where he received his Ph.D. in clinical psychology in 1971 and completed an internship in clinical psychology at Indiana University Medical Center. Professor Thompson's research interests address how biological and psychosocial processes act together in development. His primary focus has been on the adaptation of children and their families to chronic illnesses and developmental problems, including sickle cell disease, cystic fibrosis, and very low birth weight infants. He has authored more than 100 scientific publications, including a recent book entitled *Adaptation to Chronic Childhood Illness*, and has served on the editorial board for several scientific journals and as associate editor for the *Journal of Pediatric Psychology*.

Index

A

Addhealth survey, 9, 191, 194
Adipsin gene, 264
Adolescents
    datasets, 190, 191
    pregnancy, 28
    smoking, 71
Adoption studies, 29, 121, 254
ADRA1, ADRA2 genes, 305
ADRB1, ADRB2 genes, 305
Adrenergic receptors, 137
Adrenocorticotropin, 143, 298, 299
Affymetrix, 124
African Americans, 39. *See also* Race/
    ethnicity
    health disparities, 98, 102-103
    multiple jeopardy hypothesis, 103
    obesity, 241, 244, 251, 252, 253, 267
    stress, 76
    TV programming, 252
Agouti signaling proteins, 73, 257, 258
AGTR1 gene, 305
Alcohol use and alcoholism, 15, 33, 37, 78,
    79, 140, 145, 167, 172
Alpha-1 antitrypsin deficiency, 206-207,
    208
α-Melanocyte stimulating hormone, 257,
    258

Alzheimer's disease, 46, 61, 77-78, 185,
    214
American Association of Medical Colleges,
    195
American Association of Universities, 195
Americans with Disabilities Act, 207
Anger and hostility, 57, 298
Animal research
    autoimmune disease, 132, 142-144
    biomedical research, 133-134
    cancer, 149
    causality in, 132, 137
    chickens, 144
    context, pleitropy, and lifetime fitness,
        137-139
    criteria for suitable models, 149-150
    deer mice (*Peromyscus maniculatus*),
        134
    definitions from, 136-139
    early life experience, 139-140, 145
    eating behavior, 135
    ecological context, 133, 150
    epidemiological research, 330
    ethological approach, 134, 150
    experimental allergic encephalomyelitis,
        143-144
    future issues, 149-150
    gene-social environment interactions, 76,
        133, 134, 136, 137, 139-150

generalizability to humans, 133-134, 135-136
genetics, 73, 136, 142-144
house sparrows, 133
immune function, 134, 136-137, 138-139, 141, 143-144
knockout models, 142, 149, 330
limitations, 135-136
mediating variables, 137
modeling known interactions and diseases in humans, 133-134
moderating variables, 137
Morris water maze, 135
nicotine addiction, 70, 79
nonhuman primates, 133, 140, 141, 142, 145-146
nontraditional animal models, 134
obesity, 73, 267
physiology, 136-137
psychosocial traits, 133, 135-136, 144
rationale, 132-133
recommendations, 7, 150
rodent models, 133, 135, 138, 140, 141-142, 144, 145-146, 147, 149
role, 132-136
side-blotch lizards, 138-139
social affiliation and support, 141, 142
social isolation, 133, 138, 141-142
social status, 133, 144-146
social stressors, 145, 146-147
stress responsivity, 135, 137, 142-144, 145, 146-148
temperament, 140-141
Antioxidant regulatory element, 51
Apolipoprotein E protein polymorphisms, 61, 76, 214
Appetite, 257, 258, 267, 268
Arginine vasopressin, 118, 146
Arthritis, 143
Ashkenazi Jews, 38, 100, 101, 149
Asthma, 137
Ataxia telangiectasia, 149
Atherosclerosis, 118, 119, 147, 298, 302
Atopic dermatitis, 144
Attention Deficit Hyperactivity Disorder, 80
Autoimmune disease, 137, 142-144

B

Bardet-Biedel syndrome, 73, 255
Bayesian belief networks, 65

Behavioral factors. *See also* Eating behaviors; Health risk behaviors; Physical inactivity; Tobacco use
context for interaction studies, 22, 125
CVD, 57, 58, 70, 73
obesity, 245-251
peer and family influences, 71
and psychological response and physiological processes, 116-122
β_2-AR, β_3-AR genes, 137, 257, 264, 305
β-Endorphins, 301
β-Thalassemia, 54-55, 101, 284
Binge Eating Disorder, 261
Biochemical System Theory, 123
Biochemical systems and processes. *See also* Metabonomics; Proteomics
affiliative behavior and, 36, 145
genomic information embedded in, 114-116
modeling, 123
psychological response to social factors and, 21, 116-122
stress and, 117-122
Bioinformatics, 51
Biologic interaction, 311, 317, 322
Biology, hierarchical view, 123
Biomedical model, 281-282
Biomedical research, 15, 18, 133-134
Biomolecular Interaction Network Database, 124
Biopsychosocial model, 282, 283, 297
Birth defects, 47
Blood lead levels, 27
Bloom syndrome, 101
BMP6 gene, 288
Body Mass Index (BMI), 69, 72, 73, 238, 250, 252, 253-257, 263. *See also* Obesity
Bowlby, John, 34
Breast cancer, 35, 38, 46, 59, 73, 101, 141, 149, 174

C

C-reactive protein, 76, 171-172, 191
Cadmium, 114
Canavan's syndrome, 101
Cancer
animal research, 149
biomarkers, 115
depression and, 80

gene-environment interaction, 113
genetic susceptibility, 38 n.1, 45, 56,
    149
obesity and, 72
physical activity interventions, 73
smoking and, 70, 310
social environment and, 35, 38-39
stress and, 140, 146, 148
survival, 72
temperament, 140
treatment side effects, 73
Cardiovascular disease (CVD)
behavioral factors, 57, 58, 70, 73
depression and, 57, 78, 80-81, 82
gene-environment interactions, 57-58,
    75-76, 113, 137
genetic susceptibility, 45, 56, 57-58, 76
immune function and, 137
life-course perspective, 57
moderators of risk, 76, 81
obesity and, 72
personality and, 77-78
sex/gender and, 92, 96
sickle cell disease, 304-305
social environment and, 31, 35, 57, 78,
    119
stress and, 57, 78, 117, 118, 119, 298,
    304-305
Catecholamines, 117, 119, 121, 137, 298,
    301, 304
Caucasians. *See also* Race/ethnicity
A1AT deficiency, 208
insulin resistance, 251
obesity, 155, 261
stress, 76
thiopurine methyltransferase deficiency,
    59
Causality
in animal research, 132, 137
in epidemiological approaches, 311, 312,
    316, 319, 330-332
CGAP Genetic Annotation Initiative SNP
    Database, 64
Child development. *See* Child health and
    development
Child Feeding Questionnaire, 244
Child health and development
abused children, 208-209
adopted children, 29
depression, 147
eating behavior, 243-244, 247, 248

environmental influences, 268
gene-environment interactions, 53, 113,
    268, 271-272
integrative study of, 121-122
and life-course patterns of health, 39,
    139-140
obesity and overweight, 240-241, 244-
    245, 247, 248, 252, 261, 267
SES and, 27, 29, 39
stress and, 117, 118, 120-122, 139, 144,
    147, 300
Childhood acute lymphoblastic leukemia,
    59
Cholecystokinin A receptor, 73
Cholesterol, 57, 75-76, 78, 119, 263, 302
Clinical and Translational Science Award,
    195
Clinical Laboratory Improvement
    Amendments, 213
Cluster analysis, 112
Cocaine- and amphetamine-stimulated
    transcript peptide (CART), 258
Colon cancer, 73, 101
Communities
involvement in research protocols, 214-
    215
SES/health associations, 18, 26, 30, 31,
    241
social capital, 36-37
Comparative sequence analysis, 51
Computer Retrieval of Information on
    Scientific Projects, 195
Cooperative Study of Sickle Cell Disease,
    293, 294
Coping Strategies Questionnaire, 295
Coronary artery disease, 77, 101, 305
Coronary Artery Risk Development in
    Young Adults (CARDIA) study, 251
Coronary heart disease, 39, 61, 92, 98, 298,
    304
Corticosterone, 142, 143, 146
Corticotropin releasing factor-like proteins,
    79
Corticotropin releasing hormone, 117, 146,
    298
Cortisol, 72, 117, 118, 137, 147, 287-288,
    299, 300, 301, 303
CpG array-based technology, 114
Crohn's disease, 101
CRP gene, 171-172
CYP11B2 gene, 305

Cystic fibrosis, 101, 209, 219
Cytochrome P450 enzymes, 59
Cytochrome P450 genes, 59
Cytokines, proinflammatory, 299, 302

D

Data and databases. *See also individual databases*
  biological specimens, 192
  collection and analysis for this report, 223-231
  commercial databases, 124
  creating new datasets, 9, 192-193
  gene expression, 124
  guidance on data collection, 192
  guide to measures of key concepts, 191-192
  informed consent, 191
  infrastructure for transdisciplinary research, 19, 187-194, 220
  merging and integrating, 112
  metabolic pathways, 124
  pooling samples, 174
  privacy and confidentiality issues, 191
  proteomic, 112, 124
  recommendation, 9-10, 194
  replication of, 193-194
  review of existing datasets, 188-192
  security, 211-212, 220-221
  sharing, 112, 188, 191, 212, 215-216
  SNP databases, 64, 110
  for systems modeling, 124
  transcriptomic, 112, 124
  use agreements, 211
Data Quality Act, 212
Database of Interacting Proteins, 124
Department of Housing and Urban Development, 31
Depression
  and cancer, 80
  children, 147
  and CVD, 57, 78, 80-81, 82
  gene-environment interactions, 18, 80-82, 328-330
  heritability, 81
  personality and, 78
  sickle cell disease and, 292, 298
  and smoking, 71
  social environment and, 18, 31, 32, 35, 147
    social supports and, 147
    stress and, 81-82, 298, 301, 302, 313-314, 328-330
Developmental behavioral genetics, 271-272. *See also* Child health and development
Dexamethasone, 143
Diabetes, 27, 72, 80, 98, 101, 113, 236, 263
Diethylstilbestrol, 114
DNA
  damage and repair, 50
  methylation, 113-114, 139
  microarray technology, 111-112
  sequencing technologies, 49-50
Dopamine
  DRD2 gene, 73, 172, 264-265
  DRD4 gene, 80
  and eating behavior, 249
  and nicotine addiction, 71
  transporter, 73
Drug abuse, 39
Duke University, 290
Duncan Socioeconomic Index, 31, 32
Durkheim, Emile, 34

E

Earnings, 28
Eating behaviors (unhealthy)
  access to energy-dense foods and, 240, 241, 242, 246, 250-251, 252, 269
  animal research, 135
  appetite and, 257, 258, 267, 268
  and child health and development, 243-244, 247, 248
  disinhibition trait, 246, 261, 267
  epidemiology, 72
  external cues and, 244, 246
  fast-food consumption, 72, 241, 242, 250-251, 270-271
  food advertising and, 252
  gene-environment interaction, 72-73, 259-260
  genetic influences, 72-73, 243, 257-262, 264-265, 267-268
  health consequences, 72; *see also* Obesity
  heritability, 259-260, 267
  hypothalamic response and, 135, 247
  and mortality, 15

parent-child feeding dynamics and, 239,
243-245, 269, 271
portion size and, 247, 251, 269
reinforcing value of food and, 70, 75,
76, 172, 248-249, 264-265, 267
research opportunities, 267-268
restrictive feeding practices, 244-245, 271
satiation impairment and, 246-248, 267
satiety impairment and, 70, 73, 75, 135,
244, 248
self-regulation of intake, 244
smoking and, 172
social facilitation of, 72, 239, 242-243,
259
socioeconomic status and, 242
stress and, 301
style of eating, 135, 249-250
taste preferences and, 70, 75, 135, 172,
250, 257, 259, 268, 269
television viewing and, 252
time-extension mechanism and, 243
traits associated with obesity, 245-251
Eating Inventory, 261
EcoCyc, 124
Ecological model of health determinants,
18-19
EDN1 gene, 305
EDNRA gene, 305
Education and training of researchers
conferences, 182
evaluation of programs, 184
fellowships, 182, 183-184, 185, 190,
233
K-12, 181-182, 183
private support, 187, 190
professional development, 184-186
recommendation, 9, 187, 233
short course approach, 185
social sciences, 183
T90 grant, 183-184
undergraduate, 185-186
Education quality, 27
Educational attainment
and health, 27-28, 39, 103
and smoking, 70
Edwards classification, 31
Embedded information. *See* Genomic
information; Metabonomics;
Proteomics
Emotional support, 34
Endothelin-1 gene, 76

Endotoxins, 118
Energy-dense foods, 240, 241, 242, 246,
250-251, 252, 269
Energy expenditure regulation and, 257
Engel, George, 281
Enkephalin, 301
Environmental Genome Project, 49, 64
Environmental genomics, 203
Epidemiologic approaches to interactions
additive and superadditive models, 7, 8,
162, 163-164, 165-166, 168, 170,
173, 220, 313, 314, 315, 328, 334,
336
animal models, 330
biologic interaction, 311, 317, 322
biological plausibility restriction, 175,
177
Bonferroni correction, 174-175
case-control designs, 163, 168, 169,
170, 171, 174, 177, 329, 334, 335
case-only design, 49, 170, 177, 331-332
causal effect (causal contrast), 318, 319-
322, 328, 330
causal model, 315-318
causal pies heuristic (causal pathways),
316-317, 319-320, 323, 331
causality in, 311, 312, 316, 319, 330-
332
cohort designs, 163, 168, 169, 192-193,
329, 331-332, 334, 335
confidence intervals, 334-335
confidentiality issues, 171
confounding and bias, 62, 63, 65, 71,
168-169, 170, 171, 172, 174, 257,
319, 324, 330, 333, 335
counterfactual (potential outcomes)
model, 8, 316-328, 330-335, 336
definitions of interactions, 7, 161-168,
176, 220, 312-316, 317
designs of studies, 57-58, 166-173, 176-
177, 331-332
exposures, 171-172, 174-175, 311, 316,
319, 323-324
false positive test results, 175-176
family-based designs, 47, 49, 170-171,
332
framework for assessing interaction,
312-316
gene-environment interactions, 47, 49,
163, 166-173, 174, 311-312, 318,
329, 330-336

genetic context, 52, 62, 330-332, 333
genomic control methods, 169, 170
Henle-Koch principles, 310
human laboratory (intervention)
 research, 172
independent effect, 162, 165, 312
individual-level, 164-165
infectious disease models, 310
mathematical modeling, 327-330
measurement of environmental factors,
 204, 333
measurement of interaction, 161-162
Mendelian randomization, 171-172
modeling innovations, 65
molecular techniques, 46
multiple comparisons, 175-176, 177,
 334-335
multiplicative (log linear) or
 supermultiplicative model, 7, 8, 162,
 163-164, 168, 170, 173, 312-313,
 332, 334, 336
for nonbinary outcome variables, 172-
 173
permutation testing, 174, 175
population stratification, 62, 63, 65, 71,
 168-169, 170, 171, 172, 257, 335
power of, 8, 65, 166, 171, 172, 333-334
practical applications, 164, 328-330
preventive effects in, 328
primacy of genetic effect, 331
protective effects in, 168
psychologic interactions, 317, 328
recommendations, 7-8, 177
reproducibility of results, 48, 63, 175,
 177, 193-194, 204, 257
risk factor framework, 162, 310-311,
 335-336
sample size and power, 8, 65, 166, 169,
 171, 172, 173-174, 177, 192, 193,
 333-334
statistical methods and issues, 8, 161-164,
 173-175, 177, 311, 312-313, 332
sufficient-component cause model, 8,
 164-166
synergy (interaction) assessment, 165,
 311, 315, 319, 322-330, 332-333
temporality, 170
uncertainties in, 71, 332-335
validation, 174
Epigenetic phenomena, 53, 64, 112, 113-
 114, 139

Epinephrine, 117, 118, 147, 286, 298, 301,
 302, 304
ERCC2 gene, 50-51
Erythrocyte adhesion, 302-303
Ethical, legal, and social issues
 civil liberties issues, 208-209
 communicating research results, 203-205
 community involvement in research
  protocols, 214-215
 data security, 211-212, 220-221
 disclosure of research results to
  participants, 212-213
 discrimination and stigmatization, 202,
  205, 207, 209-210
 environmental regulation, 206-207
 genetic association studies, 51
 genetic screening, 209
 health care issues, 202, 207
 informed consent, 215, 216
 insurance coverage, 202, 207
 intellectual property rights, 203
 intervention policies, 205-210, 220
 lay oversight of protocols, 214-215
 "moral hazard," 206
 new hypotheses arising from research,
  214
 privacy and confidentiality concerns,
  171, 208, 211-212, 215-216
 protection of research participants, 211
 race/ethnicity as a proxy, 209
 recommendations, 11-13, 210, 215-216
Eugenics movement, 205
Experimental allergic encephalomyelitis,
 143-144

F

Familial aggregation studies, 47, 49, 170-
 171, 332
Familial dysautonomia, 101
Familial hypercholesterolemia, 45, 60-62,
 203
Familial Mediterranean Fever, 101
Family Research Consortium III, 187, 190
Fast-food consumption, 72, 241, 242, 250-
 251, 270-271
Fetal
 exposure to stressors, 139
 hemoglobin, 284, 285, 289
 nutrition, 39, 73, 113, 267

G

Gaucher disease, 101
Gel electrophoresis, 115
Gender. *See* Sex/gender
Gene-environment interactions
    animal models, 76, 133, 134, 136, 137,
        139-150
    and biological characteristics, 64, 76
    biological plausibility, 175, 177
    cancer, 113
    correlation model, 270-271
    counterfactual model, 8, 316-328, 330-
        335, 336
    CVD, 57-58, 75-76, 113, 137
    depression and, 18, 80-82, 328-330
    developmental consequences, 53, 113,
        268, 271-272
    eating behaviors and obesity, 72-73,
        243, 257-262, 264-265, 267-268
    epidemiologic approaches, 47, 49, 163,
        166-173, 174, 311-312, 318, 329,
        330-336
    gene expression mechanisms and, 52-53,
        113, 122
    intermediate phenotypes as measures of,
        74-82, 173
    life-course perspective, 39, 47, 113-114
    models of, 166-168
    moderating, 76, 251, 262, 263-266,
        269-270
    molecular mechanisms, 116-122
    for nonbinary outcome variables, 172-
        173
    in obesity, 76, 113, 236, 237-238, 251,
        260, 262-272
    personality and, 35, 37, 78-79
    physical inactivity, 74
    and proteomics, 122
    psychosocial work environment and
        health and, 37-38
    racial/ethnic disparities in health, 99-
        101, 102, 104-105
    research challenges and opportunities,
        39, 40, 64, 266-272
    responsivity to environmental factors,
        50-52, 56, 139
    social isolation, 36, 121, 141-142
    statistical modeling, 65
    study design and analysis, 57-58, 166-
        173, 176-177, 331-332

    systems approach to modeling health,
        123-126, 139
    temperament and, 35, 37-38, 79, 140-
        141, 328
    tobacco use, 50-51, 70-71
    "toxic environments," 269-270
Gene expression
    databases, 124
    defined, 287
    developmental control of, 53, 54-55
    DNA methylation and, 114
    environmental influences, 52, 116-117,
        122, 137
    epigenetic phenomena, 53, 64, 112, 113-
        114, 139
    fetal nutrition and, 73
    and gene-environment interactions, 52-
        53, 113, 122
    genetic variation in, 6, 110-114, 122
    and globin, 54-55
    heritability of, 111
    in immune response, 288
    life-course perspective, 5-6
    mechanisms, 52-56
    pharmacogenetics research, 60
    post-transcriptional control, 53
    serial analysis of, 111
    sickle cell disease, 53-56, 284, 287
    social environment and, 142
    stress and, 118, 139, 142, 148, 287-288,
        303, 305
    transcriptomics technologies, 111-112
Gene-gene interactions, 46, 51-52, 56, 62,
    236, 257
Gene Logic, 124
Gene Ontology Project, 112
Gene products, classification, 112
Genetic association studies. *See also* Gene-
    environment interactions; Gene-gene
    interactions
    Bayesian networks, 288
    BMI and fat mass, 255-257
    DNA sequencing technologies, 49-50
    eating behaviors, 261-262
    ethical issues, 51
    genome-wide approach, 111, 174
    HaloChip assay, 111
    population stratification in, 62, 63, 76,
        168-169

race/ethnicity, 99
reproducibility, 177
sickle cell disease, 288, 299
SNP profiling, 49-50, 51-52, 58, 110
transcriptomic studies, 112
Genetic imprinting, 112, 113
Genetic linkage analysis
    databases, 112
    eating behaviors, 261-262
    goal and principle, 47-48
    sibling pair method, 48
    statistical power, 48
    whole-genome, 288
Genetic susceptibility. *See also* Gene-
    environment interactions; Gene-gene
    interactions
    adverse drug reactions, 58-60
    allelic heterogeneity, 45-46
    aspects of health influenced by, 56-60
    biases in studies, 49
    cancer, 38 n.1, 45, 56, 149
    clinical variability in diseases, 53-56
    common disease, common variant
        hypothesis, 204
    CVD, 45, 56, 57-58, 76
    diabetes, 56
    differential risk, 51
    etiologic heterogeneity, 46
    founder mutations, 60-61, 100
    genetic association studies, 49-52, 255-
        257, 261; *see also* Genotype and
        genotyping
    hierarchy of causes, 311-312, 331
    life-course perspective, 57, 254
    linkage analysis, 47-48, 255, 261
    linkage disequilibrium phenomenon, 62,
        80, 175
    locus heterogeneity, 46
    mental illness, 56
    molecular epidemiology techniques, 46
    multifactorial models, 46, 48, 60, 65
    nicotine addiction, 58, 71
    obesity, 20, 72-73, 236, 248, 249, 253-
        262, 264-266
    OMIM statistics, 45
    overeating, 72-73, 243, 257, 259-262
    pathway-driven study design, 50
    penetrance of mutations, 45, 149, 312
    polygenic models, 44-45, 46, 47, 60,
        219

population-based measures, 47, 49, 63,
    71
population distribution of variations,
    60-62, 63
research approaches, 44-48
single-gene disorders, 44, 45-46, 47, 48,
    53-56, 63-64, 73, 203-204, 219,
    255, 257, 260
SNPs, 49-50, 51-52
therapeutic response to drugs, 56
to "toxic environments," 241
Genome Science Education Program, 183
Genomic information
    in biochemical systems, 114-116
    comparative genomic hybridization
        assays, 112
    epigenetic phenomena, 112, 113-114
    and gene expression, 6, 110-114, 122
    metabonomic technologies, 116
    in proteomes, 111, 115
    social and ethical implications, 203
    transcriptomics, 111-112
Genotype and genotyping
    array-based, 49, 52-53
    interactions with environmental factors,
        113-114
    pharmacogenetic research, 59-60
    moderator in gene-environment
        interaction, 269-270
    multiplexing arrays, 49
    SNP initiatives, 64
Geographic Information Systems, 269
German National Center for Environment
    and Health, 124
Glucocorticoids, 117, 118, 121, 137, 139,
    145, 146, 298, 301
Glutamate, 303
GR gene, 299
Growth hormone, 301
GSTM1 gene, 51
GSTP1 gene, 51
GSTT1 gene, 51
Guyton, Arthur, 123

H

Health determinants
    life course perspective, 5-6, 21, 22, 25-
        26, 219, 282
    systems approach to modeling, 5-6, 17-
        19, 123-126

Health Insurance Portability and Accountability Act of 1996, Privacy Rule, 211
Health literacy, 28
Health outcomes. *See* Mortality; Survival and functional recovery; *individual diseases*
Health risk behaviors. *See also* Eating behaviors; Physical inactivity; Tobacco use
 definitions, 69-70
 intermediate phenotypes, 70
 outcome pathways, 68, 69
 personality and, 70, 75, 77-79
 social and cultural environment and, 30, 37, 71, 237-238
 temperament and, 70, 75, 79-80
Hemochromatosis, 101
Hemoglobin gene expression, 53, 54-55
HER2 gene, 59
Heritability
 of BMI and fat mass, 253-255
 defined, 46-47. *See also* Genetic susceptibility
 of depression, 81
 of eating behaviors, 259-260, 267
 of gene expression, 111
 narrow sense vs. broad sense, 253
High school dropouts, 28
High/Scope Perry Preschool Project, 28
Hispanics, 103. *See also* Race/ethnicity
Histone acetylation, 114
HIV infection, 35, 77, 91-92
Home Observation for Measurement of the Environment (HOME) system, 268-269
Human DNA Polymorphism Discovery Program, 49
Human Genome Project, 3, 15, 64, 90, 109, 110, 123, 203
Human Obesity Gene Map, 255, 256
Hypertension, 27, 37, 94, 98, 102, 298, 299, 304
Hypospadias, 92
Hypothalamic-pituitary-adrenal (HPA) reactivity, 35, 81, 117, 118, 121, 135, 136, 139, 140, 143, 144, 145, 247, 297, 298-300, 301, 302, 303, 304
Hypothalamic-pituitary-ovarian axis, 301

I

Immune function
 animal research, 134, 136-137, 138-141, 143-144
 and CVD, 137
 gene expression, 288
 genetic determinants, 136
 natural vs. specific immunity, 301, 302
 obesity and, 72
 sex/gender and, 93, 94
 sickle cell disease and, 283, 287, 300-302
 smoking and, 70
 social environment and, 35, 134, 136-137, 138-139
 stress and, 21, 117, 118, 119-120, 139-141, 145, 146-147, 148, 288, 297, 298-299, 300-302
Impulsive aggression, 18
Inclusion body myopathy, 101
Income
 absolute vs. relative, 30
 health associations, 29-31
 measurement, 28-29
 psychological and behavioral associations, 30
 race/ethnicity and, 103
 societal distribution, 31
 tests of associations, 29-30
 and weight gain, 241
Incyte, 124
Indians, Asian, 263
Infants
 breastfeeding vs. bottle feeding and, 243-244
 mortality, 27
 secure attachments, 34
 sucking rates and obesity risk, 249, 268
 weight gain, 267
Inflammatory bowel disease, 101
Informational support, 34
Informax, 124
Informed consent, 215, 216
Infrastructure for transdisciplinary research
 academic institutional structure and policies and, 10-11, 19, 181, 194-198
 data, 19, 187-194, 220
 education and training of researchers, 8-9, 19, 181-186, 193, 220

federal and industry research structures vs. academic, 197
IDR recommendations applied to, 10, 20, 186, 232-235
incentives and rewards, 10-11, 20, 193, 194-198, 220
MacArthur Network Model, 188-189
NIH support, 181, 182, 186-187, 194-198
P30 Core Grant and, 187
peer review, 10, 20, 195, 198-200
private support, 187, 188-189, 190
R01 Research Project Grant and, 186-187
recommendations, 8-11, 187, 199-200
support mechanisms, 186-187, 272
Institute for Public Health Genetics, 184
Institutional Review Boards, 210-211
Instrumental support, 34
Insulin-like growth factor, 74
Insulin resistance, 70, 72, 251
Interactions. *See also* Gene-environment interactions; Gene-gene interactions
definitions of, 7, 161-168, 176, 220, 312-316, 317
framework for assessing, 312-316
measurement of, 161-162
Interdisciplinary research (IDR)
academic institutional structure, 235
academic institutions' policies, 233-234
defined, 3, 19
educators, 233
evaluation of programs, 235
funding organizations, 234
journal editors, 234
key conditions for effective programs, 194
MacArthur Network Model, 188-189
postdoctoral scholars, 190, 233
professional societies, 234
recommendations for facilitating, 10, 20, 186, 232-235
researchers and faculty members, 233
students, 232-233
team leaders, 234
Interferon γ, 301
INTERHEART Study, 78
Interleukin-1α, 143
Interleukin II, 288
Interleukin-6, 302
Intermediate phenotypes
and biological characteristics, 76

defined, 74-75
emotional/motivational states as, 80-82
measuring, 75-76
for obesity, 70, 73, 75, 78, 237
personality as, 75, 76, 77-79
for sickle cell disease, 288, 297
temperament as, 75, 76, 79-80
for tobacco use, 70, 75, 76
International HapMap Consortium, 64
International HapMap Project, 64, 190
Inuit of Greenland, 75-76

J

Job stress
and CVD, 57
effort-reward imbalance model, 37
job demand-control model, 37

K

Karolinska Institute Human Gene Bi-Allelic Sequences Database, 64
Kleinfelter Syndrome, 94
Kyoto Encyclopedia of Genes and Genomes, 112, 124

L

Laminin, 286
Leptin and leptin receptors, 73, 94, 134, 135
Life-course patterns of health
animal models, 139-140, 145
Barker hypothesis, 39
CVD, 57
determinants, 5-6, 21, 22, 25-26, 219, 282
early life experience, 39, 139-140
in gene-environment interactions, 39, 47, 113-114
in gene expression, 5-6
genetic susceptibility to disease, 57, 254
obesity, 74, 253, 266-267
physical inactivity and, 74
poverty, 4-5, 39
race/ethnicity and, 102-103
sex/gender and, 94
social-environment associations, 25-26, 27, 31, 32, 34, 39, 138

stress and, 118, 145, 300
in study design and analysis, 21, 22
Lipid metabolism, 119
Lung cancer, 50-51, 310
Lutheran blood group anitgens, 286

M

MacArthur Foundation, 187, 188
MacArthur Network Model, 188-189
Machine learning algorithms, 65, 112
Major histocompatibility complex, 138
Mass spectrometry, 116
Maternal
  attachment, 36
  education, 27
  exposure to endocrine disrupting agents, 92
  obesity, 267
  separation, 35, 118, 121, 139
Matrix-assisted laser desorption ionization-time of flight mass, 115
MC4R (melanocortin-4 receptor) gene, 257, 258, 261
Media campaigns, 71
Mediation and mediating variables
  in animal research, 137
  in models of interactions, 167
  of social environment, 147-148
  of stress reactivity, 139, 147-148, 304
Melanoma, 38
Melatonin, 301
Metabolic Control Theory, 123
Metabolic syndrome, 236
Metabolism
  gut flora and, 116
  sex/gender and, 93-94
  social stressors and, 145, 298
Metabonomics, 6, 112, 116, 124
Microarray Gene Expression Data Society, 112
Midlife health, 27
Migration, and genome selection, 109-110, 263
Minimum Information About a Microarray Experiment Guidelines, 112
Mitochondrial upcoupling proteins, 257
Modeling strategies
  recommendations, 5-7
Moderation and moderating variables
  animal research, 137

CVD risk, 76, 81
in gene-environment interactions, 76, 251, 262, 263-266, 269-270
of obesity, 76, 251, 262, 263-266
social environment, 262, 263
in stress response, 76, 285, 287-289, 290, 300
Monamine oxidase A deficiency, 208-209
Mortality
  eating behavior and, 15
  infant, 27
  obesity and, 72
  race/ethnicity, 98, 104
  SES and, 27, 29, 104
  smoking and, 70
  social and behavioral factors and, 15, 27, 35
Moving to Opportunity study, 31
mRNA transcripts, 111-112, 118, 124, 146, 149, 287
Multidisciplinary research, defined, 3-4, 19
Multivariate statistical analysis, 137
Munich Information Center for Protein Sequences, 124
Myelosuppression, 59
Myocardial infarction, 35, 39, 70, 76, 78, 81, 82, 119

N

National Association of State Universities and Land Grant Colleges, 195
National Bioethics Advisory Commission, 213
National Cancer Institute
  Breast and Prostate Cancer and Hormone-Related Cohort Consortium, 174
  SNP database, 64
National Center for Biotechnology Information, dbSNP database, 64, 110
National Health Interview Surveys, 29
National Heart, Lung, and Blood Institute Programs in Genomic Applications, 49
National Human Genome Research Institute, 15-16, 50
National Institute of Environmental Health Sciences, 49
National Institute of General Medical Sciences, 16

National Institutes of Health (NIH), 97-98, 236
Certificates of Confidentiality, 211
Office of Behavioral and Social Sciences Research, 15-16
and transdisciplinary research, 181, 182, 186-187, 194-198
National Longitudinal Survey of Youth, 241 Supplement, 9, 190, 194
National Opinion Research Center Study, 31
National Research Service Act, 182
National Science Foundation, 181, 182
Native Americans, research protocols, 214-215
NEO-Personality Inventory, 77
Neophobia, 140
Network models, 65, 112
Network on Socioeconomic Status and Health, 188-189
Network theory, 204
Neurocognitive functioning, 283, 287, 293, 294-295, 297, 303
Neuroeconomics, 36
Neuroendocrine regulation, 35
Neuropeptide-Y, 257, 258
NGFI-A transcription factor, 139
Nickel, 114
Nicotine
  addiction, 70, 79
  and cardiovascular function, 70
  cognitive and autonomic effects, 75
  patch, 82
  reward value, 70, 75, 76, 79
  tolerance and deprivation, 70, 75
  withdrawal and relapse, 79
Niemann-Pick disease, 101
NMB (Neuromedian-β) gene, 262
Norepinephrine, 147, 298, 301, 304
North Carolina Sickle Cell Center, 290
NOS1, NOS2, NOS3 genes, 305
NR3C1 gene, 137
Nuclear magnetic resonance spectroscopy, 116

O

Obesity. See also Eating behaviors; Physical inactivity
  animal research, 73, 267
  behavioral traits associated with, 245-251

BMI and fat mass, 69, 72, 73, 238, 253-255
breastfeeding vs. bottle feeding and, 243-244
children, 240-241, 244-245, 247, 248, 252, 261, 267
conceptual model of interactions, 237-238
critical growth periods for, 266-267
energy expenditure regulation and, 257
epidemiology, 72, 236
gene-environment interactions, 76, 113, 236, 237-238, 251, 260, 262-272
genetic influences, 20, 72-73, 236, 248, 249, 253-262, 264-266
health consequences, 72, 236
hypothalamic response and, 247
intermediate phenotypes, 70, 73, 75, 78, 237
life-course perspective, 74, 253, 266-267
macroenvironmental influences, 240-242, 268, 269-270
maternal, 267
microenvironmental influences, 242-245
moderating variables, 76, 251, 262, 263-266
personality and, 78
physical inactivity and, 74, 240, 251-253
race/ethnicity and, 155, 241, 244, 251, 252, 253, 261, 267
research opportunities, 266-272
restrictive feeding practices and, 244-245, 271
social environmental factors and, 20, 27, 72, 236-237, 238-245, 253-255, 263-266
socioeconomic status and, 239, 241-242, 264
"toxic environments," 237, 239, 240-241, 251, 269-270
and underreporting of intakes, 247
Occupational status, 28
  child health and, 39
  and health, 31
  measures of, 31-32
  and weight gain, 241
Office of Management and Budget, 97
Online Mendelian Inheritance in Man (OMIM) statistics, 45
Osteoarthritis, 72
Ovarian cancer, 149
Oxytocin, 36, 145

P

Pain, 286-287, 289, 295-297, 299, 302
Pair bonding, 36
Panel Study of Income Dynamics, 29, 31
Parasitic infections, 137
Parent-child feeding dynamics, 239, 243-245, 269, 271
Path analysis, 123
Pattern recognition methods, 65, 112, 116
Peptidoglycan polysaccharide, 143
Persian Jews, 100, 101
Personality
  and alcohol consumption, 78
  "Big Five" model, 77-78
  and CVD, 77-78
  defined, 77
  and depression, 78, 298
  gene-environment interaction, 35, 37, 78-79
  genetic factors, 78
  and health risk behaviors, 70, 75, 77-79
  as intermediate phenotype, 75, 76, 77-79
  and obesity, 78
  race/ethnicity and, 80
  sex/gender and, 80
  and social-environmental influences, 144-145
  and stress response, 145, 298
  and tobacco use, 78
Pew Charitable Trust, 182
Pharmacogenetics, 56, 58-60, 203
Phenylketonuria, 45, 167, 318
Physical activity
  defined, 69
  protective effects, 71, 73, 119
  recommended, 69, 73
Physical inactivity
  animal studies, 74
  epidemiology, 73-74
  gene-environment interactions, 74
  genetic influences, 74
  health consequences, 15, 73-74
  life-course perspective, 74
  nonexercise activity thermogenesis, 252-253, 268
  and obesity, 74, 240, 251-253
  research opportunities, 268
  social/cultural environment and, 27, 72, 74, 240, 241, 242
  television, video recorders and computers and, 240, 251-252

Pima Indians, 263-266
Pituitary function, 70
Polycystic kidney disease, 46
Population genetics, 60-62
Pound of Prevention Study, 251
Poverty. *See also* Income; Socioeconomic status/health associations
  definitions, 30
  gender and, 92-93
  health associations, 25
  life-course perspective, 4-5, 39
  persistent vs. transient, 39
PPARγ gene, 257
Prader-Willi syndrome, 73, 255
Preschool education, 28
Privacy and confidentiality concerns, 171, 208, 211-212, 215-216
Pro-opiomelanocortin (POMC) peptides, 257, 258
Prolactin, 301
Promoter Database of *Saccharomyces cerevisiae*, 124
Prostate cancer, 174
Proteomics, 6, 60, 112, 114, 115, 122, 124
Psoriasis, 77
Psychological/psychiatric disorders and behaviors. *See also* Depression
  occupational status and, 31
  personality and temperament and, 77
  pharmacogenetics research, 59
  SES of communities and, 18
  social connectedness and, 35
Psychoneuroimmunology, 35
Psychosocial traits
  animal models, 133, 135-136, 144
Psychosocial work environment/health associations. *See also* Job stress
  biases in, 37-38
  gene-environment interactions, 37-38
  evidence of, 25, 37-38
  measures of, 31, 32, 37
  occupational status and, 31, 32
  reciprocal relationships, 37

Q

Quebec Family study, 262
Quebec Overfeeding Study, 269

R

Race/ethnicity
and ancestral origin, 98, 99
confounding issues, 98
context for research, 96-98, 99, 125-
126, 219
definition, 97
gene-environment interactions, 99-101,
102, 104-105
and genetic variation, 38 n.1, 61, 97,
99-104
health disparities and, 90-91, 98-104
life-course perspective, 102-103
and mortality, 98, 104
and personality, 80
and SES, 27, 103-104
SNP markers of phenotypic variation,
99-101
and social stressors, 39, 76
Racism and discrimination, 39
Random forest methods, 65
Renal function, 70, 305
Renin-angiotensin-aldosterone system, 299,
304
Reproductive dysfunction, 138, 118
Research. *See* Epidemiologic approach;
Infrastructure; Interdisciplinary
research; Transdisciplinary
interactions research
Rett syndrome, 113
Rheumatic diseases, 144
Risk behaviors. *See* Health risk behaviors;
specific behaviors
Risk prevention programs, 82
Rosetta, 124

S

Satiation, 246-248, 267
Satiety, 70, 73, 75, 135, 244, 248
Scale-free networks, 65
Science Education Partnership Awards, 182-
183
Sedentary lifestyle. *See* Physical inactivity
Selective serotonin re-uptake inhibitors
(SSRIs), 59
SELP gene, 288
Sephardi Jews, 100, 101
Serotonin
central nervous system responsivity, 18

pathways, 71
stress response, 298, 303
transporter (5HTTLPR) gene, 59, 76,
78, 80, 81-82, 145, 146, 147, 261,
305, 328-330
Sex/gender
animal research, 93
and body weight and fat stores, 93-94
context for interactions research, 94,
125-126, 219
and CVD, 92, 96
definitions, 91
and delivery of health care, 92
and energy metabolism, 93-94
and health, 90, 91-96
and heritability of food intake, 259
and immune response, 93, 94
independent dimensions in humans, 95
and life-course patterns of health, 94
and personality, 80
and poverty, 92-93
and social environment, 92-93
and stress, 93
synergistic effects of biology and gender
relations, 91-92
variants, 95, 96
Sexual dimorphism, 94
Shotgun proteomic analysis, 115
Sickle cell disease
α-thalassemia, 285, 288, 289-290
ancestral origin and, 100
β-thalassemia, 54-55, 101, 284, 291,
292, 295, 296
cardiovascular and renal response, 286,
287, 304-305
clinical manifestations, 56, 286-287
definitions, 284
epistatic or modifier genes, 285, 287-
289, 290, 300
erythrocyte adhesion to endothelial cells,
285, 286, 302-303
etiology, 283-284
and family functioning, 293
fetal hemoglobin, 284, 285, 289
gene expression, 45, 53-56, 284, 287
HPA activation and, 298-300
hydroxyurea treatment, 289
and immune response, 283, 287, 300-
302
individual differences in severity, 282-
283, 287-289

intermediate phenotype, 288, 297
maternal adjustment, 291-292
and neurocognitive functioning, 283,
  287, 293, 294-295, 297, 303
pain, 286-287, 289, 295-297, 299, 302
pathogenesis, 284
pathophysiology, 55-56, 283, 285-286
psychological adjustment to, 287, 290-
  294, 295
resistance to malarial infection, 61, 100
severity, 54-55, 284-285, 286, 287, 289,
  295
stress and, 20, 283, 287-288, 292, 297-
  305
and stroke, 286, 287, 288-289, 294, 305
sympathetic nervous system activation
  and, 298, 302
trait, 53-54, 284
Sickness behavior, 301
Silicon Genetics, 124
Simian immunodeficiency virus, 141
Single nucleotide polymorphisms (SNPs),
  49-50, 51-52, 58, 62, 64, 99-101,
  110, 168-169
SLC6A3 gene, 264-265
Sleep patterns, 301
Social and cultural environment. *See also*
  Psychosocial work environment;
  Social networks; Social support;
  Socioeconomic status
  animal research, 36, 139-147
  aspects of health influenced, 38-39, 57, 78
  and biological processes, 36, 116-122
  and cancer, 35, 38-39
  context for interactions research, 5, 6-7,
    21, 25-26, 125
  cumulative effects, 39
  and CVD, 31, 35, 57, 78, 119
  defining, 21, 25-26, 238-239
  and depression, 18, 31, 32, 35, 147
  dynamic trajectories, 39
  early life experience, 139-140
  evidence of health associations, 26-38
  and gene expression, 142
  generational effects, 39
  genetic selection of, 113-114
  and health risk behaviors, 30, 37, 71,
    237-238
  and immune response, 35, 134, 136-
    137, 138-139
  life-course, multilevel perspective, 25-26,
    27, 31, 32, 34, 39, 138

macroenvironmental variables, 240-242,
  268, 269-270
measures of, 204, 268-271
mediators of effects of, 147-148
microenvironmental variables, 242-245
moderating effects, 262, 263
and mortality, 15, 27, 35
natural policy experiments, 27-28
and obesity, 20, 27, 72, 236-237, 238-
  245, 253-255, 263-266
and personality, 144-145
population density, 138
research opportunities, 39, 268-269
and sedentary lifestyle, 27, 72, 74, 240,
  241, 242
sex/gender and, 92-93
shared vs. unshared, 253-255, 267, 269
stressful, 20, 25, 30-31, 32, 39, 57, 78,
  117-122, 138-139, 145, 146-147,
  300, 312-313, 146-147
and temperament, 140-141
"toxic environments," 237, 239, 240-
  241, 251, 269-270
variables affecting health, 26, 240-245
Social attachment, 79
Social capital, 36-37
Social control, 37
Social isolation, 36, 121, 133, 138, 141-142
Social mobility, 32
Social networks/social supports and health
  animal research, 141, 142, 147
  bidirectional relationships, 35, 147
  biological basis, 36
  causality, 35-36
  community-level, 36-37
  confounding bias, 35
  definitions and types, 34
  and depression, 147
  evidence of health effects, 25, 34-37
  and health outcomes, 39
  and immune function, 35
  and incidence of disease, 39
  measures of, 34-35
  naturally occurring vs. strangers, 36
  negative, 35
  status and, 32, 302
  temperament or personality and, 35
Social status. *See also* Occupational status;
  Socioeconomic status
  animal research, 133, 144-146
  personality and, 144

Social support. *See* Social networks
Socioeconomic status/health associations.
    See also Social status
  and cancer, 38-39
  causal pathways, 27, 30, 32
  causation, 27, 29, 32
  and child health and development, 27,
    29, 39
  community-level, 18, 26, 30, 31, 241
  confounding bias, 27
  and eating behavior, 242
  economic reserves, 31
  education-related, 15, 25, 27-28
  flow of resources, 31
  income-related, 28-31
  inherited ability and, 29
  measures of, 26-27
  moderating effects of, 76, 104, 264
  and mortality, 27, 29, 104
  and obesity, 239, 241-242, 264
  occupational status-related, 31-33
  psychological disorders and behaviors,
    18, 30
  and psychosocial stress, 30-31, 76
  race/ethnicity and, 27, 103-104
  relative deprivations, 30-31
  research network, 188-189
  reverse causation, 27, 29, 31-32
  tests of, 29-30
  variations in associations, 38-39
  wealth, 31, 103-104
Spotfire, 124
Stanilas Family Study, 261
Statistical methods. *See also* Epidemiologic
    approaches to interactions
  Bonferroni correction, 174-175
  confidence intervals, 334-335
  gene expression profiles, 112
  inferring causality, 137
  Structural Equation Modeling, 123
Stress
  acute, 148, 302
  and alcoholism, 140, 145, 172
  allostatic load, 118
  animal research, 118, 120, 135, 137,
    142-144, 145, 146-148
  and autoimmune disease, 142-144
  behavioral response, 139, 140
  biochemical systems and processes, 117-
    122
  and cancer, 140, 146, 148

  and child development, 117, 118, 120-
    122, 139, 144, 147, 300
  chronic, 118, 139-140, 148, 298, 300,
    302, 303, 304
  and CVD, 57, 78, 117, 118, 119, 298,
    302, 304-305
  defined, 117, 147-148, 297
  deleterious coping behaviors, 32, 39
  and depression, 81-82, 298, 301, 302,
    313-314, 328-330
  early life experiences, 145, 300
  and eating behavior, 301
  epigenetic programming, 139
  and gene expression, 118, 139, 142,
    148, 287-288, 303, 305
  generational effects, 139
  genetic differences in responsivity, 20,
    142-144
  health-promoting effects, 147-148
  HPA axis activation, 297, 298-300, 301,
    302, 303, 304
  and immune function, 21, 117, 118,
    119-120, 139-141, 145, 146-147,
    148, 288, 297, 298-299, 300-302
  as intermediate phenotype, 297
  job-related, 25
  life-course perspective, 118, 145, 300
  and longevity, 118
  magnitude of, 148
  mediating variables, 139, 147-148, 304
  moderators of, 76, 285, 287-289, 290,
    300
  nonsocial factors, 120
  obesity and, 72
  oxytocin and, 36, 145
  and pain response, 301, 302
  personality and, 145, 298
  physiological responses, 72, 76, 81-82,
    117-122, 147-148, 287-288
  psychological response to, 21, 116-122,
    302
  race/ethnicity and, 39, 76
  research opportunities, 39
  SES and, 30-31, 76
  sex/gender and, 93
  and sickle cell disease, 20, 283, 287-288,
    292, 297-305
  social situations, 20, 25, 30-31, 32, 39,
    57, 78, 117-122, 138-139, 145, 146-
    147, 300, 312-313

sympathetic nervous system activation, 297, 298, 301, 302
temperament and, 120, 140-141
Stroke, 35, 39, 82, 98, 119, 135, 286, 287, 288-289, 294, 305
Structural Equation Modeling, 123
Study design and analysis. *See also* Epidemiologic approaches to interactions
life-course perspective, 21, 22
Suicide, 18, 81
Survival and functional recovery
cancer, 72, 73
obesity and, 72
personality and, 79
social environment and, 38, 39
Sympathetic nervous system reactivity, 35, 81, 118, 119, 144, 297, 298, 301, 302
Systems approach to interactions modeling, 123-124
biobehavioral model, 304
challenges in development and practice, 124-125
databases and systems analysis software, 124
gene-environment interactions, 123-126, 139
health determinants, 5-6, 17-19, 123-126
life-course perspective, 282
recommendations, 125-126
Systems biology, 123
System theory, 123

T

Tay-Sachs disease, 45, 101, 219
Television viewing
advertising, 252
and eating behavior, 252
and physical inactivity, 240, 251-252
Temperament
and ADHD, 80
animal research, 140-141
assessment, 79
and cancer, 140
defined, 77
gene-social environment interactions, 35, 37-38, 79, 140-141, 328
and health risk behaviors, 70, 75, 79-80

and immune function, 120
as intermediate phenotype, 75, 76, 79-80
and social and cultural environment, 140-141
and stress response, 120, 140-141
and tobacco use, 77, 79, 80
traits, 79
Temperament and Character Inventory, 79
Testicular cancer, 96
TGFBR2, TGFBR3 genes, 288
Thiocyanate, 70
Thiopurine methyltransferase deficiency, 59
Thiopurines, 59
$1,000 Genome Project, 50
Thyroid function, 70
TNFα gene, 76
Tobacco use
advertising and, 79
bupropion treatment, 172
cessation, 82, 172
and CVD, 57, 58, 70
definition of smoking, 69
depression and, 71
epidemiology, 70
gene-environment interactions, 50-51, 70-71
genetic susceptibility to nicotine addiction, 58, 70, 71
health consequences, 15, 57, 58, 70
intermediate phenotypes, 70, 75, 76
life-course perspective, 74
and lung cancer, 70, 310
nicotine addiction, 69, 70
personality and, 78
risk factors for initiation, 71
smoking persistence, 69
social and cultural environment and, 27, 30, 31, 32, 37, 70-71, 79
and social policy, 207
temperament and, 77, 79, 80
Transcription Factors Database, 124
Transcriptomics, 6, 111-112, 124
Transdisciplinary interactions research. *See also* Epidemiologic approaches to interactions; Infrastructure for transdisciplinary research
barriers to, 19-20
behavioral and psychological variables, 4, 5, 125
biological signatures, 126

defined, 4, 19, 220
diversity of groups and settings, 126, 220
funding, 272
genetic factors, 4, 125
hiring, promotion, and tenure policies and, 195-196
life-course perspective, 4-5, 94, 125
minority subjects in, 97-98
physiological measures and pathways, 4, 22, 125
race/ethnicity context, 4, 96-98, 125-126
recommendations, 3-5, 125-126
sex/gender context, 4, 94, 125-126, 219
social variables, 4, 21, 25-26, 125
systems modeling approach, 3, 5-6, 123-124, 126
team approach, 186-187, 195-196
Trust, 36
Tumor necrosis factor-alpha, 147
Tumor suppressor gene, 149
Turner's syndrome, 91, 95
Twin studies, 47, 71, 81, 254, 255, 258, 264, 266-267, 269-270, 299

U

UCP1, UCP2, UCP3 genes, 257
Ulcerative colitis, 77, 101
University of Washington, Seattle, 184
U.S. Department of Health and Human Services, 211

V

Vasopressin 1a receptor gene transfer, 142

W

What Is There, 124
Wisconsin Longitudinal Survey, 32
Work environment. *See* Psychosocial work environment
World Health Organization, 69, 72
Wright, Sewall, 123

X

Xenobiotics, 114
XRCCI gene, 50-51